Research Objects in their Technological Setting

What kind of stuff is the world made of? What is the nature or substance of things? These are ontological questions, and they are usually answered with respect to the objects of science. The objects of technoscience tell a different story that concerns the power, promise and potential of things – not what they are but what they can be. Seventeen scholars from history and philosophy of science, epistemology, social anthropology, cultural studies and ethics each explore a research object in its technological setting, ranging from carbon to cardboard, from arctic ice cores to nuclear waste, from wetlands to GMO seeds, from fuel cells to the great Pacific garbage patch. Together they offer fascinating stories and novel analytic concepts, all the while opening up a space for reflecting on the specific character of technoscientific objects. With their promise of sustainable innovation and a technologically transformed future, these objects are highly charged with values and design expectations. By clarifying their mode of existence, we are learning to come to terms more generally with the furniture of the technoscientific world – where, for example, the 'dead matter' of classical physics is becoming the 'smart material' of emerging and converging technologies.

Bernadette Bensaude Vincent is Emeritus Professor of Philosophy at University Paris 1 Panthéon-Sorbonne.

Sacha Loeve is Associate Professor of Philosophy of Technology at University Jean Moulin Lyon 3.

Alfred Nordmann is Professor of Philosophy at Darmstadt Technical University.

Astrid Schwarz holds the chair of General Science of Technology at Brandenburg Technische Universität Cottbus-Senftenberg.

History and Philosophy of Technoscience
Series Editor: Alfred Nordmann

Titles in this Series

1 **Error and Uncertainty in Scientific Practice**
Marcel Boumans, Giora Hon and Arthur C. Petersen (eds)

2 **Experiments in Practice**
Astrid Schwarz

3 **Philosophy, Computing and Information Science**
Ruth Hagengruber and Uwe Riss (eds)

4 **Spaceship Earth in the Environmental Age, 1960–1990**
Sabine Höhler

5 **The Future of Scientific Practice 'Bio-Techno-Logos'**
Marta Bertolaso (ed.)

6 **Scientists' Expertise as Performance**
Between State and Society, 1860–1960
Joris Vandendriessche, Evert Peeters and Kaat Wils (eds)

7 **Standardization in Measurement**
Philosophical, Historical and Sociological Issues
Oliver Schlaudt and Lara Huber (eds)

8 **The Mysterious Science of the Sea, 1775–1943**
Natascha Adamowsky

9 **Reasoning in Measurement**
Nicola Mößner and Alfred Nordmann (eds)

10 **Research Objects in their Technological Setting**
Bernadette Bensaude Vincent, Sacha Loeve, Alfred Nordmann and Astrid Schwarz (eds)

Research Objects in their Technological Setting

Edited by Bernadette Bensaude Vincent, Sacha Loeve, Alfred Nordmann and Astrid Schwarz

Routledge
Taylor & Francis Group
LONDON AND NEW YORK

First published 2017
by Routledge
2 Park Square, Milton Park, Abingdon, Oxon OX14 4RN

and by Routledge
711 Third Avenue, New York, NY 10017

Routledge is an imprint of the Taylor & Francis Group, an informa business

British Library Cataloguing-in-Publication Data
A catalogue record for this book is available from the British Library

Library of Congress Cataloging-in-Publication Data
A catalog record for this book has been requested

ISBN: 978-1-848-93584-6 (hbk)
ISBN: 978-1-781-44839-7 (ebk)

Typeset in Times New Roman
by Apex CoVantage, LLC

Contents

List of figures vii
Notes on contributors ix

Introduction: The genesis and ontology of technoscientific objects 1

PART I
Horizon of possibilities 13

1 **The pyramid and the ring: A physics indifferent to ontology** 15
PETER GALISON

2 **Cancer stem cells: Ontology matters** 27
LUCIE LAPLANE

3 **Robots behaving badly: Simulation and participation in the study of life** 42
CHRISTOPHER KELTY

4 **Vanishing friction events and the inverted Platonism of technoscience** 56
ALFRED NORDMANN

5 **From the birth of fuel cells to the utopia of the hydrogen world** 70
PIERRE TEISSIER

PART II
Arenas of contestation 87

6 **Heroin: Taming a drug and losing control** 89
JENS SOENTGEN

7 **Long live play: The PlayStation Network and technogenic life** 105
COLIN MILBURN

8 **A biography of a disorder that didn't want to be diagnosed** 123
SIMONE VAN DEN BURG

9 **The plasticity and recalcitrance of wetlands** 136
KEVIN C. ELLIOTT

10 **The life and times of transgenics** 150
HUGH LACEY

11 **Cardboard: Thinking the box** 166
CHERYCE VON XYLANDER

PART III
Multiple temporalities 183

12 **The multiple signatures of carbon** 185
SACHA LOEVE AND BERNADETTE BENSAUDE VINCENT

13 **Monitoring and remediating a garbage patch** 201
JENNIFER GABRYS

14 **Polar ice cores: Climate change messengers** 215
AANT ELZINGA

15 **Nuclear waste: An untreatable technoscientific product** 232
SOPHIE POIROT-DELPECH

16 **Biography of a 'sand heap': Staging the beginnings of nature** 245
ASTRID SCHWARZ

Index 261

Figures

2.1	Off with their heads! The Queen of Hearts running after cancer stem cells. Illustration by the author.	30
3.1	Scale of situatedness.	46
5.1	Advertisement for the hydrogen economy, UTC Power (2013).	71
5.2	First published picture of the gaseous voltaic battery, W. Grove (1842). Source: Wikimedia Commons.	75
5.3	1952 patent GB667298 (A) by F. T. Bacon. 'Improvements relating to galvanic cells and batteries'. Source: European Patent Office online database.	79
5.4	1.5 kW fuel cell power plant for Apollo Program, Pratt and Whitney (1967). Source: NASA SP-5115, Crowe (1973).	80
6.1	Fake heroin in the hand of the author (replica of a Bayer-Heroin-Flask). Photo: Author.	90
7.1	Folding@home client for the PlayStation 3.	109
7.2	Anonymous message to Sony, April 12, 2011.	115
10.1	Golden rice. Courtesy of the International Rice Research Institute (RRI).	150
11.1	Advertising poster, Container Corporation of America: "Safe delivery in paper all over the world," Egbert Jacobsen (1943). Herbert Bayer: © 2015 Artists Rights Society (ARS), New York / VG Bild-Kunst, Bonn.	172
11.2	Advertising poster, Container Corporation of America: "All three in one [. . .] all in one organization!" Matthew Leibowitz (1943). Herbert Bayer: © 2015 Artists Rights Society (ARS), New York / VG Bild-Kunst, Bonn.	179
12.1	The 'coal tree'. By-products obtained from bituminous coal degasification. Courtesy of the Pharmazie-Historisches Museum, University of Basel.	185
14.1	Working in the science trench. Courtesy of the University of Copenhagen.	223

15.1 Projects of creating warning markers to deter inadvertent human intrusion into underground waste burial sites. 1991 Sandia National Laboratories report. Concept by Michael Brill, art by Safdar Abidi. 239
16.1 Construction plan of Chicken Creek. 246
16.2a and b Aerial photographs taken in 2009 and 2013. Courtesy of BTU Cottbus-Senftenberg, FZLB. 246

Contributors

Bernadette Bensaude Vincent is Emeritus Professor of Philosophy at University Paris 1 Panthéon-Sorbonne. Her research topics span from the history and philosophy of chemistry to materials science and nanotechnology with a continuous interest in science and the public issues. Her recent publications include *Chemistry. The Impure Science* (co-author J. Simon, 2008); *Matière à penser* (2008); *Bionanoéthique* (2008); *Les Vertiges de la technocience* (2009); and *Fabriquer la vie. Où va la biologie de synthèse* (co-author D. Benoit-Browayes, 2011).

Kevin C. Elliott is an Associate Professor at Michigan State University with joint appointments in Lyman Briggs College, the Department of Fisheries & Wildlife, and the Department of Philosophy. His work lies at the intersection of the philosophy of science and practical ethics, focusing especially on issues related to environmental ethics, research ethics and the role of values in science. He is the author of *Is a Little Pollution Good for You? Incorporating Societal Values in Environmental Research* (2011), as well as more than fifty journal articles and book chapters.

Aant Elzinga is Professor Emeritus at the University of Gothenburg, Sweden. As a philosopher of science he built on his studies of theoretical physics and applied mathematics to discuss the practical utopias of science. Aside from formal research training, participation in movements for the social responsibility of science continued to afford him an important alternative source of knowledge about science in context. He was President (1991–1997) of the European Association of Science and Technology Studies (EASST), and also participated in the Swedish Antarctic Expedition 1997–1998. One of the champions of Antarctic Humanities, he has been studying and shaping Antarctic research for the last 25 years.

Jennifer Gabrys is Reader in the Department of Sociology at Goldsmiths, University of London, and Principal Investigator on the ERC-funded project, "Citizen Sensing and Environmental Practice: Assessing Participatory Engagements with Environments through Sensor Technologies." She is author of a study on electronic waste, *Digital Rubbish: A Natural History of Electronics* (2011), and a forthcoming study on environmental sensing, *Program Earth: Environment as Experiment in Sensing Technology* (University of Minnesota Press). Her work can be found at http://citizensense.net and http://jennifergabrys.net.

Peter Galison is the Joseph Pellegrino University Professor of the History of Science and of Physics at Harvard University. His work explores the interaction between the subcultures of physics – experimentation, instrumentation, and theory, and their embedding in politics and materiality. Among his books are *How Experiments End* (1987), *Image and Logic: A Material Culture of Microphysics* (1997), *Einstein's Clocks, Poincaré's Maps* (2003), and *Objectivity* (2007), with Lorraine Daston. He has also made two documentary films: *Ultimate Weapon: The H-bomb Dilemma* (2000), and with Robb Moss, *Secrecy* (2008).

Christopher Kelty is an Associate Professor at the University of California, Los Angeles. He has a joint appointment in the Institute for Society and Genetics, the Department of Information Studies and the Department of Anthropology. His research focuses on the cultural significance of information technology, especially in science and engineering. He is the author of *Two Bits: The Cultural Significance of Free Software* (2008), as well as numerous articles on open source and free software, including its impact on education, nanotechnology, the life sciences, and issues of peer review and research process in the sciences and in the humanities. He is trained in science studies (history and anthropology) and has also written about methodological issues facing anthropology today.

Hugh Lacey is Scheuer Family Professor of Philosophy Emeritus at Swarthmore College in Swarthmore, Pennsylvania, and Research Collaborator in a project, "The Origins and Meaning of Technoscience: On relations between science, technology, values and society," in the Institute of Advanced Studies at the University of São Paulo in São Paulo, Brazil. He is the author of *Is Science Value Free?* (1999), *Values and Objectivity in Science* (2005) and *A Controvérsia Sobre os Transgênicos: questões científicas e éticas* (2006).

Lucie Laplane is a philosopher of biology and medicine. She is a researcher at the CNRS (UMR8590-IHPST, University Paris I Panthéon-Sorbonne and UMR1170. While still working on the identity of stem cells, she extends her research to the study of (1) induced pluripotent stem cells as a new model in oncology, and (2) the clonal evolution theory.

Sacha Loeve is Associate Professor of Philosophy of Technology at University Lyon 3. Since his PhD on molecular nanomachines (2009), he practices a 'fieldwork philosophy' centered on objects and primarily interested in rethinking technology as a narrative logos of technics. His publications include works on Gilbert Simondon, nano-objects, images and technology, pharmacotechnology, synthetic biology, Moore's Law and electronic waste.

Colin Milburn holds the Gary Snyder Chair in Science and the Humanities at the University of California, Davis. He is a professor in the departments of English, Science and Technology Studies, and Cinema and Technocultural Studies. He is the author of *Nanovision: Engineering the Future* (2008) and *Mondo Nano: Fun and Games in the World of Digital Matter* (2015).

Alfred Nordmann is Professor of Philosophy at Darmstadt Technical University, after receiving his PhD in Hamburg (1986) and several years on the faculty at the University of South Carolina (1988–2002). In the field of History and Philosophy of Science and of Technoscience he is interested primarily in the questions of knowledge and objectivity, explanation and understanding under representational and technological conditions of knowledge production; see *Science Transformed* (co-edited with Hans Radder and Gregor Schiemann, 2011) and *Science in the Context Application* (co-edited with Martin Carrier, 2011).

Sophie Poirot-Delpech is Associate Professor at University Paris 1, Panthéon-Sorbonne, where she teaches sociology, and founding member of the CETCOPRA (Center for Study of Technology, Knowledge & Practices). She has conducted field research on aeronautics and more recently on nuclear waste. She is currently exploring how technologies have been taken into account in the French sociological tradition. In all research areas she is primarily concerned with the memorial dimension of technological objects. Here publications include *Mémoire et histoires de l'automatisation du contrôle aérien. Sociobiographie du CAUTRA* (2009).

Jens Soentgen has been scientific director of the Environmental Science Centre of Augsburg University since 2002. Born 1967 in Bensberg, North-Rhine Westfalia (Germany), he studied chemistry and sociology and wrote his PhD thesis in philosophy (on the notion of substance, 1996). In 1999 and 2001 he worked in Brazil as guest professor of philosophy. He realized three interactive science exhibitions (on dust, on CO_2 and on nitrogen), and publishes philosophical papers as well as popular books. Two of them have been nominated for the German Youth Literature Prize (2004, 2011). He is co-editor of the *Stoffgeschichten* book series, which publishes historical accounts of politically relevant substances. His research interests are in the field of science and technology studies, science communication, substance stories and the history and philosophy of chemistry.

Astrid Schwarz holds the chair of General Science of Technology at Brandenburg Technische Universität Cottbus-Senftenberg. In her research she asks questions about the status and power of concepts, models, and objects in the process of generating, stabilizing and demarcating scientific knowledge. Recently, she focuses on debates around green cultures and ecotechnology, as well as on gardens as models to act in the Anthropocene. Her last book was *Experiments in Practice* (2014).

Pierre Teissier is Associate Professor at the Centre François Viète of Epistemology, History of Sciences and Techniques from the University of Nantes in France. He investigates two topics of contemporary history of sciences and technology: materials chemistry between academy, industry and society in Europe and the United States; and alternative energies in their social and cultural dimensions, especially fuel cells, electric cars and solar energy. He has published a monograph entitled *Une histoire de la chimie du solide. Synthèse, formes, identities* (2014).

Simone van der Burg is senior researcher and theme leader of Responsible Innovation at Radboud University Medical Center (Nijmegen, the Netherlands). Her research projects primarily focus on the development and use of stakeholder involvement methods to conduct ethical reflection on the future of genomics, newborn screening and personalized medicine. She is member of Cogem, an advisory board on societal and ethical aspects of genetic modification for the Dutch government, and she is co-founder and associate editor of the *Journal for Responsible Innovation.*

Cheryce von Xylander currently teaches intellectual history and philosophy at the Technical University Darmstadt and lives in Berlin. American by birth, German by upbringing and English by the wiles of the Norns, she studies the aesthetic politics of knowledge transmission. Taking cognition to be a form of situated practice, she explores the madeness of mind. Her work, cross-cultural and trans-disciplinary, clusters around themes of applied popular philosophy: aesthetic re-education inside and outside the asylum, bracketed interconnections between the natural sciences and liberal arts, comparative global imaginaries.

Introduction

The genesis and ontology of technoscientific objects

This book seeks to contribute toward the philosophy of technoscience and a better understanding of its significance. It proposes to do so by following objects that occupy a place of prominence in contemporary science and technology. Objects of advanced research such as smart materials, brain-machine interfaces, stem cells and artificial organisms tend to capture the popular imagination. They are attractive not necessarily for what they are but for what they promise. A carbon nanotube is just a tiny bit of soot, but – rather like in the Cinderella fairy tale – we think it can do miraculous things if only we can fit it with a glass slipper. The 'dead matter' of classical physics becomes 'smart material' in the hands of materials scientists and other nanotechnology researchers, DNA molecules become a wonderful 'molecular machine' in the hands of bioengineers. Where does the seductive power of such things come from? Do they fit the mould of innovation that would change the world? And to what extent does technoscientific research redefine our relationship to things in the world?

Attractive and mundane objects

While the epistemology and ethics of science are focused mainly on the subjects of knowledge, we propose to shift attention from the subjects to the objects of knowledge. This volume displays a collection of diverse objects: polar ice cores, the oceanic garbage patch, graphene as the latest incarnation of carbon, video game networks, cardboard, GMO seeds, 'social' robots, stem cells, 'natural' wetlands, frictionless surfaces, nuclear waste, fuel cells, heroin and a giant heap of 'sand' that simulates the genesis of an ecosystem. Taken together, these objects map a large part of the bizarre landscape of the technoscientific world. Many of them are frequently discussed in newspaper columns. Unlike scientific objects constructed in the closed world of laboratories that need to be popularized, technoscientific objects have a propensity to attract public attention. They prove to be attractive in more than one sense.

First, they act as 'attractors' of young talent, experts, funds and lobbies from all over the world. As their design requires a coming together of various specialists from established disciplines such as mathematics, physics, chemistry, biology and computer science, they prompt the creation of new fields of research, new journals and new institutions boasting technical platforms and considerable financial investments. By bringing people together and being a magnet for resources, they reorganize research activities and orchestrate networks of both people and resources.

Second, as knowing and making become ever more entangled, researchers' attention shifts from the characterization of things to how they perform, from structures to functions, from properties to processes. Objects of scientific research are prepared intellectually and technically in such a way that they cease to be mundane things; they are purified of all kinds of interests and values. They are considered to be nothing but bearers of properties and are recorded as matters of fact or states of affairs. By contrast, objects of technoscientific research are attractive in a mundane and familiar way, an observation which challenges Bachelard's epistemology of a break from ordinary experience (Bachelard 2002). They can be as simple as cardboard or rubber or beer. But whether 'hi-tech' or 'lo-tech', they are always value laden.

For a familiar illustration of this point, consider the history of the gene. The gene was introduced as a hypothetical substrate of inheritable traits and thereby served an explanatory and theoretical purpose. Encapsulated in 'the book of life', it raised questions about human nature and about the structure of reality to which each of us owes our phenotypic appearance. The confirmation of genetic theories of inheritance went hand in hand with the experimental manifestation or 'realization' of the gene as, for example, the molecular structure of DNA and mechanisms of replication were elucidated. As it was the referent of concerns about racial purity, about defective and healthy bodies, and about the myth of biological destiny, the gene signified something immutable and inescapable – all we could do was trace ourselves back to it and acknowledge its supreme importance. Nowadays, by contrast, genes are malleable or plastic, and as research objects they are hypothesized not as real entities that have an explanatory function but as a potent material that affords many kinds of manipulation. Researchers take an interest in them as instruments, materials and tools. According to the rhetoric of the day, genes are at last being released from the straightjacket of (deterministic) evolution: they are no longer features of an exclusively natural history but have become part of cultural history too, in that they are used in medical therapies and devices or as an agrotechnical object for the improvement of crops, farm animals and human beings. Once the gene had thus proved interesting for what it could do in the creation of new products and procedures, for new ways of acting in and relating to the world, it became what we call an attractive object of technoscientific research.

Third, technoscientific objects are also 'attractive' in that they imply a kind of 'Copernican revolution'. They shift the focus from subjects to objects and display the wide spectrum of interactions between these objects and their surroundings. In becoming technoscientific objects, such things embody new relationships, values, more and less intense concerns that change their modes of existence. As Aant Elzinga shows in his contribution to this volume, polar ice has switched from being an archive of natural history to becoming a messenger of the future enrolled in urgent calls for political action to mitigate climate change. Of course, arctic ice has been 'sitting there' for millions of years, but it came into being as a technoscientific object by being transplanted into new material setups and practices (from ice drilling, coring and trenching to global climate modeling) as well as in new discursive settings – a fragile object that has become a witness to the fragility of the human condition.

Accordingly, these attractive objects are of interest well beyond the limits of research laboratories. It is an important feature of technoscientific research that it operates in the open field of society and the environment, where its objects propagate, colonize and transform their social and natural surroundings by (re-)shaping them and by imposing their own temporal dynamics upon them. Whether genetically modified organisms, nuclear waste or stem cells – all of them alike shape the outer and inner environment of the human body, all of them exert an effect on the material and imaginative level, and all of them deeply penetrate mundane practices and habits in almost all spheres of society.

But what exactly is meant by technoscientific research and how can the outcomes of this research be adequately conceptualized? What is so special about technoscientific objects? And does it really make sense to distinguish between science and technoscience? Why do we need to compound the familiar notions of 'science' and 'technique' or 'technology'?

Why technoscience?

Technoscience is a rather suspect term indeed. At first glance, it connotes power and money, politics and capitalism rather than particular objects. Taking a closer look at the development of this concept, it turns out that 'technoscience' has been subjected to three kinds of uses since the 1980s. The Belgian philosopher Gilbert Hottois used the term 'techno-science' (written with a hyphen) as a provocative wake-up call to rouse philosophers from their dogmatic slumber (Hottois 1979, 1981).[1] Hottois contended that most philosophers were mainly concerned with linguistic issues in science, delegating the study of reality to the techno-sciences and thereby condemning philosophy to a condition of what he called 'secondarity'. Hottois switched from the philosophy of language to the philosophy of 'technoscience', referring to a new epoch in which technology is at once the milieu, the driver and the final goal of research (Hottois 1984). Our use of 'technoscience' comes close to that of Hottois, but our approach is slightly different: while science and technology are still considered as distinct but interacting spheres from the perspective of science, they become indistinguishable from the perspective of technoscience. Indeed science appears as a mode of knowledge production in which technology is often used to create phenomena and to make representations such as models or diagrams. In the technoscientific mode of knowledge production, theories, models and algorithms become tools by which to achieve predictive and technical control and to make things work.

Moreover, unlike historian Paul Forman, who situated technoscience in the transition from modernism to postmodernism (Forman 2007), we do not claim that technoscience represents a historical break – that would require us first to posit an age of science and Enlightenment and then, from this vantage point only, an age of technoscience (Nordmann et al. 2011). Rather than claim such a break, we argue that technoscience denotes a certain mode of existence of research objects, one which is now dominant, although it can certainly be found in past centuries as well. Examples are plentiful: mechanical arts, alchemy and pharmacy, pneumatic chemistry, agronomy, automated calculation, zootechnics and so on. The historical depth of technoscience shows clearly that the technoscientific relation to research objects is not new, albeit it is strikingly prevalent in contemporary culture.

A second – quite distinct – meaning of the term 'technoscience' has been disseminated by Bruno Latour, Donna Haraway and other scholars in the science and technology studies (STS) movement and in feminist studies. For Latour, technoscience is nothing but 'science in the making', the true expression of the real, impure and mixed practices of the sciences as they are made (Latour 1987). 'Technoscience' has become a polemic term used to debunk the myth of pure and autonomous science. It is also a heuristic tool for rendering visible the various mediations and hybrid alliances (instrumental, technical, rhetorical, political . . .) required to construct scientific facts. This is most often done by means of a process of 'purification', at the end of which the contributions of humans and non-humans can be sorted into two opposite categories: 'nature' on the one side and 'society' on the other. In the STS world, science has always been technoscience, and today's explicitly impure technosciences (nanotechnology or synthetic biology) are nothing more than the 'truth speaking' of science, a sign that we are beginning to stop believing that we have ever been 'modern'. Unlike the STS concept of technoscience, we do not think that all of science can be exposed as technoscience. Neither do we claim that we are going to unmask the true face of scientific activity or that the work of purification is a futile and meaningless pursuit. What we contend instead is that a technoscientific object is encountered whenever the scientific work of 'purification' proves impossible, undesirable or unnecessary.

Finally, the phrase 'technoscience' is widely used to denounce the corrosion of the scientific ethos and the contamination of scientific research by economic interests, the pressure of funding applications and consumerist society. This has also been noticed by Hottois, who criticized that 'technoscience' is used rather vaguely alongside fashionable terms like technocracy, capitalism, neoliberalism, globalization and so on (Hottois 2006). What is more, technoscience carries a dubious 'postmodernist' flavor that celebrates the alleged surrender of the grand narratives of modernity (Lyotard 1979). All these connotations may be one reason why the scientists and engineers who are the actors of technoscience never adopted the term. In not doing so, however, they deny its positive and heuristic power to reconceive not only science and technology but also the role of scientific research in society.

This book is not interested in arguing *pro* or *contra* technoscience. It does not even take the dichotomy between science and technoscience as given. Far from assuming that science and technoscience are two distinct entities with stable, transhistorical identities, we consider them as 'ideal types' forged against specific historical and cultural backgrounds. They are supported by more or less explicit grand narratives of progress that guide research policy. The ideal type of technoscience is associated with notions like 'responsible research and innovation' or 'sustainable development', which are thought to orient the future direction of research and development (Bensaude Vincent 2014). It encourages a mode of research that is open to the world, which takes on board all kinds of societal, environmental and economic concerns. Research institutions or individual researchers who have been socialized in a very traditional way are now engaged in a diverse range of academic and entrepreneurial pursuits. The activities of understanding and making, knowing and producing are becoming indistinguishable from one another. In this context of the hybridization of science and technology, epistemic values (such as truth, simplicity, etc.) are displaced by non-epistemic values such as social robustness and environmental and economic sustainability (Carrier et al. 2008).

Consequently, technoscientific research can be neither appropriately characterized using the oddly established terms of 'applied' and 'pure' science nor considered as being research conducted in a trans- or interdisciplinary context. Technoscientific research explicitly involves problem solving in many different societal domains, including personalized medicine, climate change and alternative energy, and it does so by mobilizing everyday practices as well, such as game playing or consuming. Two examples are discussed in this book, the 'Sony PlayStation', presented by Colin Milburn, and social robots, discussed by Christopher Kelty. At the same time, consumers are confronted with genetically modified (GM) products on the grocery shelves while biomedical research opens up a market in human enhancement in sports, cosmetics and medicine, for instance. Accordingly, technoscientific objects are heavily invested with a variety of interests. Far from being neutral, they are to serve societal interests such as durability, human empowerment, beauty, user-friendliness, efficiency or solidarity. To some extent, the research conducted in nanotechnology or biotechnology, or in information technology or cognitive science, takes society as its laboratory (Schwarz 2014).

Technoscientific objects are either already deeply rooted in the daily routines of societies or else they are at least virtually present by virtue of embodying a blueprint of future everyday life. Above all, they raise great expectations – and fears. By the same token, to behold them is to participate in their power, potential and promise. Some of them, such as stem cells (presented by Lucie Laplane) and fuel cells (presented by Pierre Teissier), bear the promise of a cure for cancer, or of the hydrogen economy, and open up a world of unbounded possibilities. Other objects, by contrast, such as nuclear waste (Sophie Poirot-Delpech) and oceanic plastic garbage (Jennifer Gabrys), are fascinating because they confront us with vexing issues that have no immediate solutions. All of them point to the future. They become matters of interest or matters of concern; they fascinate and engage because of the futures embedded in them.

The furniture of the technoscientific world

Like the curiosities of an early modern *Wunderkammer*, the collection of technoscientific objects displayed in this volume blurs the boundaries between the natural and the artificial as well as, in some cases, between the living and non-living and between subject and object. They are often presented as hybrids, and everyone seems to agree on their hybrid nature. In this respect, the archetypical technoscientific research object is still the oncomouse, as introduced and discussed by Donna Haraway (Haraway 1997). This genetically engineered laboratory animal is at once a living thing and an artificial organism and, as such, is a hybrid of technology and nature. However, 'hybrids' wouldn't exist if it were not for our belief that some things are clearly natural, others artificial, and hybrids an entanglement of the two.

However, it is characteristic of technoscientific objects that the question of their hybridity does not even arise. Instead of confounding nature and artifice, they are simply impure and not in need of purification. As mundane objects, they are as simple and straightforward as a wooden chair or an aspirin tablet – since it would not occur to anyone to ask whether and to what extent these things are natural or artificial, there is also no reason to consider them 'hybrids'. The great divides that served as the pillars of modern culture are not ones that structure the technoscientific world.

As things of the world (or mundane things), technoscientific objects are so ordinary that we view them with ontological indifference. However, it is not for science to be ontologically indifferent when its business is described as identifying an immutable substrate of reality, as Peter Galison argues in this volume. Technoscientific research, by contrast, is mainly interested in making everyday things that are meant to require no special conditions for their existence, even though it goes to nanoscale extremes and has the audacity to create biological organisms. Technoscientific objects are as richly ordinary as Martin Heidegger's 'things'. Like the jug that he takes as his example (Heidegger 1971), they gather within themselves (*sammeln*) the ways and conditions of the world; they grant us (*schenken*) new ways of doing things in the world.

For all their straightforward simplicity, the mode of existence of such worldly things deserves our attention. As objects of technoscientific research they replace the paradigm of objects that are believed to be grounded in immutable facts of nature. Now, terms like 'furniture of the world' or 'mode of existence' evoke the philosophical field of ontology. Typically, it encompasses a type of categorical analysis that focuses on substance and accidents, on manifest and dispositional properties, on primary and secondary qualities, and the like. Technoscientific objects – and this is one of the main theses of this book – require distinctively different ontological categories to understand them as things of the world.

They need to be characterized by their performance rather than by their structure or composition. They undermine the traditional distinction between *substantia* (what it is) and *potentia* (what it can do or become). More precisely, their *potentia* relies on affordances rather than on dispositional properties. The concept of affordance plays a central role here, in that it points to a relationship and not primarily to the intrinsic properties of a structure. Affordances accord something to an organism in its environment that wouldn't otherwise be possible. What James Gibson, who first established the concept of affordance, sought to suggest with it is a particular perspective on the furniture of the world (Gibson 1997). A vertical, flat, extended and rigid surface such as a wall or a cliff face affords falling – it is fall-off-able – but might also become climb-up-able if we are attentive, and so on. Similarly, the design of technoscientific objects relies on what materials and natural phenomena can afford in certain circumstances or in relation to specific conditions. The design process does not consist in the application of general laws. Turning molecules into machines or using proteins built up through the evolution of life on earth as tools is all about seizing opportunities to provide abilities. The activity of design is organized so as to develop an ability in material entities to do something (switch a signal, store information) or to produce something (a toxin, a fuel, a material). The ultimate goal of research is to make things 'actionable' or to 'enable' certain actions – two key words in the language of contemporary nano- and biotechnology.

Such terms encourage discourses about 'shaping the world' or 'redesigning life'. Common phrases such as 'materials by design' or 'smart technology' often suggest a Promethean view of designers projecting form and information on a passive matter. But the practices of technoscientific design do not fit this 'hylomorphic' conception of artifacts.[2] Technoscientific designers are not dealing with matter in general. They use materials, working with their singular qualities, propensities, or with their surfaces, their interactions, and with the environment (Bensaude Vincent 1998).

More importantly, when it comes to technoscientific objects, genesis and ontology are intimately linked. Typically, such objects come into being as proofs of principle – they neither constitute evidence nor are they demonstrations. Rather, they are exhibit(ion)s of a prototype that promise indefinite possibilities. They are designed to become; their mode of existence is simultaneously one of 'no longer' and 'not yet'. Genes are no longer an inexorable destiny; they can be redesigned, re-engineered and reassembled to create new forms of life in the future as well as to serve as materials that encode their own construction plan. Matter is no longer a rigid constraint, a pole of resistance or recalcitrance. It is plastic and malleable as soon as one takes materials or cells in their nascent state and works with them to generate a material or a tissue by design. Technoscientists are taking advantage of the intrinsic dynamic of material entities and their collective behaviors. In thus playing with the spontaneous dispositions of things – self-assembly, self-organization or self-repair – they address nature as *natura naturans* rather than *natura naturata*. In their efforts to seize all opportunities, they co-operate with natural phenomena and use molecules, bacteria and viruses as partners. Nature is neither to be commanded nor obeyed; it is an ally of technoscientific adventures.

Such alliances determine the ambiguous modes of existence of technoscientific objects. They are designed with an eye to the future as a door opening onto a new set of possibilities. Whether material or merely virtual, whether wetware or software, they are perched between promise and reality. They are 'always within reach' (if only the proper technical conditions were realized) and at the same time 'always just slightly beyond reach'. As a consequence, they call for action, for more research, more investment, more regulations. Technoscientific objects, in contrast to well-defined scientific objects, are never stabilized; they are always 'works in progress'. They are endless processes and as such continuously require the mobilization of a variety of human and non-human resources to keep moving.

Finally, they are torn between the utopia of emancipation from matter and the affirmation of unrestricted potentials lying in material structures. On the one hand, objects such as frictionless surfaces or graphene undermine the notion of materiality as a constraint. They thereby revive the grand narratives about mankind being in control of nature, and the modern ideal of emancipation of mankind from the constraints of matter and the limits of life. On the other hand, the same objects may also be seen as the manifestation of the potentials hidden in matter or even as an enhancement of matter. As new material creatures they have a life of their own, which far exceeds that of their designers and users.[3] Technoscientific objects come to life; they show resilience and a certain obstinacy; they interfere with evolutionary processes. It is thus possible to characterize their temporal dynamics in terms of their life cycles, from cradle to grave. Plastics or microchips have their own times, which are hardly commensurable with the timescales of their ephemeral use.

Featuring technoscientific objects

Rather than including a general philosophical essay on the ontology of technoscientific objects, this volume contains *stories* about the genesis and life of a selection of such objects. Narratives are a powerful tool that serves to throw a spotlight on objects as the central characters of a philosophy and history of technoscience focused on processes rather than on substances, on individual modes of existence rather than on essences, while paying attention to the various timescales implied in their existence. The narrative approach is an alternative way to address philosophical issues while mobilizing the traditions of philosophy and of science studies.

A first set of papers – grouped under the heading 'horizon of possibilities' – explores the ontological commitments present in technoscientific research process. Peter Galison's simulations, nano-objects and strings inspire a new map of knowledge that profoundly changes the identity of disciplines. Their 'ontological indifference' comes into stark contrast with the ontological anxiety expressed in twentieth-century debates that pit realists against anti-realists in theoretical physics.

However Lucie Laplane's analysis of cancer stem cells research suggests that ontological indifference is actually a problem for technoscience. Depending on the ontological status of stem cells, the therapeutic strategies will be conceived quite differently. She argues that ontology matters for very practical purposes.

The next two chapters explore the relations between technoscientific objects and modes of theorizing. Whether physical or simulated, the robots presented by Christopher Kelty are used as a kind of model organism that affords a way of accelerating the timescales involved in biological evolution. These robots, which subvert the ontological divide between reality and copies, exemplify the productivity of technoscientific simulacra. Subverting the ontological divide between reality and copies proves to be a heuristic epistemic gesture.

The frictionless surfaces pursued by nanotechnologists and presented by Alfred Nordmann exemplify the quest for an ideal world that lies not *behind* the appearance, as in Platonism, but *beyond* the appearances as the vanishing point of engineering. In spite of a marked ontological indifference, they hint at a possible new world, an ideal world emancipated from the constraints of matter – the dream world of perpetual machines that modern thermodynamics had repudiated.

To counterbalance the dreams of dematerialization, the fuel cell, to which Pierre Teissier gives a voice, instantiates the recalcitrance not only of nature but of technology as well. Fuel cells embody the promise of an eco-friendly hydrogen economy capable of overcoming the finitude of a world with limited and depleted resources. They show that not every object can be turned into affordances.

The objects displayed in the second set of papers – 'arenas of contestation' – raise the issue of the interplay between political and economic values and scientific norms. Jens Soentgen's *longue durée* history of a controversial chemical substance known as heroin exhibits the tensions between the scientific and the legal control of this drug, and calls attention to the struggle between gaining and losing control.

The story of the outage of Sony PlayStation Network (PSN) told by Colin Milburn questions the distinction between experts and amateurs – hackers – as well as between the individual and the collective. The PSN is simultaneously a technical assemblage, an experimental platform and an affective space of pleasure and devotion for eighty million players. As a corporate mechanism that redistributes fun as well as capital, knowledge as well as risk, the PSN presents itself as a 'quasi-object' that brings into existence a new way of 'technogenic' life.

Writing the biography of a mysterious disease affecting a young boy, Simone van den Burgh explores the precarious instability of a disorder as it is negotiated between the patient's family, health professionals and medical technosciences. She moves from the patient's family to the physician's cabinet to emphasize the changing identity of the disorder, and she argues that this shifting meaning deserves attention in translational medicine.

Kevin Elliot's story of swamps reconceptualized as wetlands draws attention to the combination of traditional 'scientific' values with 'social' values. Environmental science interferes with conservation and regulatory policy in the design of an object, which also blurs the boundary between naturally grown and artificially designed pieces of nature.

Hugh Lacey's genetically modified seeds illustrate the same elusive boundary between nature and artifice. While GM crops exemplify the potentials and the powers of technoscientific objects, they also emphasize the complex environments they require as well as the problematic character of their impacts on nature and society.

The chapters gathered in this group strongly suggest that technoscientific objects cannot be grasped through laboratory inquiry alone because they reconfigure the social fabric. This is particularly clear in the case of the mundane cardboard boxes presented by Cheryce von Xylander, which have been a key factor in reframing our ways of mass consumption in the twentieth century. She argues that cardboard as a material has deeply changed the function of paper in culture with the shift from a literacy of signs toward a literacy of things.

The objects assembled in the third batch of chapters – 'multiple temporalities' – raise the issue of the persistence of the ephemeral, an issue that might yet become a central concern for a philosophy of technoscience. Following the traces and footprints of carbon, Sacha Loeve and Bernadette Bensaude Vincent explore the multiple modes of existence of this common object in a tentative 'ontography'. Carbon has displayed so many profiles over the course of the centuries that it seems to create various '*personae*' that move with a momentum of their own and weave amongst a wide spectrum of timescales and timescapes.

The story of ice core drilling told by Aant Elzinga also displays a wide range of temporalities in the context of climate research. The compressed snowflakes and air bubbles of each individual ice core literally encase its temporal states. It needs to be disclosed by bringing the durable international standards international standards of measurement and data collection to its ephemeral physical condition.

Recounted in a different tone, Jennifer Gabrys's geomythological narrative about the Great Pacific Garbage Patch is also a story of inextricable entanglements of actors and timescales. Whereas the ice cores originally have a solid aggregate state, the garbage patch, a floating soup of small and medium sized debris, is fluid with no firm or permanent boundary. A crucial question raised here is how to discern and identify such a radically plastic object, how to monitor its becoming.

An even more intractable polychronic and technopolitical object is nuclear waste. Sophie Poirot-Delpech portrays a 'monster', a thing that is neither natural nor technological (a product of human industry and a source of artificial nuclei) since it has no function and no potential use. With its period of decay exceeding the limits of human experience and intellectual imagination, it raises anxieties about the future and doubts about recent technological choices. The ontology of nuclear waste radically questions the limits of technoscientific power to remake the world.

Astrid Schwarz's artificial heap sand is the drama of an emerging ecosystem featuring the malleability of nature. The artificial water catchment 'Chicken Creek' was designed in order to study the initial states of natural systems. It works both as a *real-world simulation* in that it simulates the dynamic properties of the genesis of ecosystems and simultaneously as a real-world system that embodies them in real time. This research object gathers together theories, skills and interests and, in doing so, affords the proof of concept of the relation between the epistemic and the ontological: theories and concepts *about* things are inevitably linked with a theory *of* things.

In medias res

The papers gathered together in this volume were conceived and discussed in the context of the project *Biographies of Technoscientific Objects* funded by the German and French Research Councils (ANR-09-FASHS-036–01 / DFG NO 492/5–1). The 'GOTO project' (as it has come to be known[4]) has fostered a dialogue between the French and German philosophical thinkers who begin *in medias res*.

Whereas classical scientists step back and view the facts from a distance in order to construct an image of the world, technoscientific researchers behold their objects from the midst of things by participating in the natural, technical and social processes that make these things salient in their world. These objects of research are prototypes for the furniture of a world populated by things that gather within themselves not only found and processed materials or products of human crafts, not only the values of an experimental society but also the knowledge and power acquired by humans in the history of science and technology.

This characterization resonates with Heidegger's view of things as they appear beyond modern metaphysics and modern science and beyond the Cartesian subject that pictures a world of objects. Heidegger's is one of many philosophical avenues, however, toward an understanding of technoscientific objects, their attractive and quotidian mode of existence and their particular ontology irrespective of the distinction between the natural and artificial.

Simondon's philosophy of technology with its notion of concrete objects offers another avenue. It also begins *in medias res*, as it strongly rejects the abstract notions of technical objects as the materialization of a design project or as means toward an end. Whether natural or artificial, objects become concrete when they display functions that integrate the environment where they operate and turn it into the object's 'associated milieu'. Simondon considers technical objects not as static things, definable by their here and now. He describes them in terms of individuation as a process, a sequence of stages that makes up an interesting story rather than a stable entity.

Finally, a qualified extension of Gaston Bachelard's conception of metachemistry (Bachelard 1968, 45) replaces the Kantian question as to the preconditions for the possibility of true representations by attempting to identify the preconditions for the possibility of knowing through making. This allows us to consider the attractive objects of technoscience as objects of a peculiar kind of knowledge (Nordmann 2013).

From the midst of things in the world, of concrete objects known through the making of them, objectivity is the agreement among objects and not the agreement of subjects. From the midst of things, limits of knowledge and limits of technical control are not discernible but can only be probed by running up against and perhaps transgressing them. They are therefore not absolute or even – in the Kantian, transcendental sense – constitutive, but appear only as erstwhile failures to surmount them. And from the midst of things, it is a challenge to gain critical distance on attractive objects and the peculiar social or economic values that make them so attractive. It means that grand historical narratives, ambitious social and political agendas are not in sight, but that there are many revealing stories to be told and philosophical issues addressed within the midst of things. As we come to terms with the specific relations between people and things in our contemporary world, we hope that this book will stimulate further revelations about the unsettling attractiveness of objects.

Notes

1 A few occurrences of the adjective 'techno-scientific' can also be found in the science policy literature before the 1980s.
2 Hylomorphism means 'something = matter + form'. For a critics of hylomorphic dualism in technology as well as in ontology, see Simondon (1989, 2005).
3 It is not just a matter of re-scription or of co-construction of users and technology; see Akrich (1992) and Oudshoorn and Pinch (2003).
4 See our collaborative paper: Bensaude Vincent, Loeve, Nordmann and Schwarz (2011).

References

Akrich, M. 1992. 'The Description of Technical Objects.' In *Shaping Technology/Building Society: Studies in Sociotechnical Change*, edited by Bijker, W. and Law, J., 205–224. Cambridge, MA: MIT Press.

Bachelard, G. 1968. *The Philosophy of No: A Philosophy of the Scientific Mind.* New York: Orion Press.

———. 2002. *The Formation of the Scientific Mind: A Contribution to a Psychoanalysis of Objective Knowledge.* Bolton: Clinamen.

Bensaude Vincent, B. 1998. *Eloge du mixte. Matériaux nouveaux et philosophie ancienne.* Paris: Hachette Littératures.

———. 2014. 'The politics of buzzwords at the interface of technoscience, market and society: The case of "public engagement in science".' *Public Understanding of Science* 23 (3): 238–253.

Bensaude Vincent, B., Loeve, S., Nordmann, A. and Schwarz, A. 2011. 'Matters of interest: The objects of research in science and technoscience.' *Journal for General Philosophy of Science* 42 (2): 365–383.

Carrier, M., Howard, D. and Kourany, J. (eds.) 2008. *The Challenge of the Social and the Pressure of Practice: Science and Values Revisited.* Pittsburgh, PA: University of Pittsburgh Press.

Forman, P. 2007. 'The primacy of science in modernity, of technology in postmodernity and of ideology in the history of technology.' *History and Technology* 23 (1/2): 1–152.

Gibson, J. J. 1997. *The Ecological Approach to Visual Perception.* Boston, MA: Houghton Mifflin.

Haraway, D. 1997. *Modest_Witness@Second_Millenium. FemaleMan__Meets_Onco-MouseTM: Feminism and Technoscience.* New York: Routledge.

Heidegger, M. 1971 [1951]. *The Thing.* In Lecture to the Bayerische Akademie der Schönen Künste, 1950, translated by A. Hofstadter. New York: Harper.

Hottois, G. 1979. *L'inflation du langage dans la philosophie contemporaine.* Bruxelles: Editions de l'Université de Bruxelles.

———. 1981. *Pour une métaphilosophie du langage.* Paris: Vrin.

———. 1984. *Le signe et la technique: La philosophie à l'épreuve de la technique.* Paris: Aubier.

———. 2006. 'La technoscience de l'origine du mot à son usage actuel.' In *Regards sur les technosciences*, edited by Goffi, J. Y., 21–38. Paris: Vrin.

Latour, B. 1987. *Science in Action.* Cambridge: Harvard University Press.

Lyotard, J.-F. 1979. *La condition postmoderne: Rapport sur le savoir.* Paris: Minuit.

Nordmann, A. 2013. 'Metachemistry.' In *The Philosophy of Chemistry: Practices, Methodologies, and Concepts*, edited by Llored, J.-P., 725–743. Newcastle: Cambridge Scholars.

Nordmann, A., Radder, H. and Schiemann, G. (eds.). 2011. *Science Transformed? Debating Claims of Epochal Break.* Pittsburgh: University of Pittsburgh Press.

Oudshoorn, N. and Pinch, T. 2003. *How Users Matter: The Co-Construction of Users and Technologies.* Cambridge, MA: MIT Press.

Simondon, G. 1989. *Du mode d'existence des objets techniques*, 1958, translation in progress. Paris: Aubier, https://www.academia.edu/4184556/Gilbert_Simondon_On_the_Mode_of_Existence_of_Technical_Objects.

———. 2005. *L'individuation à la lumière des notions de forme et d'information*, 1964/1989, translation of the introduction. Paris: Million, https://english.duke.edu/uploads/assets/Simondon%20Genesis%20of%20the%20Individual.PDF.

Schwarz, A. 2014. *Experiments in Practice.* London: Pickering and Chatto.

Part I

Horizon of possibilities

1 The pyramid and the ring

A physics indifferent to ontology

Peter Galison

The credo of fundamental explanations

In the 1970s there was a confidence among particle physicists, a sense that they were on the verge of cracking the ultimate code. It is easy to understand why. Two dramatically distinct theories – one of ordinary electricity and magnetism (quantum electrodynamics), the other one of the nuclear forces (weak interaction theory) – could be subsumed under a single integrated structure. The photon carrier of light and the carriers of this nuclear force appeared to be different versions of the same thing; the new theory avoided disastrous infinities, and the first experimental confirmations were coming in right on target. Soon, the particle physicists extended this kind of theory to include what used to be called the strong nuclear force, and theorists' experimental predictions began rolling in at a fast and furious pace, on both sides of the Atlantic.

The sense that unified field theory was at last on the right track led to a breathless enthusiasm throughout the field. Within a few years, certainly by 1983, the Standard Model, as it had come to be called, was so well matched to observation that it became the background knowledge against which all other particle experiments were calibrated. At the core of this theoretical assemblage lay a few structureless, fundamental particles that carried mass, alongside a few particles that carried force. Aside from gravity, all the forces were really versions of the same thing – and the objects of the world, its fundamental constituents, could be listed in short order: a few particles like the photon, Z boson or gluon carried force, and a small number of other particles like the electron or quarks carried mass. The then elusive but now observed Higgs particle was responsible for splitting the forces at the low energies of our everyday life – cleaving electrodynamics from the weak force, for example.

It made sense for both physicists and science journalists to produce books with titles like *From Atoms to Quarks* (1980) or *Inward Bound* (1986). It became plausible to insist, as Steven Weinberg did, that even if physicists could not calculate from the fundamental theories, by dint of too much complexity, the everyday properties of biological or even ordinary materials, all questions eventually would come back to the fundamental entities and the laws that governed their interaction (Trefil 1980; Pais 1986; Weinberg 1992).Why is glass clear? Because in the visible spectrum, there are no atomic transitions to allow absorption. Why are electron orbitals what they are? Because quantum mechanics, applied to the interaction between electrons and nuclei in silica, forbid such transitions. What determines these basic interactions between electrons and nuclei? They are determined by quantum electrodynamics, a specification of the electroweak theory. Pick a problem of ordinary matter and, it appeared, a similar train of rainy-day questions will eventually drive us back to the unified quantum field theory: radioactivity, nuclear structure, chemistry, biology. This was a fundamentalism not of prediction but of explanation.

The pyramid and the ring

True, with Weinberg's interesting take on reduction, different questions would pick out different paths for the explanatory arrows, but in the end they all would converge at the pinnacle of a vast explanatory pyramid of knowledge. And if, in the 1980s, the precise form of the generalization of the electroweak theory to include the strong interactions (the totality known as the Grand Unified Theory, GUT) wasn't clear, it was imagined that it soon would be – maybe the final form would be the simplest solution of all, the 'minimal SU(5)', maybe a bit more complicated variant. If not quite in view, the peak of a single vast structure seemed to loom near, its peak hidden by clouds for now but not for long. Ending physics with a set of fundamental entities or fundamental laws or fundamental explanations sealed the deal: one more generalization of gauge theory, one more step toward finding the first of all particles, and we would be there.

Within microphysics, at least, that meant a full list of entities, an ontology coupled together with a set of laws that described the entities' fundamental interactions. By the mid-1970s, neutrinos and positrons were not only universally accepted as members of the narrow set of truly elementary particles; they had become tools. Physicists routinely used neutrino beams as probes in experiments; medical technicians used positrons to diagnose and treat disease. Sure, there were debates, but they took place at a moving edge of high energy: Was there a particle like the electron only many times heavier? Yes. Then came a third quark, then a fourth, a fifth and finally a sixth. These formed the building blocks of every nucleus: up, down, strange, charm, top and bottom. There was grumbling, of course. Condensed matter physicists, foremost Phillip Anderson, militated against the overbearing wealth and power of elementary particle physics – he disliked what he considered the disproportionate draw the particle physicists exerted over graduate students and government funding. He opposed the founding of the accelerator laboratory (Fermilab) in the early 1970s, and objected to the Superconducting Supercollider in the 1980s and 1990s. He understood early and deeply that particle physics joined its claim to temporal authority to its claim to be the branch of physics uniquely charged with the pursuit of the fundamental (Galison 1993).

Anderson's was a strong voice – a Nobel Prize winner, he eloquently and powerfully testified to Congress against the Superconducting Supercollider. His countervailing watchword was made famous in an essay of that name: 'More Is

Different' (Anderson 1972).[1] For Anderson, there are emergent phenomena, phenomena not contained in the properties of individual particles taken one by one (or one against one) altogether as 'fundamental' as anything offered by particle physics. Genuinely new things, he argued, are to be found in complex configurations of matter. Superconductivity – current flowing without any resistance in very cold conductors – was in Anderson's view a good example of something involving bulk matter that was altogether as fundamental as the way an electron scattered from a photon. For many condensed matter physicists and their philosophical allies, the universe was more quilt than quark. Fundamentality existed; it was just that fundamental fruit did not only grow in the garden of particle physics.

Looked at from afar – as is perhaps useful here – the drive to ontology has been a governing matter for a good long time. James Clerk Maxwell's followers were persuaded that the world was nothing more than a world-ether, a universe-spanning substance whose states made up all other objects (Galison 1983). Twisting, spinning, smoke-ring-like vortices of ether were really what made up charged particles. Electrical and magnetic forces, matter itself – subtract ether and you had nothing at all. Then came the Dutch theorist, Hendrik Lorentz, in many ways the first of a long sequence of charismatic theoretical physicists, who pushed for a dualistic world – ether, yes of course, but also fundamental charged particles that were not explained or built from ether. Charged particles altered the ether; the ether, in turn, told the charged particles how to move.

One or many, no- or anti-ontology?

Now, within science and within philosophy too, there has long been tradition of anti-ontology, a view that science should be more about prediction than the establishment of fundamental entities. The sixteenth-century theologian, Andreas Osiander, whether he personally believed his preface to Copernicus's *De Revolutionibus* or not, used his introduction to shift attention away from the reality of heliocentrism. The scientist-philosophers of the Vienna Circle fulminated against ontology. In fact, most of the Vienna Circle avoided any talk of ontology at all (an exception comes in the essay "Empiricism, Semantics, and Ontology" by Carnap in 1950, but that is another story). Henri Poincaré, the turn of the twentieth century French scientist-philosopher, militated for a science predicated on real relations ('*les rapports vrais*') rather than the 'truth' of objects in and of themselves. More recent physicists too have had moments where they wanted to hold ontology at arm's length: Murray Gell-Mann, one of the physicists who introduced the world to the idea of quarks, at first clearly held them to be calculational, classificatory entities, not to be taken as physically really real; later he became persuaded that they were to be taken seriously as physical entities (Pickering 1984; Pais 1986; Gell-Mann 1994).

But whether one is aiming toward the ontological or pointing one's arrow away to avoid it, ontology has been taken to organize scientific reason. The history of physics – and its philosophical re-telling – has often been structured around an historical ontology: vacuum against plenum, phlogiston against oxygen. It has been considered necessary, even central, to pick out the building-block objects of a theory. So too is it in the social sciences. Are there groups, or are they explicable only as aggregated individuals? Are there socioeconomic classes as such that merit being considered as constituent, even fundamental objects of political economic inquiry? What, we ask when we encounter a new people or new scientific arena, is the roster of allowable entities?

I ask such questions not from some extra-historical standpoint – many of us in the broad set of science and technology studies fields have been engaged in this project. Thomas Kuhn took the paradigm to be an exemplary way of showing how ontology, nomology and epistemology were to be prosecuted by way of an example. Each paradigm carried its own ontology: Newton and the Newtonians had one set of objects; Einstein and the Einsteinians another. Indeed, nothing exemplified the paradigmatic switch better than the switch in ontology (time as absolute, for example, to relativistic time). Ian Hacking has used his notion of historical ontology to good effect. Philippe Descola returns again and again to the ontology of different cultures – of which he takes there to be four: animism, totemism, analogism and naturalism (2013). My own work on inter-languages and trading zones depends on the tracking of shared objects in the local configurations of language and action: Two scientific cultures might disagree on fundamentals, but still come to accord about a shared set of properties and dispositions for certain objects (Galison 1997).[2]

But I want here to talk about something that is happening now, something that in my view is quite unprecedented in the history of the physics: a shift not from one set of objects to another – not even a switch from the ontological to the anti-ontological (this would not be news, and in fact the anti-ontological is, of course, always ontological). No, over the last few years, something has begun to change in the sciences themselves, not toward a by-now familiar positivism, but toward what one might call the anontological: an indifference to the ontological. My claim is that a broadening set of scientific activities aims not to find out the most basic entities in the world; nor does it struggle to formulate a way to avoid such a roster of fundamental things. Not one ontology, not many ontologies, not an ontology of no ontology. Not atheism, not even agnosticism toward ontology – instead, a thoroughgoing unconcern about existence as an attribute. For emphasis, surely most circuit and app designers using global positioning technology know that their work depends on quantum mechanics and even general relativity. But if you asked them what they thought about debates over how to reconcile quantum mechanics and relativity at the edge of a black hole, I would venture that most would be, understandably, utterly indifferent. They are not withholding judgment – struggles over the 'information paradox' are, quite simply, skew to the sum total of their hardware, software, theoretical and even financial work.

I want to describe a world in which key branches of physics and related sciences no longer see the objects of their concern as a pyramid, but instead see their universe as a ring – a connected space of knowledge, but one that is connected without center.

In the middle, nanofacture – ethos of science, ethos of engineering

The mesoscopic world of nanoscience is one caught between poles. It is in just that region of sizes and energies that puts it at the boundary of macroscopic, classical physics on one side, and a small scale in which quantum physics holds well. It is a region touching the scientific at one extreme (atomic physics, surface chemistry, virology), and the much more applied (chemical, electrical and mechanical engineering) on the other. In different ways and idioms, if one distilled a common exchange, it might look like this:

CRITIC: "What you are doing is engineering not physics, drilling holes, making devices. . .these are jobs for commercial technicians not scientists."

NANO: "I want to make things, to construct devices at the smallest possible scale – a manufactured atom, an atomic-sized transistor, an instrument-probe at the atomic scale. The behavior, the manufacture of matter at the meso-scale, between quantum and classical realms, is the most interesting of all structures. I am concerned about robustness and scalability – not whether something is 'real'. Ontology is simply irrelevant to me."

For a particle physicist of the 1960s or 1970s, nothing was more important than existence questions: Did Omega Minus really exist? If it did, that meant the quark classification system was on the right track. Were weak neutral currents real, or were they nothing more than an artifact produced by ordinary and uninteresting neutrons? That early 1970s question was crucial: If they were real (produced by a heavier version of the photon), the new unified electroweak theory was truly a harbinger of a new epoch of physics. Images – bubble chamber images most strikingly – could establish existence. The omega minus bubble chamber image is an icon of physics; a senior German bubble chamber physicist waved the first image of a candidate weak neutral current electron over his head as he joined his colleagues.

Intriguingly, in the nanoscientific domain, images are still, perhaps more important than ever in the history of science. But very often the plethora of image stations in a nano-lab are not there to look, after the fact, at images in order to establish whether a new kind of entity is real. Instead, the images function in real time as part of what one might call 'nanofacture': the cutting of DNA strands or carbon nanotubes, the attachment of nanodots to other circuit elements. The nano-lab is making, altering, combining things, and imaging is a way of getting the process done. Here a conjoint ethos: making objects of genuine scientific concern and doing so within a frame of engineering.[3]

In the end there is nothing behind the screen

One sign and instance of indifference toward ontology is in the world of simulations. In the early 1950s, physicists struggled to figure out what these 'Monte Carlos' were.

Consider a simple example: To model a marble bouncing through a pinball machine, start with a position and velocity and, at regular intervals, throw a pair of dice. If a one, scatter at 60 degrees; if a 2, scatter at 120; a 3, at 180; and so on. By modeling such random events, a huge amount can be predicted about the natural world, even where no overarching theory exists.

If we numerically simulated the dispersion of a gas, was it because the random numbers used in simulations were 'like' the random collisions of molecules? That is, were numerical simulations a deeper replication of stochastic phenomena in the world than deterministic theory could ever be? Or, as others argued, were these simulations like experiments, where different tries would never quite generate the same result? For quite some time, many physicists and astrophysicists thought of these simulations as something preliminary, a surface solution that would only be completed through the deeper considerations of analytic theory that alone could truly bring 'understanding' (Galison 1997). Such Monte Carlo methods spread (e.g., calculating integrals by randomly sampling the value of a function) to a very wide range of domains – from chemistry and nuclear physics to the calculation of galaxy formation and fluid dynamics.

But the intrusion of such numerical simulations into the heartland of the physical sciences also carried with it a long-term battle, in part generational, that marked a shift from seeing simulations as an indication but not true science – to the pragmatic attitude that counts simulation as a way of reaching a goal. Again, if one imagined the following idealized exchange between critic and defender, it might go like this:

CRITIC: "All you simulators have done is preliminary – real physics is to show just the term in the physical law that explains what has happened. Only then can we point to a term in the equation and say *that* is the real cause of the phenomenon in question."

SIMULATOR: "If we can start with Newton's inverse square law and some simple assumptions about the distribution of matter, and if we can show that the galaxies in fact do form into a spiral or lens shape by means of a simulation using a Monte Carlo, then we are done! We have shown that some basic processes, repeated over and over, produce one of the great and dominant visible structures of our universe. All your analytic techniques are just antiques, a poor man's crutch useful before computers, but hardly the sacred core of science. We simulators are interested in prediction, in deriving structures through calculation. But we are utterly indifferent to whether the Monte Carlo calculational processes capture the reality of the phenomena. I want the results to match what we see through our telescopes. Whether the world is random in just the way that my pseudo-random number generator is random simply does not matter. I am indifferent to the ontology of the elementary bits along the way. I do not care at all if it is possible, in some labored and far future mathematical inquiry, to find an analytical solution. Ontology is irrelevant.

In the beginning, there is no beginning

At first, string theory seemed to promise the pyramid of all pyramids: an account that would unify beyond the wildest expectations of the most ambitious particle physicists. The latter had wanted to show that the weak, the strong and the electrodynamics forces all derived from a single 'ur-force'. The photon, gluons, the W and the Z were just low energy recombinations of a more originary set of force-carrying particles. String theory promised more. Instead of thirty or so free parameters, string theory would offer none – a theory so constrained by mathematical consistency that there was nothing left to be fixed by experimental knowledge. String theory would join gravity to particle physics and end the historical project of inquiry into the building blocks nature. In the beginning, so the answer would go, there were extended, one-dimensional objects held under a tension producing load of some 10*39 tons, whose excited states constituted the particles that for decades we have taken as fundamental.

For this project to work, there would have to be a law of physics, *the* law of physics. And for a time – from what physicists called the first string revolution in 1984 through the second revolution of 1995 and even a few years beyond – it seemed possible. As Weinberg insisted in many places, one day physicists could wake up to find that someone had actually written down the right Lagrangian – the law governing the physics of strings. And from that law (it was hoped) would come all else: the masses, charges and other characteristics of the electron and its heavier relatives, neutrinos, quarks and the particles that bound them together. From those objects could, in principle (as the saying goes), be derived the features of chemistry, biology, planets, solar systems and galaxies.

It is not that string theory had been unopposed. Experimentalists had long derided it for its lack of contact with laboratory results. 'Theology', Sheldon Glashow called it, better studied on Divinity Avenue than in the Jefferson Physics Laboratory. String theorists responded that this was in fact science, constrained it was true not by accelerator results but by mathematical self-consistency. Even theoretical particle physics blasted the theory – for people like Howard Georgi, string theory had lost the back and forth between theoretical ideas and experimental results that had made gauge theory so successful back in the 1970s. No physical theory – no scientific theory – had ever achieved the remarkable correspondence between prediction and result that gauge theories had provided. Why give up this happy collaboration? Ah, the string theorists responded, good science was not always done in this way – think of Einstein ferreting out the equations of relativistic gravity with no more to go on than the precession of the perihelion of Mercury. Should Einstein have given up because he couldn't match results with his pals in the precision laboratories?

But now, in these last few years, comes a third stance toward string theory because it has become clear that even if there is a single governing set of equations – the world equation – there are a lot of solutions. A vast amount of solutions – on the order, some estimate of 10^{100} solutions (a googol) maybe more. The question then arises: Which one is ours? Imagine a hugely complex system of mountains where every valley, peak, or inflection point represents a solution. Every one picks out a particular set of particles and forces. Which is our universe? Maybe there's a principle that would pick out ours, but maybe there isn't. It is this latter possibility that has split the string theory community itself right down the middle – right down the Continental Divide in the United States, with many of the West Coast physicists opting for what has been called the landscape and the East Coast lobbying hard against it.

Here's the idea

String theorists hoped to have one equation and one solution; instead they got many with no selection principle on the basis to choose our real universe's actual values of things like the ratio of the electron to the proton. Maybe, say the anthroposophs, all of the possible solutions are, in fact, realized. Now most all of them could not support galaxies, could not make higher elements and in particular could not generate life – and us. "So what?" they say. We have the right values of the ratio of the mass of the electron to the mass of the proton (e/p) not because anyone or anything made them for us but because all the possible values of e/p are in fact in use, and our presence in the universe means we live in a universe where a value of e/p allows life. After all, are you surprised that your great grandparents all survived the flu pandemic of 1918? You shouldn't be. If they hadn't, you wouldn't be here. Are you surprised that we live in one of the solution spaces that include electrons and protons that are compatible with life? *Mutatis mutandis* – you shouldn't be.

Now the anthroposophs have enemies. The East Coast of the Continental Divide, with some West Coast allies, detests this form of us-centered argumentation. Ed Witten of the Princeton Institute for Advanced Study finds the prospect of such anthropic account exceedingly depressing; so does his colleague Juan Maldecena, Harvard's Andrew Strominger and many others. Stanford, by contrast, is anthropic HQ: Leonard Susskind, also one of the founders of string theory, along with his colleagues Andre Linde (known for his work on cosmic inflation) and particle phenomenologist and model builder Savas Dimopoulos, all see the anthropic turn as an incipient scientific revolution of the first order.

David Gross, director of the Kavli Institute for Theoretical Physics (Santa Barbara) puts himself into the Pacific Time Zone's column of resistance: he sees the effort as, in the first instance, not physics. "Look," he says, "imagine we were back in 1940, facing the theoretical mess and sprawling disconnected data associated with nuclear physics." We could well be thinking that the myriad nuclear structure relations were miraculously tuned to make life possible. Someone clever could easily have invoked the anthropic principle to explain these fortuitous relations. But they aren't fortuitous at all, at least not at the level of nuclear physics; they are the explicable consequence of a deeper, underlying theory that explains what the protons and neutrons are and how they bind together. That theory – the theory of quarks and their glue, the 'gluons' – arose thirty or so years later: quantum chromodynamics. Why not expect something of this order with string theory? Sure, we do not know what it might be yet, but why throw in the towel because we can't solve the problem (Gross 2008)?

In July 2005, all the key leaders of string theory gathered for a major meeting (Shenker 2005), and no topic was as hotly contested as the anthropic principle. One after another, the lead panel weighed in – and split their judgment down the middle, four against four. Then Steve Shenker addressed the question to the very large audience. "I want to ask you I guess as a snap question: Do you think that the smallness of the cosmological constant will be explained by the anthropic principle or by a physical principle? That there will be a large landscape and . . . it will rely on some sort of environmental variable?" (Andy Strominger interjected: "We're talking about the year 3000!"). The audience voted, and Shenker exclaimed, "Wow, holy shit! 5:1 for a physical explanation! . . . That's it. The anthropic principle is out of office!" He then asked, "How many people think God made the universe just for us?" Very few hands went up. Then someone from the audience said, "I think God made the landscape" (Shenker 2005).

On the anthropic side there is clearly a putting aside of the original hope for a theory that would offer the one true answer to what there was in the universe. Intriguingly, even among those who are not particularly sympathetic to the anthropic program and who are optimistic about string theory, there is a growing sense that the claims for a 'theory of everything' were overblown. Strominger, for example, has great hopes for string theory – he thinks that we will, in fact, come to understand many of the current puzzles about the theory. But he does not think that string theory now – or in the future – is all of physics. Instead, he sees it as a fascinating corner of the field, one that will productively deepen both our understanding of quantum field theory and of the deepest mathematical structures.[4]

Strominger's view – that string theory was a generative 'corner of the field', but hardly the end of physics, has become increasingly a view even of those who reject the anthropic assault on the fundamentalist position. John McGreevy, a theorist at MIT, takes as his starting point the idea that there are two ways to advance and connect string theory with the rest of physics. One is to use string theory to look at the smallest possible distances, and then to use the results to explain very basic problems of physics (e.g., Does the topology of spacetime change?). When advanced students at MIT come to his class on 'applied string theory', however, they learn through another approach, one increasingly in favor among physicists: use of a relation discovered in string theory, called the 'holographic principle' (or more technically AdS/CFT), that shows two theories can be equivalent – one in a higher dimension and the other essentially defined as a kind of projection onto its boundary.

Here's how McGreevy's class began in the fall of 2008:

> So here's my crazy plan: we will study the AdS/CFT correspondence and its applications and generalizations, without relying on string perturbation theory.
>
> Why should we do this? You may have heard that string theory promises to put an end once and for all to that pesky business of physical science. Maybe

something like it unifies particle physics and gravity and cooks your breakfast. Frankly, in this capacity, it is at best an idea machine at the moment.

But this AdS/CFT correspondence, whereby the string theory under discussion lives not in the space in which our quantum fields are local, but in an auxiliary curved extra-dimensional space . . . is where string theory comes the closest to physics. . . .The role of string theory in our discussion will be like its role in the lives of practitioners of the subject: a source of power, a source of inspiration, a source of mystery and a source of vexation.

(McGreevey 2008)[5]

Actually doing string theory, in McGreevy's view, is not about the hunt that once seemed to be closing in on its quarry: a single, final equation that provided the last station of the physicists' two thousand-year-old voyage toward the innermost, deepest fundamental level of physical reality. Instead, the view that is emerging is that string theory has led to a series of mathematical clues. Among these are the linking principles, like the 'holographic principle', which says that theories can be equivalent, even if one looks enormously different from the other. Here is a theory that says we can learn about gravity, black hole entropy and much else besides by looking at the very 'ordinary', experimental phenomena of condensed matter physics, or the hitting of one big fat ion against another in accelerators like the one at Brookhaven – seemingly far from the cutting edge of CERN, much less the staggering energies one would think necessary to look at string theory directly.

Applied string theory is an account of physics that links domains, but not by solving things at the 'most fundamental' and then deriving reality down the food chain: quarks and leptons, nuclei, atoms, molecules, ordinary matter, people, worlds and galaxies. Our slogan recurs: fields linked, but not hierarchically. What is surprising is that it recurs at the heart of a long tradition that just about defined the ontological lineage of modern science: molecules, atoms, nuclei, nucleons, quarks, strings. So here again, imagine an idealized exchange of the type we fastened on earlier. The first, without exaggeration, is precisely the kind of exchange that occurred in the mid-1980s; the others draw from more recent struggles (Galison 1995).

CRITIC: "Physics without experiment is theology, not science."

STRINGER (CA. 1985): "No! Physics without experiment is constrained by mathematical consistency. Maths is the new experiment."

CRITIC: "Physics that justifies itself by mathematical yield is not science; it has abandoned the historical mission of hunting for fundamental entities."

STRINGER (2010S): "No! String theory is one science among others, with the mission of building structures that offer insight into both physics (at a variety of scales) and a wide swath of mathematics from algebraic geometry to knot theory."

CRITIC: "Physics that does not predict, that relies on evolutionary explanation is not physics."

STRINGER (LANDSCAPE VARIETY): "No! Like Darwin, this is a new and revolutionary modesty, knowing when not to explain is just as important as knowing what to explain. And even if we want physical laws all the way down, so to speak, they bear relations of duality, not hierarchical ordering to a range of other parts of physics, from condensed matter to heavy ions smashing into a quark soup at accelerators around the world."

An engineering way of being within the sciences

We are used to thinking of a fight to the finish between two positions grounded at the intersection of physics and philosophy. On the one side is an absolute fundamentalism that grounds the connectedness (unity) of science in a hierarchically centered world that imagined a pyramid, with atoms or elementary particles or strings at the high center. This view captures very well the enormous weight put upon the discovery of specific objects – indeed we celebrate and periodize the discovery of the electron, the neutron, the quarks, the particulate photon and a host of other entities just because the building-block picture of the sciences is so alluring.

On the other side stands a rebellious alternative that sets many disciplinary domains as equal, each just as valid as the other, but disconnected, more a quilt with weak stitching than a pyramid. Here, scientists and philosophers emphasize emergence not reduction; they seek to establish the ontological autonomy of different domains based on complexity, life or laboratory methods.

But what may be emerging in the twenty-first century – and I am by no means insisting that everyone subscribes to this view – is an image of science skew to the perpetual oscillation between fundamentality and autonomy. It is a view that differs from fundamental ontology because it refuses a center. But it simultaneously differs from the autonomy (quilt or island) view because it holds a wide variety of sciences to be connected, though without a governing core. A focus on novel effects, materials, and objects, but constructed through an engineering way of being that values the making and linking of structures with little regard for the older fascination with existence for its own sake. We may be witnessing the arrival of a different kind of science, inflected by making but deeply imbricated in the sciences: linked sciences, but formed into a ring by a broad and expanding consensus leaves the pyramidical hierarchy in the sands of Giza.

Notes

1 See also U.S. Congress, Senate, *Joint Hearing before the Committee on Energy and Natural Resources and the Subcommittee on Energy and Water Development, Importance and Status of the Superconducting Supercollider*, 102nd Cong., 2nd Sess., June 30, 1992.
2 For more on trading zones, see, e.g., Strübing et al. (2004), Gorman (2010) and Balducci and Mäntysalo (2013).
3 For more on this argument, see Daston and Galison (2007), especially ch. 7.
4 Overbye (2005) reports on theorists trying democracy by voting with a 4:4 split on the panel and the audience voting overwhelmingly for the non-anthropic principle. On the panel were Raphael Bousso (UC Berkeley), Shamit Kachru (SLAC & Stanford), Ashok Sen (Harish-Chandra Research Institute), Juan Maldacena (IAS, Princeton), Andrew Strominger (Harvard), Joseph Polchinski (KITP & UC Santa Barbara), Eva Silverstein (SLAC & Stanford) and Nathan Seiberg (IAS, Princeton).
5 The reason: it offers otherwise-unavailable insight into strongly coupled field theories (examples include QCD in the infrared, high-temperature superconductors, cold atoms at

unitarity) and into quantum gravity (questions about which include the black-hole information paradox and the resolution of singularities), and through this correspondence, gauge theories provide a better description of string theory than the perturbative one.

References

Anderson, P. W. 1972. 'More is different: Broken symmetry and the nature of the hierarchical structure of science.' *Science* 177 (4047): 393–396.

Balducci, A. and Mäntysalo, R. (eds.). 2013. *Urban Planning as a Trading Zone*. Heidelberg, New York and London: Springer.

Carnap, R. 1950. 'Empiricism, semantics, and ontology.' *Revue Internationale de Philosophie* 4: 20–40.

Daston, L. and Galison, P. 2007. *Objectivity*. New York: Zone Books.

Descola, P. 2013. *Beyond Nature and Culture*. Chicago: University of Chicago Press.

Galison, P. 1983. 'Re-Reading the Past from the End of Physics: Maxwell's Equations in Retrospect.' *Functions and Uses of Disciplinary Histories*, edited by Graham, L., Lepenies, W. and Weingart, P., 35–51. Dordrecht: Kluwer.

———. 1993. 'Metaphysics and Texas.' Review of *Dreams of a Final Theory* by S. Weinberg, *The New Republic* 6 (September): 41–42.

———. 1995. 'Theory Bound and Unbound: Superstrings and Experiment.' In *Laws of Nature: Essays on the Philosophical, Scientific and Historical Dimensions*, edited by Weinert, F., 369–408. Berlin and New York: Walter de Gruyter.

———. 1997. *Image and Logic: A Material Culture of Microphysics*. Chicago: University of Chicago Press.

Gell-Mann, M. 1994. *The Quark and the Jaguar: Adventures in the Simple and the Complex*. New York: W.H. Freeman and Co.

Gorman, M. E. (ed.). 2010. *Trading Zones and Interactional Expertise: Creating New Kinds of Collaboration*. Cambridge: MIT Press.

Gross, D. 2008. 'Einstein and the Quest for a Unified Theory.' In *Einstein for the 21st Century: His Legacy in Science, Art, and Modern Culture*, edited by Galison, P., Gerald Holton, L. and Schweber, S. S., 287–298. Princeton: Princeton University Press.

McGreevey, J. 2008. Notes from class at MIT: 8.821 (Fall), http://web.mit.edu/~mcgreevy/www/fall08/index.html.

Overbye, D. 2005. 'Lacking hard data, theorists try democracy.' *New York Times*, 2 August, http://www.nytimes.com/2005/08/02/science/lacking-hard-data-theorists-try-democracy.html (accessed October 22, 2016).

Pais, A. 1986. *Inward Bound: Of Matter and Forces in the Physical World*. Oxford: Clarendon Press.

Pickering, A. 1984. *Constructing Quarks: A Sociological History of Particle Physics*. Chicago: University of Chicago Press.

Shenker, S. 2005. The Next Superstring Revolution Panel Discussion. Toronto Strings 12 July 2005, http://www.fields.utoronto.ca/audio/05–06/strings/discussion.

Strübing, J., Schulz-Schaeffer, I., Meister, M. and Gläser, J. (eds.). 2004. *Kooperation im Niemandsland: Neue Perspektiven auf Zusammenarbeit in Wissenschaft und Technik*. Opladen: Leske & Budrich.

Trefil, J. S. 1980. *From Atoms to Quarks: An Introduction to the Strange World of Particle Physics*. New York: Scribner.

Weinberg, S. 1992. *Dreams of a Final Theory: The Scientist's Search for the Ultimate Laws of Nature*. New York: Pantheon.

2 Cancer stem cells

Ontology matters

Lucie Laplane

Cancer stem cells are the new targets of the 'war on cancer' launched by Richard Nixon in the 1970s (Lenz 2008). After forty years and the $105 billion spent by the National Cancer Institute alone, cancers are still there, killing thousands of people every day.[1] Despite undeniable progress, the comparative failure of oncology has put the whole therapeutic classical enterprise under suspicion. In this context, any new therapeutic strategy would be welcome and revive the expectation of 'breaking the cancer war stalemate' (Haber et al. 2011). Thus, when the role and existence of the so-called cancer stem cells was revealed to the public as a potential therapeutic strategy against cancers, it quickly gained attention from researchers and the public. Cancer stem cells (CSCs) are cancerous cells with 'stemness' properties. According to the cancer stem cell model of carcinogenesis based on accumulating evidence, CSCs are the only tumorigenic cells of cancers. In this view cancer is initiated, developed and spread by the CSCs and never by any other cancerous cells. From this model of carcinogenesis, biologists draw two inferences: first, the relapses after apparently successful therapies are due to resistance of few CSCs; second, it would be necessary and sufficient to kill those CSC in order to cure cancer (Reya et al. 2001; Clarke et al. 2006).

Thus cancer stem cells appear as ideal targets for launching a second war on cancer. But precisely aren't they too ideal to be true? This chapter will first describe how cancer stem cells came into existence and briefly outline the general theory that supports the promise of targeted cures of cancer. Then it will question the ontological assumption underlying the prospect of such targeted cures. The view that cancer stem cells are a special category of discrete entities can be challenged for a number of reasons. Alternative modes of existence for cancer stem cells can be – and are increasingly being – envisaged. There is a great variety of stem cell types, and subtypes have proliferated over the past decades: toti-, pluri-, multi- and unipotent stem cells, followed by the technological reproduction of in vitro stem cells, such as 'induced pluripotent stem cell' (iPS cells) or 'embryonic stem cells' (ES cells). To what extent these proliferating varieties fit in the standard definition of stem cells? I will argue that the ontological status of stem cells is a crucial issue that deserves research attention. How to define 'stemness' is not just a speculative question of interest for philosophers, but also a decisive conceptual tool for shaping therapeutic strategies against cancers.

The elusive identity of cancer stem cells

The biology of cancer is highly complex. It gathers together heterogeneous diseases affecting various parts of the organism at various levels of organization (genetic and epigenetic, molecular and cellular) and involving a multitude of genes and molecular pathways. Because of this diversity, oncologists had to give up the expectation of designing *the one* miracle cancer cure. Brain cancers are very distinct from blood cancers, among which acute myeloid leukemia is different from chronic lymphoid leukemia, which themselves can result from various deregulations of normal functions. There is thus very little hope to find a universal cure.

The standard therapeutic protocols, surgery apart, involve anti-mitotic agents. These agents kill cells in division by inducing damages in their DNA (which result in various side effects like hair loss, digestive disorders and blood aplasia). This therapeutic strategy came from the apparently common problem of proliferation involved in every cancer. Indeed, cancers cells are assumed to escape the steady state of the organism and outgrow. Thus, cancer cells have been thought to divide a lot. But this was proven wrong in the 1950s: cell division appeared equivalent in cancers and in steady-state tissues. Many cancer cells were found to be not dividing (Astaldi et al. 1947; Killmann et al. 1963). This observation has been explained by two rival hypotheses: either all cancer cells have the ability to divide but do so at different stochastic rates ('self-maintaining system hypothesis') or most cancer cells do not divide but there is a pool of stem cells that feeds the cancer cells population ('stem cells hypothesis') (Gavosto et al. 1967; Clarkson 1969). In the case of leukemia, the latter hypothesis has gained support from clonogenic studies.

The model role of leukemic and hematopoietic stem cells

The emergence of the cancer stem cells hypothesis is not alien to the growing public fear of nuclear power. Although radioactivity has been used for therapeutic purpose with the creation of radium institutes all over the world, the damages caused by radiation became a serious concern after World War II. In the nuclear era, radiations caused by nuclear wars, by fallout from nuclear weapon testing or by accidents in nuclear power stations generated public fear. The first cause of death after non-directly lethal radiations was blood depletion. Thus, a number of biologists actively searched for a way to rescue the hematopoietic system by transplantations of bone marrow cells (Lorenz et al. 1951; Ford et al. 1956).[2] In vitro and in vivo assays were designed to determine the capacity of a single cell to produce an entire population of cells, called a 'clone' (Till and McCulloch 1961; Becker et al. 1963). Later on, such assays resulted in the description of a hierarchy of cells with distinct clonogenic abilities in both normal blood system and leukemia (Till and McCulloch 1980; cf. McCulloch and Till 1981; Griffin and Lowenberg 1986). At the top of the hierarchy are the so-called 'hematopoietic stem cells' and 'leukemic stem cells', respectively. These cells are highly able to produce clones containing different kind of cells. But it rapidly appeared that the most immature cells detectable by these assays were not the true hematopoietic and leukemic stem cells.[3]

The identification of hematopoietic stem cells became even more important in the 1990s, with the expectations raised by gene therapies to cure blood genetic diseases caused by single-gene defects. Indeed, the aim of gene therapies is the 'permanent correction of genetic deficiencies of the hematopoietic system' (Larochelle et al. 1995, 163). It requires transducing the right target, which is the hematopoietic stem cell. Otherwise, the disease is likely to reappear (Larochelle et al. 1996). The development of three technologies played a major role in this race to the identification of hematopoietic stem cells: production of strains of immunodeficient mice in which xeno-transplantations of human cells were possible, FACS technologies (Fluorescence-Activated Cell Sorting) and production of monoclonal antibodies. The latter two, once combined, allow cell sorting according to their cell surface antigens, without killing them. Immunodeficient mice played a central role in the process of modeling and studying blood cancers, HIV and immune disorders (Shultz et al. 2007; Belizário 2009). A lot of strains were rapidly produced, among which the NOD-SCID mice (non-obese diabetic-severe combined immunodeficiency) played a major role in the identification of human hematopoietic and leukemic stem cells (Dick 1996).

FACS technology is a smart technology developed in partnership with Becton Dickinson Electronics on the model of ink jet printing technology, which is able to generate droplets and change their direction (Herzenberg et al. 1976). Its development has benefited from two impetuses of automation. The first one was the need of automation to standardize clinical-diagnostic practice. The second one came from the NASA research in exobiology (Cambrosio and Keating 1992b; Keating and Cambrosio 1994). In combination with the production of monoclonal antibodies from the hybridoma technology developed by immunologists (Cambrosio and Keating 1992a; Keating and Cambrosio 1994), FACS can sort hematopoietic and leukemic cells according to their antigens. Immunologists (e.g., Irving Weissman's group at Stanford)[4] and hematologists (e.g., John Dick's group at Toronto) used it for the research of the hematopoietic stem cell. They sorted the cells, transplanted them into lethally irradiated immunodeficient mice and then studied their ability to rescue the hematopoietic system. From these studies emerged the first phenotypic characterizations of murine and human hematopoietic stem cells (Visser et al. 1984).[5] Identification of leukemic stem cells with the same phenotypic characteristics ($CD34^{+}CD38^{-}$) followed rapidly (Bonnet and Dick 1997). The protocol for cell sorting and transplantation in immunodeficient mice promptly became the gold standard for the isolation and characterization of any kind of 'adult' or 'somatic' stem cells.

The cancer stem cell theory

Leukemic and hematopoietic cells helped shape a general cancer stem cell theory based on three distinct and complementary assumptions. First, cancers are hierarchically organized: they are initiated, developed and propagated exclusively by a specific sub-population of cancer cells, the so-called 'cancer stem cells' (CSCs); second, CSCs explain relapses after apparent successful cure because they can escape and/or resist therapies, and they are able to initiate a relapse; third, targeting CSCs is necessary and sufficient to permanently cure cancers.[6]

These assumptions renewed the promise to cure cancers, especially when a lot of data in favor of the existence of CSCs were accumulated in various cancers such as leukemia, breast cancers, brain tumors, prostate cancers, ovarian cancers, liver cancers, colorectal cancers, pancreatic cancers, head and neck squamous cell carcinomas, lung cancers and many others.[7] Nowadays, a lot of public and private funding is devoted to the research and development of drugs specifically targeting CSCs. Biotech companies specialized in the R&D of CSC targeting therapeutics have bloomed in recent years, and the big pharmaceutical companies invested a lot in this field. Initial public offerings are extremely successful, showing a real optimism for CSC targeting therapeutic strategy. Conversely, the significant funding of research on CSCs strengthened the field and attracted a lot of biologists, a phenomenon illustrated by the exponential number of publications on CSCs (beginning

Figure 2.1 Off with their heads! The Queen of Hearts running after cancer stem cells. Illustration by the author.

with a few publications per year in the early 1990s and increasing to more than 1000 publications per year after 2005, according to the Web of Science database[8]). Funders and researchers, but also patients and their families in demand of targets to shoot so that the cancer war can be won, could together express their conviction that killing CSCs might solve the cancer problem through the famous words of the Queen of Hearts in *Alice's Adventures in Wonderland*: "Off with their heads" But are they right? Or are they just acting in 'blind fury' like the Queen of Hearts?

Ontology matters

Although CSCs have been an object of research for decades,[9] and were established matters of fact in the late twentieth century, they came into existence as a matter of public concern around the early twenty-first century with other kinds of stem cells like embryonic stem cells (ES), induced pluripotent stem cells (iPS), totipotent, pluripotent and multipotent stem cells. The growing population of stem cells of various kinds raise vexing questions of definition.

Stem cells are classically defined in textbooks as cells that have two properties: the capacity to self-renew and the ability to differentiate into two or more cell types. This definition is meeting growing problems with the profusion of types of stem cells. Several biologists had criticized it for being either too inclusive or too exclusive. Both criteria appeared too inclusive because a number of non-stem cells like lymphocytes can also self-renew (Mikkers and Frisen 2005; cf. Younes et al. 2003; Zipori 2004), and some of them like many progenitors (or transit-amplifying cells) can also differentiate into two or more different cell types (Lander 2009). A possible (and common) response to the problem of inclusivity is to consider that stem cells possess the two properties, whereas progenitors do not self-renew (or not for long anyway), and lymphocytes do not differentiate into different cell types (see, e.g., Seaberg and van der Kooy 2003). But then, we face an alternative problem: Together, these two properties, used to define stem cells, have been criticized for being too exclusive. Some stem cells, such as mammal embryonic stem cells, do not self-renew in vivo (Lander 2009); indeed they belong to transient 'emergent tissues' and are part of the organism for a short period of time (Shostak 2006). Additionally, "there are unipotent self-renewing cells, most notably germ-line stem cells, which most scientists would argue are obvious stem cells" (Mikkers and Frisen 2005, 2715). To overcome this difficulty, one can assume that germinal stem cells, embryonic stem cells and all the cells that do not share the two properties are not real stem cells. However, this position fails to reach a consensus among stem cell biologists.

Another critical issue concerning the ontology of stem cells has been discussed over the past decade. As stem cells nurture the promises of cancer cures and regenerative medicine, it becomes vital to come up with reliable phenotypic characterizations of the various stem cells. The search for a 'stemness signature' is based on two major assumptions. First, the two defining properties of stem cells (self-renewal and differentiation potency) allow a qualitative distinction between stem cells and non-stem cells. This presupposition has met the critics we highlighted previously. Second, the two properties (self-renewal and differentiation potency) are reducible to molecular characteristics. This presupposition relies on a rather simple idea: "Because all SCs share fundamental biological properties, they may share a core set of molecular regulatory pathways" (Ivanova et al. 2002, 601). On the basis of this second assumption, three research groups started a race for the genetic characterization of *stemness*. Ihor Lemischka's group at Princeton compared the genetic profiles of human and mice hematopoietic stem cells and found 283 shared highly expressed genes. They consider some of these genes as constitutive of the stemness 'genetic program' (Ivanova et al. 2002, 604). Douglas Melton's group at Harvard has compared transcription profiles of embryonic, neural and hematopoietic mouse stem cells and found a list of 216 highly expressed genes (Ramalho-Santos et al. 2002). Finally, Bing Lim's team from the Genome Institute of Singapore compared gene expression profiling of embryonic, neural and retinal stem cells. They identified 385 common genes. The latter group compared their data with the two previous groups and found only one common gene: integrin-alpha-6 (ITGA6) (Fortunel et al. 2003). ITGA6 gene codes for the α6 subunit of the α6β4 transmembrane protein, which is by no means specific to stem cells. Indeed, it is primarily found in epithelial differentiated cells (see the Genetics Home Reference of NIH).

Such disappointing results have prompted debates in the stem cell research community. Some researchers questioned the experimental settings without questioning the hypothesis (Burns and Zon 2002; cf. Ivanova et al. 2003; Vogel 2003). In their views, scientists should work harder to turn the "current impressionistic portrait of a stem cell into a more realistic one" (Burns and Zon 2002, 613). In contrast, few members of this community dare venture that maybe "there is no such thing as intrinsic stemness at the molecular level, such that perhaps stemness should be understood as a relational property between cells and their microenvironment generating the functionality of stem cells" (Robert 2004, 1007). To investigate this hypothesis, a number of scientists try to understand the role of the microenvironment in the fate of stem cells. To what extent do environmental factors determine self-renewal or differentiation (Hackney et al. 2002)?

The very project of reducing *stemness* to genetic features raises an additional issue. The two distinctive properties used to delineate stem cells from other cells are not actual properties because stem cells seldom divide; they are often quiescent. Thus, supposing that those two 'fundamental biological properties [. . .] share a core set of molecular regulatory pathways' as suggested by Ivanova et al., and supposing that those molecular characteristic were identified, they would not be sufficient to define or portray stem cells because such a characterization would miss all the non-dividing stem cells. This question calls for further discussion among stem cell scientists.

Last but not least, the standard view of differentiation as a one-way process going from stem cells to differentiated cells, with no way back, has been challenged by cloning first and more by iPS cells technology. The nuclei of mature cells can be 'reprogrammed' by injecting them within enucleated oocytes, from where they could generate an entire organism, like Dolly the sheep and many other cloned animals (Wilmut et al. 1997).[10] The differentiated cells themselves can be turned into pluripotent stem cells by forcing the expression of a few genes through transfection by diverse types of viral vectors (Takahashi and Yamanaka 2006; Takahashi et al. 2007), plasmids (Okita et al. 2008) or merely proteins (Zhou et al. 2009). However, such transformation requires sophisticated experiments and the data are restricted to in vitro conclusions. They do not challenge the one-way differentiation in vivo. But there are now accumulating data challenging this dogma in vivo, too. Some appealing data from the regeneration field demonstrates processes of dedifferentiation before regeneration, although these are restricted to few species (see, e.g., Lo, Allen, and Brockes 1993). In drosophila and mice testis and ovary, spermatogonia and cytocytes do dedifferentiate into germline stem cells under certain circumstances (Brawley and Matunis 2004; cf. Kai and Spradling 2004; Barroca et al. 2009). Chaffer and collaborators have showed that non-stem epithelial cells of the breast can spontaneously dedifferentiate to a stem state (Chaffer et al. 2011). They have also highlighted that this phenomenon is increased in cancers, where non-stem cancer cells can give rise to cancer stem cells in vitro as well as in vivo. The observation of epithelial-mesenchymal transitions has been followed by the observation that they might cause a *de novo* acquisition of 'stemness' (Mani et al. 2008). Breast CSCs have been generated by this process (Morel et al. 2008). In fact, far from being restricted to laboratory experiments, dedifferentiation might occur naturally and be particularly efficient in the context of cancers. It follows from all these discussions about the origin of CSCs that at least some CSCs might come from non-stem cells (Passegue et al. 2003; Krivtsov et al. 2006). This means that non-stem cells could become stem cells, while other investigations suggest that 'stemness' could be the outcome of the cancer process (Rapp et al. 2008; Thirant et al. 2011).

Up to now, the doubts raised by such results have not really undermined the standard view of stem cells. The central dogma of stem cell biology, considering stem cells as stable discrete entities at the top of hierarchical ladder of cells, seems resilient. But altogether these objections suggest that cancer research can no longer dispense with addressing ontological questions, such as determining the type of property that can be categorized as 'stemness'.

A few biologists tackle this sort of issue. Loeffler and Roeder suggest to revise the classical definition: they advocate a shift from a cellular view to a system view, "emphasizing stemness as a capability rather than as a cellular property" (Loeffler and Roeder 2002, 13). In considering stemness as a capability, biologists gradually pay more attention to the environment, which seems to play a major role in the expression or non-expression of the capability. Since the microenvironment has an impact on the action of stem cells (quiescence, asymmetric division, self-renewal symmetric division or differentiation symmetric division), it has to be integral part of a definition of stemness. Lander claims that the standard definition of stemness as self-renewal and differentiation "is not really a property that a cell has independent of its environment." He discusses the example of glucose that can be converted in quite distinct substances, depending on enzymes, temperature, pH, and so on (Lander 2009, 70.2). This led him to the idea that stemness should be considered as a system-level property:

> Stemness is a property of systems, rather than cells, with the relevant system being, at minimum, a cell lineage, and more likely a lineage plus an environment. A system with stemness is typically one that can achieve a controlled size, maintain itself homeostatically, and regenerate when necessary.
>
> (Lander 2009, 70.5)

In a similar vein, a few biologists argue that stemness refers to a (set of) function(s) instead of to an entity. This view is consistent with a very original and interesting thesis that stemness "can be induced in many distinct types of cells, even differentiated cells" (Blau et al. 2001, 829).[11] These new perspectives on stemness – known as the 'cell state' hypothesis – constitute an alternative to the established view designated as the 'entity' theory (Fagan 2013; Robert 2004; Leychkis et al. 2009). In fact, the two rival theories are based on four different metaphysical views of stemness.[12]

1. In the standard established view underlying the entity theory, stemness is the biological fundamental property of particular cell types (the stem cells). Hence stemness is a 'categorical property'.
2. In a slightly different version, stemness could be conceived of a distinctive property of a number of cells. But the property itself is just a *disposition*, which might not be expressed unless the right conditions are met. The microenvironment has to be taken into account to define the property itself. As a result of this 'dispositional ontology', one cannot investigate stemness without studying the extracellular factors involved in its expression and regulation.
3. In a radically different metaphysical assumption, stemness can be considered as the result of the interactions between a cell and its specific microenvironment. In this 'relational ontology', stem or non-stem are states of cells depending on interactions. In other terms, the 'niche' can induce the stemness property in non-stem cells.
4. Finally the cell development could be conceived as a continuum, in which cells can occupy different functional roles depending on their particular states. In this 'system ontology', every cell in a multicellular organism can go through all states, including the stem state. The probability of being in the stem state declines as cells differentiate, but there is no sharp boundary between the states. In this perspective, stemness is a systemic function of some cells in a given system. Stemness is maintained at the system level. That is, at any time, some cells of the system express stemness, but cells that express stemness can change through time. Stemness can be maintained through different kind of factors such as stochastic events affecting gene expression, or cell-population level regulations.

All three tentative propositions for revising our ontological understanding of stemness share the presupposition that stem cells are not entities like neurons or erythrocytes.

Cutting heads an endless task?

Whether stemness refers to a categorical property, a disposition, a relational property or a system property remains a question opened to debate. What might be the consequences of any change in ontological perspective on stemness? Such theoretical considerations about the ontological status of stemness have a practical impact on therapeutic strategies. The model of anti-CSC therapeutic strategy is based on the categorical ontology of stemness: If CSCs are somehow qualitatively distinct and distinguishable from other cancer cells, stemness is a stable property and differentiation is a one-way irreversible process. Accordingly the elimination of the CSCs could definitely eradicate a cancer. But would this therapeutic strategy still be adapted if stemness were a disposition, a relational property, or a system property?

What if stemness were a disposition?

If stemness is a disposition, then the CSCs can be distinguished from non-CSCs and specifically targeted in order to cure cancer. If successful, the outcome would predictably be the same as in the case if stemness were a categorical property: without CSCs, cancers that respond to the CSC model will not be able to maintain themselves and will regress and disappear. This would ultimately lead to an effective cure.

However this ontological perspective emphasizing the behavior of stem cells stumbles on a difficulty for the elimination of the CSCs: any CSCs targeting therapy that would focus on the stemness properties of CSCs would require a genetic portrait of stemness. But the population of active CSCs and the population of quiescent CSCs may have a different signature. This heterogeneity should be taken into account in the design of new drugs. Targeting CSCs with a functional signature could miss the quiescent CSCs. Only the elimination of all CSCs could achieve the goal of an effective and definitive cure.

The dispositional view also suggests another plausible therapeutic strategy. If stemness is a disposition, then cancer stem cells activities rely on the conditions of expression of stemness. Destroying these conditions should lead to the maintenance of the CSCs out of the stem state. This should, in turn, lead to the same result as the elimination of the CSCs (i.e., to the disappearance of the cancer cells population, which would no longer be fed by the CSCs).

What if stemness were a relational property?

If stemness is a relational property, targeting CSCs might be useless, as niches can transform cancer non-stem cells in CSCs at anytime. In this case, CSCs targeting therapies would either never end or likely lead to relapse. If niche is necessary to the acquisition of stemness, then the niche-targeting therapeutic strategy is a plausible alternative. Breaking the cell-niche relationship would bring the loss of stemness and get the same result as the direct elimination of CSCs in the classical conception: the cancer cells population no longer fed by any CSCs would disappear. Notably, the niche targeting strategy fits both relational and dispositional ontology but not the categorical one.

What if stemness were a system property?

Finally, if stemness is a system property, targeting CSCs might also be endless, as cancer non-stem cells can become CSCs at anytime. This 'system conception' of stemness subverts the plausible therapeutic strategies against cancers. Neither CSCs targeting nor niche targeting can guarantee the success of the cure if stemness is a system property. Of the two types of factors potentially involved in inducing stemness, only the population-level regulations can be targeted, and the lack of knowledge about such regulation makes such targeting therapy quite hypothetical. Indeed, it should be possible, as in the case of the niche hypothesis, to intervene on key molecular receptors in order to prevent or inhibit the induction of stemness by extrinsic factors. However the efficiency of such a therapeutic strategy would depend on the possibility of inducing stemness by intrinsic stochastic events (Adler and Sánchez Alvarado 2015). The higher the probability that dedifferentiation occurs in a stochastic way, the less efficient the therapy would be.

Stem cell system biologist Sui Huang uses a concept of landscape to represent cell states (Huang 2011). With the assumption that stemness is only one of the possible states of a cell, states like valleys figures, or 'attractors', are stable, whereas others like hills figures, or 'repellors', are unstable. The state of cells is determined by internal and external stimuli that modify genes expression. If the system view of attractor states is confirmed, some states might be less easily reversible than others. A red cell, for example, is definitely unable to turn back to the stem state, given that it no longer has a nucleus. Other differentiation states might also be highly stable, and dedifferentiation might also be negatively correlated with differentiation. This alternative ontology is worthy of exploration because, if true, then differentiation therapies might turn out to be able to cure cancers more efficiently than any other CSC specific therapies.

To conclude, cancer stem cells are an exemplar case demonstrating the significance of ontological reflections on technoscientific objects. The common view of stem cells as a category of cells has undeniably inspired the recent research efforts on cancer stem cells. The conventional ontological category of 'natural species' allows locating the source of cancer as an identifiable and actionable object. It is an ideal motivation to revive the hope of curing cancers with surgical strikes on specific targets. At the same time it rejuvenates the old alchemical dream to get a hand on natural entities in their nascent state in order to reshape or destroy them for human practical purposes. However, the intensification of research efforts resulted in a growing and extremely diverse population of stem cells that challenges the dream of a radical elimination of the roots of cancer.

More broadly, this case study questions the traditional ontological assumption that refers diseases to specific localized discrete entities. The complexity of carcinogenesis and cancer cures requires new ontological frameworks that take into account multiple agencies such as dispositions and niches, and multiple modes of existence such as state in a system or relational identities. Most importantly, this case study illustrates that ontological clarifications are badly needed to develop more effective cures for cancer.

Notes

1 There has been a less than 5 percent drop in the death rate, while other important causes of death like stroke and cardiovascular diseases have substantially declined (improving by more than 64 percent). These numbers are from Kolata (2009); see also the *Facts & Figures* of the American Cancer Society.
2 For an historical analysis, see Kraft (2009).
3 Reviewed in Dick (2008).
4 For an historical account, see Fagan (2007, 2010).
5 See also Spangrude et al. (1988), Morrison and Weissman (1994) and Larochelle et al. (1996).
6 For a precise analysis of the structure and content of the CSC theory, see Laplane (2014).
7 Reviewed in Kreso and Dick (2014).
8 I searched for the occurrence of the expression 'cancer stem cell*' in the 'Topic' research category, which includes titles, abstracts and key words of all the articles referenced in Web of Science. The research was undertaken in June 2013.
9 In fact, the hypothesis of stem cells at the origin of cancers presupposes the nineteenth-century cell theory. Rudolph Virchow, who formulated the second principle of the cell theory, '*omnis cellula e cellula*' (the first one being that organisms are made of cells), also claimed that tumors are developed from the transformation of normal immature cells, as did Remak and Cohnheim with their embryonic rest theory. For a historical perspective, see Duchesneau (1987) and Laplane (2013).
10 For a history, see Maienschein (2003).
11 Zipori is a major advocate of this view; see Zipori (2004, 2009).
12 For more detailed explanations and descriptions, see Laplane (2013).

References

Adler, C. and Sánchez Alvarado, A. 2015. 'Types or States? Cellular Dynamics and Regenerative Potential.' *Trends in Cell Biology* 11 (25): 687–696.

Astaldi, G., Allegri, A. and Mauri, C. 1947. 'Experimental investigations of the proliferative activity of erythroblasts in their different stages of maturation.' *Experientia* 3 (12): 499–500.

Barroca, V., Lassalle, B., Coureuil, M., Louis, J. P., Le Page, F., Testart, F. J., Allemand, I., Riou, L. and Fouchet, P. 2009. 'Mouse differentiating spermatogonia can generate germinal stem cells in vivo.' *Nature Cell Biology* 11 (2): 190–196.

Becker, A. J., McCulloch, E. A. and Till, J. E. 1963. 'Cytological demonstration of the clonal nature of spleen colonies derived from transplanted mouse marrow cells.' *Nature* 197: 452–454.

Belizário, J. E. 2009. 'Immunodeficient mouse models: An overview.' *The Open Immunology Journal* 2: 79–85.

Blau, H. M., Brazelton, T. R. and Weimann, J. M. 2001. 'The evolving concept of a stem cell: Entity or function?' *Cell* 105 (7): 829–841.

Bonnet, D. and Dick, J. E. 1997. 'Human acute myeloid leukemia is organized as a hierarchy that originates from a primitive hematopoietic cell.' *Nature Medicine* 3 (7): 730–737.

Brawley, C. and Matunis, E. 2004. 'Regeneration of male germline stem cells by spermatogonial dedifferentiation in vivo.' *Science* 304 (5675): 1331–1334.

Burns, C. E. and Zon, L. I. 2002. 'Portrait of a stem cell.' *Developmental Cell* 3 (5): 612–613.

Cambrosio, A. and Keating, P. 1992a. 'Between fact and technique: The beginnings of hybridoma technology.' *Journal of the History of Biology* 25 (2): 175–230.

———. 1992b. 'A matter of FACS: Constituting novel entities in immunology.' *Medical Anthropology Quarterly, New series* 6 (4): 362–384.

Chaffer, C. L., Brueckmann, I., Scheel, C., Kaestli, A. J., Wiggins, P. A., Rodrigues, L. O., Brooks, M., Reinhardt, F., Su, Y., Polyak, K., Arendt, L. M., Kuperwasser, C., Bierie, B. and Weinberg, R. A. 2011. 'Normal and neoplastic nonstem cells can spontaneously convert to a stem-like state.' *Proceedings of the National Academy of Sciences* 108 (19): 7950–7955.

Clarke, M. F., Dick, J. E., Dirks, P. B., Eaves, C. J., Jamieson, C. H., Jones, D. L., Visvader, J., Weissman, I. L. and Wahl, G. M. 2006. 'Cancer stem cells—perspectives on current status and future directions: Aacr workshop on cancer stem cells.' *Cancer Research* 66: 9339–9344.

Clarkson, B. D. 1969. 'Review of recent studies of cellular proliferation in acute leukemia.' *Journal of the National Cancer Institute Monographs* 30: 81–120.

Dick, J. E. 1996. 'Normal and leukemic human stem cells assayed in SCID mice.' *Seminars in Immunology* 8 (4): 197–206.

———. 2008. 'Stem cell concepts renew cancer research.' *Blood* 112 (13): 4793–4807.

Duchesneau, F. 1987. *Genèse de la théorie cellulaire*. Paris: Vrin.

Fagan, M. B. 2007. 'The search for the hematopoietic stem cell: Social interaction and epistemic success in immunology.' *Studies in History and Philosophy of Biological and Biomedical Sciences* 38 (1): 217–237.

———. 2010. 'Stems and standards: Social interaction in the search for blood stem cells.' *Journal of the History of Biology* 43 (1): 67–109.

———. 2013. *Philosophy of Stem Cell Biology: Knowledge in Flesh and Blood*. London: Palgrave-Macmillan.

Ford, C. E., Hamerton, J. L., Barnes, D. W. and Loutit, J. F. 1956. 'Cytological identification of radiation-chimaeras.' *Nature* 177 (4506): 452–454.

Fortunel, N. O., Otu, H. H., Ng, H. H., Chen, J., Mu, X., Chevassut, T., Li, X., Joseph, M., Bailey, C., Hatzfeld, J. A., Hatzfeld, A., Usta, F., Vega, V. B., Long, P. M., Libermann, T. A. and Lim, B. 2003. 'Comment on "'Stemness': Transcriptional profiling of embryonic and adult stem cells" and "A stem cell molecular signature".' *Science* 302 (5644): 393.

Gavosto, F., Pileri, A., Gabutti, V. and Masera, P. 1967. 'Non-self-maintaining kinetics of proliferating blasts in human acute leukaemia.' *Nature* 216 (5111): 188–189.

Griffin, J. D. and Lowenberg, B. 1986, 'Clonogenic cells in acute myeloblastic leukemia.' *Blood* 68 (6): 1185–1195.

Haber, D. A., Gray, N. S. and Baselga, J. 2011. 'The evolving war on cancer.' *Cell* 145 (1): 19–24.

Hackney, J. A., Charbord, P., Brunk, B. P., Stoeckert, C. J., Lemischka, I. R. and Moore, K. A. 2002. 'A molecular profile of a hematopoietic stem cell niche.' *Proceedings of the National Academy of Sciences* 99 (20): 13061–13066.

Herzenberg, L. A., Sweet, R. G. and Herzenberg, L. A. 1976. 'Fluorescence activated cell sorting.' *Scientific American* 224: 108–117.

Huang, S. 2011. 'On the intrinsic inevitability of cancer: From foetal to fatal attraction.' *Seminars in Cancer Biology* 21 (3): 183–199.

Ivanova, N. B., Dimos, J. T., Schaniel, C., Hackney, J. A., Moore, K. A. and Lemischka, I. R. 2002. 'A stem cell molecular signature.' *Science* 298 (5593): 601–604.

Ivanova, N. B., Dimos, J. T., Schaniel, C., Hackney, J. A., Moore, K. A., Ramalho-Santos, M., Yoon, S., Matsuzaki, Y., Mulligan, R. C., Melton, D. A. and Lemischka, I. R. 2003. 'Response to comments on "'Stemness': Transcriptional profiling of embryonic and adult stem cells" and "A stem cell molecular signature".' *Science* 302 (5644): 393d.

Kai, T. and Spradling, A. 2004. 'Differentiating germ cells can revert into functional stem cells in drosophila melanogaster ovaries.' *Nature* 428 (6982): 564–569.

Keating, P. and Cambrosio, A. 1994. 'Ours is an engineering approach': Flow cytometry and the constitution of human T-cell subsets.' *Journal of the History of Biology* 27: 449–479.

Killmann, S. A., Cronkite, E. P., Robertson, J. S., Fliedner, T. M. and Bond, V. P. 1963. 'Estimation of phases of the life cycle of leukemic cells from labeling in human beings in vivo with titriated thymidine.' *Laboratory Investigation* 12: 671–684.

Kolata, G. 2009. 'Advances elusive in the drive to cure cancer.' *New York Times*, 23 April, http://www.nytimes.com/2009/04/24/health/policy/24cancer.html?pagewanted=all (accessed April 11, 2016).

Kraft, A. 2009. 'Manhattan transfer: Lethal radiation, bone marrow transplantation, and the birth of stem cell biology, ca. 1942–1961.' *Historical Studies in the Natural Sciences* 39 (2): 171–218.

Kreso, A. and Dick, J. E. 2014. 'Evolution of the cancer stem cell model.' *Cell Stem Cell* 14 (3): 275–291.

Krivtsov, A. V., Twomey, D., Feng, Z., Stubbs, M. C., Wang, Y., Faber, J., Levine, J. E., Wang, J., Hahn, W. C., Gilliland, D. G., Golub, T. R. and Armstrong, S. A. 2006. 'Transformation from committed progenitor to leukaemia stem cell initiated by MLL-AF9.' *Nature* 442 (7104): 818–822.

Lander, A. D. 2009. 'The "stem cell" concept: Is it holding us back?' *Journal of Biology* 8 (8): 70.

Laplane, L. 2013. 'Cellules souches cancéreuses: ontologie et thérapies.' PhD dissertation, Université Paris Ouest Nanterre La Défense, https://bdr.u-paris10.fr/theses/internet/2013PA100119.pdf (accessed October 20, 2016).

———. 2014. 'Identifying Some Theories in Developmental Biology: The Case of the Cancer Stem Cell Theory.' In *Toward a Theory of Development*, edited by Minelli, A. and Pradeu, T., 246–259. Oxford: Oxford University Press.

Larochelle, A., Vormoor, J., Hanenberg, H., Wang, J. C., Bhatia, M., Lapidot, T., Moritz, T., Murdoch, B., Xiao, X. L., Kato, I., Williams, D. A. and Dick, J. E. 1996. 'Identification of primitive human hematopoietic cells capable of repopulating NOD/SCID mouse bone marrow: Implications for gene therapy.' *Nature Medicine* 2 (12): 1329–1337.

Larochelle, A., Vormoor, J., Lapidot, T., Sher, G., Furukawa, T., Li, Q., Shultz, L. D., Olivieri, N. F., Stamatoyannopoulos, G. and Dick, J. E. 1995. 'Engraftment of immune-deficient mice with primitive hematopoietic cells from beta-thalassemia and sickle cell anemia patients: Implications for evaluating human gene therapy protocols.' *Human Molecular Genetics* 4 (2): 163–172.

Lenz, H. J. 2008. 'Colon cancer stem cells: A new target in the war against cancer.' *Gastro-intestestinal Cancer Research* 2 (4): 203–204.

Leychkis, Y., Munzer, S. R. and Richardson, J. L. 2009. 'What is stemness?' *Studies in History and Philosophy of Biological and Biomedical Sciences* 40 (4): 312–320.

Lo, D. C., Allen, F. and Brockes, J. P. 1993. 'Reversal of muscle differentiation during urodele limb regeneration.' *Proceedings of the National Academy of Sciences* 90 (15): 7230–7234.

Loeffler, M. and Roeder, I. 2002. 'Tissue stem cells: Definition, plasticity, heterogeneity, self-organization and models—a conceptual approach.' *Cells Tissues Organs* 171 (1): 8–26.

Lorenz, E., Uphoff, D., Reid, T. R. and Shelton, E. 1951. 'Modification of irradiation injury in mice and guinea pigs by bone marrow injections.' *Journal of the National Cancer Institute* 12 (1): 197–201.

Maienschein, J. 2003. *Whose View of Life? Embryos, Cloning and Stem Cells*. Cambridge and London: Harvard University Press.

Mani, S. A., Guo, W., Liao, M. J., Eaton, E. N., Ayyanan, A., Zhou, A. Y., Brooks, M., Reinhard, F., Zhang, C. C., Shipitsin, M., Campbell, L. L., Polyak, K., Brisken, C., Yang, J. and Weinberg, R. A. 2008. 'The epithelial-mesenchymal transition generates cells with properties of stem cells.' *Cell* 133 (4): 704–715.

McCulloch, E. A. and Till, J. E. 1981. 'Blast cells in acute myeloblastic leukemia: A model.' *Blood Cells* 7 (1): 63–77.

Mikkers, H. and Frisen, J. 2005. 'Deconstructing stemness.' *Journal of the European Molecular Biology Organization* 24 (15): 2715–2719.

Morel, A. P., Lievre, M., Thomas, C., Hinkal, G., Ansieau, S. and Puisieux, A. 2008. 'Generation of breast cancer stem cells through epithelial-mesenchymal transition.' *PLoS One* 3 (8): e2888.

Morrison, S. J. and Weissman, I. L. 1994. 'The long-term repopulating subset of hematopoietic stem cells is deterministic and isolatable by phenotype.' *Immunity* 1 (8): 661–673.

Okita, K., Nakagawa, M., Hyenjong, H., Ichisaka, T. and Yamanaka, S. 2008. 'Generation of mouse induced pluripotent stem cells without viral vectors.' *Science* 322 (5903): 949–953.

Passegue, E., Jamieson, C. H., Ailles, L. E. and Weissman, I. L. 2003. 'Normal and leukemic hematopoiesis: Are leukemias a stem cell disorder or a reacquisition of stem cell characteristics?' *Proceedings of the National Academy of Sciences* 100 (Suppl. 1): 11842–11849.

Ramalho-Santos, M., Yoon, S., Matsuzaki, Y., Mulligan, R. C. and Melton, D. A. 2002. '"Stemness": Transcriptional profiling of embryonic and adult stem cells.' *Science* 298 (5593): 597–600.

Rapp, U. R., Ceteci, F. and Schreck, R. 2008. 'Oncogene-induced plasticity and cancer stem cells.' *Cell Cycle* 7 (1): 45–51.

Reya, T., Morrison, S. J., Clarke, M. F. and Weissman, I. L. 2001. 'Stem cells, cancer, and cancer stem cells.' *Nature* 414 (6859): 105–111.

Robert, J. S. 2004. 'Model systems in stem cell biology.' *Bioessays* 26 (9): 1005–1012.

Seaberg, R. M. and van der Kooy, D. 2003. 'Stem and progenitor cells: The premature desertion of rigorous definitions.' *Trends in Neurosciences* 26 (3): 125–131.

Shostak, S. 2006. '(Re)defining stem cells.' *Bioessays* 28 (3): 301–308.

Shultz, L. D., Ishikawa, F. and Greiner, D. L. 2007. 'Humanized mice in translational biomedical research.' *Nature Reviews Immunology* 7 (2): 118–130.

Spangrude, G. J., Heimfeld, S. and Weissman, I. L. 1988. 'Purification and characterization of mouse hematopoietic stem cells.' *Science* 241 (4861): 58–62.

Takahashi, K., Tanabe, K., Ohnuki, M., Narita, M., Ichisaka, T., Tomoda, K. and Yamanaka, S. 2007. 'Induction of pluripotent stem cells from adult human fibroblasts by defined factors.' *Cell* 131 (5): 861–872.

Takahashi, K. and Yamanaka, S. 2006. 'Induction of pluripotent stem cells from mouse embryonic and adult fibroblast cultures by defined factors.' *Cell* 126 (4): 663–676.

Thirant, C., Bessette, B., Varlet, P., Puget, S., Cadusseau, J., Tavares Sdos, R., Studler, J. M., Silvestre, D. C., Susini, A., Villa, C., Miquel, C., Bogeas, A., Surena, A. L., Dias-Morais, A., Leonard, N., Pflumio, F., Bieche, I., Boussin, F. D., Sainte-Rose, C., Grill, J., Daumas-Duport, C., Chneiweiss, H. and Junier, M. P. 2011. 'Clinical relevance of tumor cells with stem-like properties in pediatric brain tumors.' *PLoS One* 6 (1): e16375.

Till, J. E. and McCulloch, E. A. 1961. 'A direct measurement of the radiation sensitivity of normal mouse bone marrow cells.' *Radiation Research* 14: 213–222.

———. 1980. 'Hemopoietic stem cell differentiation.' *Biochimica et Biophysica Acta* 605 (4): 431–459.

Visser, J. W., Bauman, J. G., Mulder, A. H., Eliason, J. F. and de Leeuw, A. M. 1984. 'Isolation of murine pluripotent hemopoietic stem cells.' *The Journal of Experimental Medicine* 159 (6): 1576–1590.

Vogel, G. 2003. 'Stem cells: "Stemness" genes still elusive.' *Science* 302 (5644): 371.

Wilmut, I., Schnieke, A. E., McWhir, J., Kind, A. J. and Campbell, K. H. 1997. 'Viable offspring derived from fetal and adult mammalian cells.' *Nature* 385 (6619): 810–813.

Younes, S. A., Yassine-Diab, B., Dumont, A. R., Boulassel, M. R., Grossman, Z., Routy, J. P. and Sekaly, R. P. 2003. 'HIV-1 viremia prevents the establishment of interleukin 2-producing HIV-specific memory CD4+ T cells endowed with proliferative capacity.' *The Journal of Experimental Medicine* 198 (12): 1909–1922.

Zhou, H., Wu, S., Joo, J. Y., Zhu, S., Han, D. W., Lin, T., Trauger, S., Bien, G., Yao, S., Zhu, Y., Siuzdak, G., Schöler, H. R., Duan, L. and Ding, S. 2009. 'Generation of induced pluripotent stem cells using recombinant proteins.' *Cell Stem Cell* 4 (5): 381–384.

Zipori, D. 2004. 'The nature of stem cells: State rather than entity.' *Nature Reviews Genetics* 5 (11): 873–878.

———. 2009. *Biology of Stem Cells and the Molecular Basis of the Stem State*. Dordrecht and New York: Humana Press.

3 Robots behaving badly

Simulation and participation in the study of life

Christopher Kelty

A 2011 paper in the journal *PLOS Biology* described an experimental test of Hamilton's Rule; the authors report success (Waibel et al. 2011). It is, they claim, "the first quantitative test of Hamilton's rule in a system with a complex mapping between genotype and phenotype," and they "demonstrate the wide applicability of kin selection theory."

This is surprising for many reasons. First, Hamilton's rule is one of the more controversial features of evolutionary biology in the last half-century.[1] Second, the authors refer to their experiment as one in 'artificial evolution'. This is obvious enough, given that experimenting on evolution is not easy to do with any but the most short-lived organisms like fruit flies and microbes, except that, third, this experiment did not use social organisms of any predictable kind such as wasps, ants, or even microbes, but tiny mobile robots named *Alice*.

These robots raise a set of confusing questions that this article will address. The first concerns the role of the robots in biological research: Do they *simulate* something (life, evolution, sociality) or do they *participate* in something? The second question concerns the physicality of the robots: What difference does embodiment make to the role of the robot in these experiments, where, as we will see, there are some subtle distinctions between what is abstract, what is digital, what is simulated and what is physical? Thirdly, how do life, embodiment and social behavior relate in contemporary biology, and why is it possible for robots to illuminate this relation? These questions are provoked by a strange similarity that has not been noted before: between the problem of simulation in philosophy of science, and Deleuze's reading of Plato on the relationship of ideas, copies and simulacra. Whereas robot scientists, biologists and some philosophers of science may argue that robots are an 'object' on which one can do precise experiments about a 'target' (life, social behavior, evolution), Deleuze might counsel instead that we do not treat robots enough like robots, nor take our curiosity about them quite seriously enough – for they themselves might be the object that we are studying, or should be.

The robot invasion begins . . .

Using robots to study evolution might seem at first extremely unlikely. The practice has steadily grown more common over the last twenty years, owing in large part to the manifest enthusiasm that computer scientists and engineers have shown for evolutionary theory as both a theory and as an engineering principle. Recently, however, biologists have also started to take such work a bit more seriously. Several reviews have appeared recently, aimed at enlightening biologists about this work (Garnier 2011; Krause et al. 2011; Mitri et al. 2013). And indeed, one of the authors of the paper about Hamilton's rule, Laurent Keller (University of Lausanne), is a widely lauded biologist of social evolution (primarily in real ants) who is thus lending a modicum of legitimacy to a style of experiment that might otherwise be dismissed by his colleagues.

Since about 2005, Keller and collaborator Dario Floreano, a robot scientist at École Polytechnique Fédérale de Lausanne (EPFL), have done a number of experiments with robots that explore different aspects of evolution. In most of them, tiny robots move around on motorized wheels in an enclosed area and interact with each other. Often there is 'food' or 'poison' (objects with blinking LEDs, for instance, that can be detected by a sensor), which the robots are programmed to forage for (or to avoid). In some, the robots are predator or prey. Properly programmed and set to work, the robots have demonstrated the evolution of communication, the evolution of altruism, the evolution of information suppression, the relationship of signal reliability and relatedness, predator/prey co-evolution, and various aspects of the morphology of a robot body (Floreano et al. 2007, 2010; Mitri et al. 2013).

In the test of Hamilton's rule, the Alice robots had infrared sensors that allowed them to differentiate the other robots from the 'food items' deposited among them, and two vision sensors that allowed them to 'see' the walls (three black and one white) of the enclosure. The robots were instructed to gather up the food (i.e., push it toward the white wall) and then decide whether or not to 'share' it with the other robots. The robots did not consume the food, nor metabolize it – they are designed only to be highly simplified phenotypic vehicles for a limited set of genes that govern their behavior. They did not reproduce, sexually or asexually, but were manually regenerated by a combination of precise but capricious software (the source of mutation) and beneficent human intervention (the robots needed to be plugged in or otherwise connected to a computer in order to download the new genome and become their own offspring).

Since these robots had neither brains nor DNA, they were equipped with a software-based 'neural network.' This simple setup allowed the experimenters to precisely vary the measure of relatedness in Hamilton's rule (r), and watch as the fitness of the robots changes over a series of generations. In each generation, the inclusive fitness of each individual was determined by how many times a robot shared the food, and how many times food was shared by another robot in the group. The robots were then 'selected' based on inclusive fitness, subjected to mutations, and then the next generation was put back into the arena with new food items. After hundreds of generations, the gene frequencies of the resulting robots confirmed just what Hamilton's rule would predict. As the experimenters put it,

> Because the 33 genes were initially set to random values, the robots' behaviors were completely arbitrary in the first generation. However, the robots'

> performance rapidly increased over the 500 generations of selection. The level of altruism also rapidly changed over generations with the final stable level of altruism varying greatly depending on within group relatedness and c/b ratio. (Waibel et al. 2011, 5)

A careful reading of the article reveals that the experiment was conducted with two hundred groups of eight robots over five hundred generations and twenty-five different treatments. That's a lot of robots. However, as the authors explain, "because of the impossibility to conduct hundreds of generations of selection with real robots, we used physics-based simulations that precisely model the dynamical and physical properties of the robots." So, to be clear, there are at least two kinds of robots in these experiments: 'real' (physical) robots and simulated robots.[2] But both kinds do 'exactly' the same thing, which raises the interesting question: Why use physical robots at all? What exactly is the difference between a physical robot and a simulated robot? Do robots *simulate* nature or *participate* in it?

Robot trouble

Certain aspects of the distinction between the natural and the artificial are reassuringly intransigent. And yet, as with anyone who studies animals, robot scientists love to chip away at those divisions. In the 1990s, Artificial Life researchers like Christopher Langton insisted that the pixelated creatures on their screens were alive. Other more jocular researchers, like Rodney Brooks, weren't so much convinced that robots or their simulations were alive – only that they can no longer be ignored, or maybe even controlled. For most participants in the fields of robotics and artificial life, the processes of biology, evolution or the meaning of 'life' were useful primarily in order to build better robots. John Holland and David Goldberg applied the ideas of evolutionary theory to the design of algorithms; researchers in robotics initiated projects in evolutionary, cognitive, developmental, affective and epigenetic computing; the 'inspiration' of nature was oft-cited among engineers as 'bio-mimesis' or 'bio-inspired' technology.[3] Robots have obvious power in understanding issues of locomotion, perception and communication, as well as, more recently, emotion and human health.

At roughly the same time, and often with very little interaction, evolutionary theory itself has seen a series of changes so that there is now a marked split between those who study *evolutionary theory* and those who study living organisms (now or in the past). The link between evolutionary theory and the study of life has steadily become less obvious, with generations of thinkers applying the concepts of evolution to everything from economic growth to epistemology to universes (Toulmin 1972; Nelson and Winter 1982; Smolin 1997). As 'evolutionary theory' has expanded, it has come to be regarded less as a theory of life and its organization and more as an incredibly powerful theory of change and diversity in any system.[4] It is clear that evolutionary theory has taken on not just a life but an ecology of its own.

Classical biologists might be justifiably skeptical: driven to extremes, evolutionary theory of this kind loses touch with the empirical – as well as with other domains of theory like those of ecology, development and physiology. They demand exploration, experimentation and the observation of living things.

But at the same time, computer simulation has risen in respectability: not only has there been a marked shift away from 'law-governed' theories to 'model-based' theories over the course of the twentieth century, but also a recognition that computer simulation might be scientifically and philosophically novel (Morgan and Morrison 1999; Creager et al. 2007).[5] Debates have emerged about the difference between simulation and experiment (Rohrlich 1990; Winsberg 2009), about the rise of 'exploratory' experiments (Steinle 1997; Franklin 2005), about the role of surprise generated by simulations (Lenhard 2006), and about the epistemological and ontological consequences of understanding based on the radically advancing computational power available to scientists (Fox Keller 2003; Humphreys 2008; Parker 2009). Robots, therefore, seem to fall into this niche of respectability, as they too become more powerful, diverse and tractable as tools of exploration and experiment.

Robots can be treated like model organisms responding to an experimental setup, or they can be used as traps, lures or decoys that provoke behavior or reaction from an animal or human. The distinction between a robot that *simulates* something else (stands in for) and one that *participates* in something is not at all clear, whether to the biologists using robots or for those observing the scientists.

Using robots (or computer simulations) to study *evolution* is apposite, though, because experiments on evolution are difficult to design in the first place. The timescales involved require experimental setups that accelerate time relative to that of humans and other animals, as well as a significant degree of inference and assumption. Today the mathematical theory of evolution allows considerable analytic and predictive power, but often requires an unsettling degree of simplification and only a tentative generality. Computer-based simulations offer a way to add 'complexity' back in, and robots, therefore, seem to be the next obvious step in such an exploration. Still they are just computer simulations. For many scientists, the question remains: How do these simulations relate to the nature they describe? Philosophers of simulation pose this question routinely as well. Eric Winsberg (among others) offers a helpful distinction by pointing out that both in simulations and in experiments there is a difference between the *object* of explanation and the *target* of explanation – a mouse model can be an object while the target is humans or human cancer, just as a simulation of fluid dynamics in a computer can be an object while the target is the center of a black hole – something that manifestly cannot be accessed by humans (Winsberg 2009). Target and object can be distinguished without depending on other spurious distinctions like digital/analogue or real/virtual and can be distinguished equally in bench experiments in a lab, field experiments, or experiments using software models of phenomena.

Such a distinction would seem to appeal to the experimental biologists who are using robots, as it implies the possibility of a continuum of relations between object and target. Indeed, in a recent review by Mitri, Floreano and Keller, the uses of robots in biological experiments concerning social behavior are surveyed; they lay out just such a continuum that they call the 'scale of situatedness' (Figure 3.1).[6]

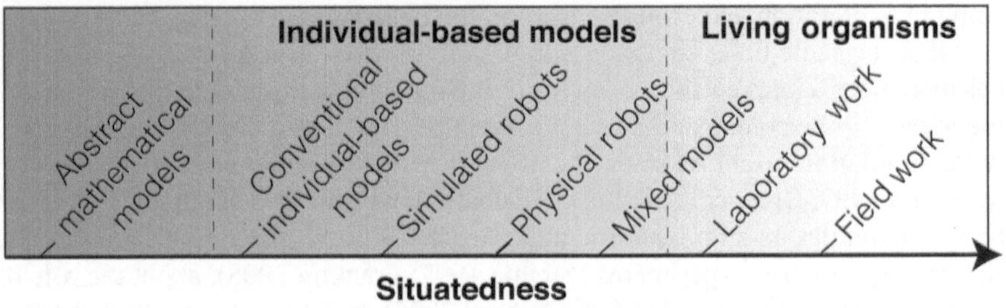

Figure 3.1 Scale of situatedness: The extent to which individuals are embedded in an environment that they can sense and modify.

On the right-hand end of the scale are versions of 'situatedness', such as 'field work' and 'laboratory work', which "include the whole complexity of the organisms and their environment . . . but rarely permit the unambiguous demonstration of causations," especially concerning social behavior – pure 'target', to be sure, but seemingly as complex and inaccessible as a black hole.

On the other end they put "abstract mathematical models" that "boil down collective systems to their minimal components." They are abstractions and formulas that model populations as a whole, rather than individuals in populations. Next on the scale are 'individual' or 'agent-based models' that can take into account the varying behavior of a potentially large population of individuals. These models necessarily inhabit a computer simulation (rather than a formula on a page or screen in the case of 'abstract mathematical models'), usually represented not graphically but as something like a database or spreadsheet of changing parameters over time. Such models depend on the ability to both compute the complexity of interacting individuals (each of which might have its own more or less complex genotype and phenotype) in a population and in most cases to display that computation in some form (though not always as a graphical visualization).

Further down the scale are robots (both physical and simulated), which are essentially agent-based models in physical, programmed robots that interact in real (or simulated) space. Physical robots are better than agent-based model simulations because they do not need to make simplifying assumptions about the physical environment or the 'physiology' of the agent/individual (i.e., physical organization of the robot). As the authors put it, realism in an agent-based model is *costly* and *complex*, whereas "the laws of physics are included 'for free' in robotic models."[7]

The scale is implicitly one of complexity of bodies and their environments. Robots straddle a boundary: they share some bodily complexity with organisms at the same time that they share putatively simpler (or just more controllable) computational and digital modes of existence with simulated organisms. That robots have bodies is not interpreted as a metaphor or a supplement, but instead as a crucial determinant of what robots are and can be. Embodiment is central to cognition not just in the sense that the robot needs to perceive and sense the world, but also implies that the kind of body it has will transform the kind of cognition of which it is capable.[8] When the authors say that the "laws of physics are included for free," they mean that all of the complexity of the robot body – it's weight, orientation, speed, the distance between its eyes and its wheels, and so on – do not need to be simulated by a digital computer; they are just there, as part of what a robot is.

I'm not a robot, but I play one on TV . . .

> In Plato's world variation is accidental, while essences record a higher reality; in Darwin's reversal, we value variation as a defining (and concrete earthly) reality, while averages (our closest operational approach to essences) become mental abstractions.
>
> Stephen Jay Gould (Gould 1996, 41)

A key moment in the work of philosopher Gilles Deleuze is his confrontation with Nietzsche's claim to be 'overturning' Plato's philosophy.[9] For most interpreters, and arguably for Nietzsche, such an inversion concerns the opposition between the transcendent realm of the Ideas, and the fallen or immanent realm of copies. To invert Platonism would mean that the material or the concrete is elevated over the realm of ideas.

However, for Deleuze it is not the case that Plato is distinguishing between a transcendent realm of Ideas and a fallen realm of copies or images, but he offers instead a tripartite relation amongst idea, true copy and false copy, or among idea, copy and *simulacrum*. Copies in Plato are not all equally subordinate to the Ideas they mimic – indeed they do not mimic at all, but instead *participate* in the Idea, some much better and more authentically than others. The sculptor captures the image of a man when she understands the internal and external resemblance of the copy she creates to the model she observes. Conversely, the 'simulacrum' or false copy bears only a semblance: it looks like the real thing but isn't the same 'on the inside' as it was. According to Deleuze, Plato observes the problem of the *rivalry* amongst copies, and the problem of deciding among them.[10] Thus the question of whether robots simulate evolution, or actually participate in it (i.e., actually evolve), might be put to the test of Deleuze's reading by asking how the scientists themselves conceive of and practice the exploration of evolution using robots.

Deleuze's reading of Plato bears a striking similarity to the problem of simulated and physical robots discussed here. Rather than a simple opposition between real (living things) and fake (robots), the 'scale of situatedness' seems to set up a scale of rivalry along which some kinds of copies are better than others. But as in the case of Plato's philosophy, the question remains: On what basis does one choose one rival over others? What troubles some biologists about using robots is that *they simply are not alive.*

It would seem therefore that the question is not 'What is the difference between an animal and a robot?' (model and copy), but 'What is the difference between a physical robot and a simulated robot?' (copy and simulacrum). For Plato, the danger of the simulacrum is that it resembles the original on the outside, but lacks internal similarity. (It is not a true copy.) This would seem to describe any robot or especially classical automata that have the appearance of being what they copy but a radically different internal organization (the story of the mechanical Turk notwithstanding). Robots with fleshy plastic masks that simulate emotionally specific expressions are troubling and uncanny because they copy (increasingly well) the outside without being the same on the inside. Whatever we think it is that 'makes us human', it is present only in the *behavior* of the robot, not in its 'essence'. So what is it then that biologists think they are seeing when they decide to use a physical, embodied robot instead of a simulated one? That is to say, on what grounds do they distinguish these claimants to participation and, in turn, place them in order on the scale of situatedness?

One robot may hide another . . .

The difference between agent-based models and robot models is frequently mentioned in experimental studies using robots. But is there also a difference between *physical* robots and *simulated* robots. In 'physics-based' robot simulations, simulations of the robots in question include more of the physical parameters of the robot than an agent-based simulation would, but presumably less that of the physical robot. In these physics-based simulations, the price of the laws of physics starts to fluctuate: How detailed must a physics-based simulated robot be to be the same as a physical robot?

There is obviously a lot of complexity to a robot simulation, and not much information is offered by the experimenters when it comes to understanding what a simulated robot is or does. The materials and methods sections in the work of Floreano and Keller, for instance, often describe in detail the robot's configuration, precise dimensions and speed, programming, and so on, as if all the robots in the study were physically real, and they rarely refer readers to the simulation tools used. They are modular, extensible software environments of a standard sort that can be used to cobble together simulated robots and run experiments using these robots. Physical and simulated robots are indistinguishable because they use the same operating systems, programming environments and software tools and libraries. It is a feature of all robots that they are first simulated in software and then, essentially, 'printed out' in physical form and set to work – and it is always the differences, breakdowns, or hardware-dependent surprises that form the core of an iterative process of inquiry in the robot sciences. The assumption that there is no difference between a simulated and a physical robot is therefore both deeply ingrained in the practice of creating robots and warranted by an understanding of what gives the robot its being: its software.

To return to Deleuze, the problem of distinguishing between false claimants is that they resemble the copy on the outside but not the inside – they do not really *participate* in the idea they copy. The simulated robots, however, offer a twist on this: they may have a visual representation (such as a 3D rendering on a screen) that resembles a physical robot, but what is indistinguishable is actually the internal aspect – the program or software governing the robot. Simulated robots have simulated bodies but their 'brains' and 'genes' are identical to those of physical robots: they are identical on the inside, but share no resemblance on the outside.

But if so, why not just run simulated robots through the motions and report on that? Implicit in the design of this experiment is that physical robots are necessary to produce that minimum of difference – that surprise – which can only come by running an experiment (Rheinberger 1997). Researchers expect, perhaps, that physical robots will confirm what the simulation demonstrates, but they must be included, observed, and the hypotheses thereby tested in order to assure them that the simulation acts as the physical robots would. In every case reported by Floreano and Keller so far, they use physical robots in the experiment – but only a few – and they do not seem to do anything other than confirm the results of the simulation. They produce no surprises but are nonetheless essential to confirming that they are, as yet, unnecessary.

These aren't the droids you are looking for . . .

There are some obvious reasons why scientists might prefer robots (whether physical or simulated) to animals. They can do all kinds of ethically suspect things with robots that they cannot do with animals. There are also obvious reasons why an experimenter would prefer *simulated* robots over physical ones: they can reproduce and mutate dramatically faster than the embodied robots, they don't break down as often, and they can be reprogrammed and reconfigured much more quickly. The trade-off is that robots are not animals, and simulated robots are not robots. On the 'scale' of situatedness, there is a clear hierarchy.

On the one hand, the troubling implication of such a spectrum is that mathematical abstractions are figured as the least appropriate object for understanding a target – as opposed to the traditional assumption that they bear a kind of representational accuracy, if not identity, with the target (mathematics as the 'language of nature'). But on the other, it seems to put mathematical abstractions, models, robots, model organisms and living organisms all on the same ontological plane – formulas are simpler versions of agent based models, which are simpler forms of robots, which are simpler forms of organisms, all of which *participate* in something not quite specified: life, evolution or social behavior. In other terms, it collapses the object/target distinction.

There appears to be an epistemological tension here: on the one hand, robots are merely physical entities that can be simulated as other physical entities (assuming we get the physical theories 'right') and what they do is ontologically indistinguishable from what mathematical formulae do or from what organisms do. But on the other hand is the assertion that *robots are different*: robots are *embodied* and so they will exhibit forms of behavior and/or cognition that can only come from being embodied (i.e., something that cannot be or does not reduce to their program, something that comes of having a 'body' not just a 'mind' [or program]).

Thus the choice to work with robots appears strictly pragmatic: robots give some kind of partial access to (more complex) nature in return for being more tractable (simpler).[11] It sets up a *rivalry* between different kinds of objects which are also targets, and at the same time implicitly orders those that are most authentic expressions of some unstated essence: life, evolution, behavior and so on.

The implicit definition of embodiment here, however, is a physical one. 'Having a body' means 'including the laws of physics for free'. By doing so we gain purchase on the physics of bodies and environments that we would otherwise have to painstakingly model in a software system. But in the case at hand, what is being simulated is something else: *social behavior*, and in particular motivated social behavior that can be labeled as either selfish or altruistic. The scientists involved are no doubt committed to the epistemological claim that all such behaviors are fundamentally physical – either in a reductionist or an emergentist sense. There are genes which are biochemical entities that interact with the physical features of an environment, which are then translated into proteins and on up the chain into organisms, whose metabolically self-sustaining whole is liable to behave in predictable ways – such as by choosing to share food with another organism (QED, Hamilton's rule).

The question the scientists do not pose, however, is whether these robots have a 'social' body (real or simulated) as well as a 'physical' one, and whether there is any difference. Is social behavior a feature of the environment or is there a kind of social behavior that is not influenced by physical environments?

Is 'social behavior' simply 'physical behavior' governed by brains and genes? Could it be something more, and could the robots reveal it? Do the scientists expect – or perhaps hope – that the robots will surprise them? Unfortunately, to make things even worse, the very concept of 'social behavior' (to say nothing of 'altruism' or 'self-interest') is so elastic and vague that it's hard to know what most scientists mean, even when humans and animals are both object and target, much less robots (Piliavin and Charng 1900; Levitis et al. 2009; Bergner 2011). One cynical answer to the question "Why use physical robots?" is simply that it's cool, or that it brings in funding, or that interdisciplinarity is (over)valued. However, the larger question concerning simulation and the 'style of reasoning' that is being developed does not thereby disappear – robots and simulations have entered the practice of science to such an extent that it is no longer possible to continue to treat them as gimmicks or illustrations.

But neither is it possible to treat these robots as less complex versions of living organisms. To do so is to misrecognize what they are, to mistake the false for the true claimant, and even more importantly, to follow Deleuze, to fail to give the false claimant a positivity of its own – to investigate, forthrightly, the internal difference that these robots harbor vis-à-vis animals and humans and even, ultimately, to argue against the existence of any model or original.

Do robots have a life of their own?

When we learn something from an experiment using robots, are we learning about robots or are we learning about something else? Floreano and Keller tell us that we are learning something about ('confirming') Hamilton's rule when we watch these robots do what they have been programmed to do. The algorithms of the robots instantiated in their physical and simulated bodies unfold in accordance with our expectations about how ants or bees or humans would do the same. We are then comparing expectations to expectations and confirming their identity – this tells us that theory is correct for robots, although does not necessarily tell us that it is true for animals.

Given these robots and their algorithms, we can assert that simulations (including robots simulating animals) *animate* theory: they bring theories to life in time. In the case at hand, the robots are animating Hamilton's Rule. Eric Winsberg's term for this is that simulations are 'downward': they draw on theory and perform it in a computer rather than being the kind of thing from which one makes observations and builds theories ('upward') (Winsberg 2009). However, the spatial metaphor is misleading, as it relies too much on a hierarchical relationship and separation between theory and observation. Yet, what simulations do is less about higher and lower, and more about static and dynamic. A simulation is preferable to an equation because it gives the equation life. Confirmation becomes something the simulation does for us by enacting the theories we devise.

Further, if equations can be observed and experienced, then they have the capacity to generate insight, surprise and difference in an experimental system. They do in fact possess the potential to be 'upward' in Winsberg's sense, but only if one approaches them differently – as that which must be observed in its own right: not a true copy but a false one, a simulacrum with its own positivity, its own internal difference as a motor of change and exploration. There are many examples of such simulations generating both surprise and new difference and understanding. Simulations therefore are obviously 'downward' when they animate a theory, but they can also be upward, with a swerve, when observed. They generate surprises and insights, and when this happens, there is no obvious commitment to realism, but there is a commitment to curiosity – the simulation does not represent something; it literally *becomes* something else. When this happens, a simulation is neither an attenuated nor a false copy of the real, but an object and process with its own complexity, an *assemblage*. Simulations therefore – and one must agree with Winsberg here, even if the metaphor is wrong – take on a life of their own; some of them even escape and find new niches in neighboring disciplines. To say that the robots *participate* in the experiment, however, implies that the robot (physical or simulated) is an image of something else – an instance of some other ideal form, in which it participates. A mouse, as a model organism, clearly participates in 'life' insofar as it is a 'copy' of human life.

To view robots as participating in life or evolution is exactly what generates the skepticism among biologists – they are so obviously false claimants to the idea of life that it is exceedingly hard to treat any experimental result as having even the faintest relevance to understanding life. These robots have 'genes' and 'brains' that can only be used in scare quotes: they lack that resemblance (merely homologous though it be) that we comfortably attribute to mice and men. However, this suspicion of robots requires a certain Platonism among the biologists – or perhaps not enough Deleuze. The robots in these studies are not model organisms; they should not be taken as simply lower down on the great chain of being, less representative models – less realistic – but still useful. They are what they are – and we ought to be honest about it: they are *robots*. They are simulacra with a power of their own: assemblages of software, hardware, theories of computation and embodiment, theories of evolution and neural processing, vision sensors and motors, human operators and software environments for testing, all cobbled together for a range of reasons like understanding, human companionship, capitalist efficiency as well as insane longings for other kinds of creation and control. Thus, they are not *copies* of a 'wild type' of robots against which we might compare our laboratory results. Neither alive nor not alive, robots give us a glimpse into the complex imbrications of knowledge, living substance, existence in time and the ability to affect or control any of these things. The more robots that are inserted into our environments and bodies, the more our sense of what it means to live will be transformed; the lines that so clearly seem to be both warrant for an experiment using robots, and argument against it, will fade.

Notes

1 Hamilton's Law (often written $rB > C$) asserts that a behavior that otherwise seems to be unselfish, perhaps even altruistic, might actually confer a fitness benefit (i.e., more success in reproduction) because it increases the chance that not only an individual's genes but those of his close kin will succeed. Hamilton gave precise form to this intuition by introducing a formalism for what is now called 'kin selection': if the organism benefiting from the altruist is sufficiently genetically related to it, then it might cancel out the reproductive cost of that act to the altruist; see Hamilton (1964). The concept has been controversial ever since – whether associated with socio-biology, evolutionary psychology or the biology of social behavior; see Kurland (1980), Okasha (2001) and Leigh (2010).
2 There is some slippage between the terms 'real' and 'physical' – often the scientific publications in question use the two terms interchangeably. I follow their usage where possible, otherwise default to 'physical' to refer to robots that are extended in space, use electricity and are built out of plastic, metal and other materials.
3 See, e.g., Holland (1973) and Goldberg and Holland (1988). For history and critical readings, see Emmenche (1996), Helmreich (1998), Bensaude Vincent (2007) and Riskin (2007).
4 See especially Daniel Dennett's claim that evolutionary theory is "substrate-neutral" (Dennett 1995).
5 On simulation, see Schweber and Wächter (2000), Humphreys (2004), Lenhard et al. (2006) and Grune-Yanoff and Weirich (2010).
6 Figure 3.1 is based on a diagram in Mitri et al. (2013).
7 Mitri et al. (2013).
8 See, e.g., Chrisley and Ziemke (2006). Mitri et al. cite two influential books in cognitive science: Varela et al. (1991) and Clark (1997). See also Pfeifer et al. (2007).

9 Key locations in the corpus include Deleuze (1990, 1983), appendix 1; (1994), conclusion.
10 I rely here on Daniel Smith's reading of Deleuze (Smith 2006).
11 A stronger assertion might be that there is 'ontological indifference' at work here (Galison 2016); perhaps the scientists are no longer concerned about the fundamental features of life, but only about the tractability or manipulability of an experimental setup and the controllability of its components – how to make life (evolutionary theory, social behavior) *do something*.

References

Bensaude Vincent, B. 2007. 'Reconfiguring nature through syntheses: From plastics to biomimetics.' In *The Artificial and the Natural: An Evolving Polarity*, edited by Bensaude Vincent, B. and Newman, W. R., 293–312. Cambridge, MA: MIT Press.

Bergner, R. M. 2011. 'What is behavior? And so what?' *New Ideas in Psychology* 29 (2): 147–155.

Chrisley, R. and Ziemke, T. 2006. 'Embodiment.' In *Encyclopedia of Cognitive Science*, edited by Nadel, L., 10.1002/0470018860.s00172. John Wiley & Sons, Ltd.

Clark, A. 1997. *Being There: Putting Brain, Body and World Together Again*. Cambridge, MA: MIT Press.

Creager, A. N. H., Lunbeck, E. and Wise, N. M. 2007. *Science without Laws: Model Systems, Cases, Exemplary Narratives*. Durham: Duke University Press.

Deleuze, G. 1983. *Nietzsche and Philosophy*. New York: Columbia University Press.

———. 1990. *The Logic of Sense*. New York: Columbia University Press.

———. 1994. *Difference and Repetition*. New York: Columbia University Press.

Dennett, D. 1995. *Darwin's Dangerous Idea: Evolution and the Meanings of Life*. New York: Simon & Schuster.

Emmeche, C. 1996. *The Garden in the Machine: The Emerging Science of Artificial Life*. Princeton, NJ: Princeton University Press.

Floreano, D., Mitri, S. and Keller, L. 2010. 'Evolution of adaptive behavior in robots by means of Darwinian selection.' *PLoS Biology* 8 (1): e1000292.

Floreano, D., Mitri, S., Magnenat, S. and Keller, L. 2007. 'Evolutionary conditions for the emergence of communication in robots.' *Current Biology* 17 (6): 514–519.

Fox Keller, E. 2003. 'Models, Simulation, and "Computer Experiments".' In *The Philosophy of Scientific Experimentation*, edited by Radder, H., 198–215. H. Pittsburgh, PA: University of Pittsburgh Press.

Franklin, L. R. 2005. 'Exploratory Experiments.' *Philosophy of Science* 72 (5): 888–899.

Galison, P. 2017. 'The Pyramid and the Ring: Physics Indifferent to Ontology.' In *Research Objects in their technological Setting*, edited by Bensaude Vincent, B., Loeve, S., Nordmann, A. and Schwarz, A. Abingdon: Routledge.

Garnier, S. 2011. 'From Ants to Robots and Back: How Robotics Can Contribute to the Study of Collective Animal Behavior.' In *Bio-Inspired Self-Organizing Robotic Systems*, edited by Meng, Y. and Jin, Y., 105–120. Berlin, Heidelberg: Springer.

Goldberg, D. and Holland, J. H. 1988. 'Genetic algorithms and machine learning.' *Machine Learning* 3 (2–3): 95–99.

Gould, S. J. 1996. *Full House: The Spread of Excellence from Plato to Darwin*. New York: Random House.

Grune-Yanoff, T. and Weirich, P. 2010. 'The philosophy and epistemology of simulation: A review.' *Simulation & Gaming* 41 (1): 20–50.
Hamilton, W. D. 1964. 'The genetical evolution of social behavior: I.' *Journal of Theoretical Biology* 7 (1): 1–16.
Helmreich, S. 1998. 'Recombination, rationality, reductionism and romantic reactions: Culture, computers, and the genetic algorithm.' *Social Studies of Science* 28 (1): 39–71.
Holland, J. H. 1973. 'Genetic algorithms and the optimal allocation of trials.' *SIAM Journal on Computing* 2 (2): 88–105.
Humphreys, P. 2004. *Extending Ourselves Computfational Science, Empiricism, and Scientific Method.* New York : Oxford University Press.
———. 2008. 'Computational and conceptual emergence.' *Philosophy of Science* 75 (5): 584–594.
Krause, J., Winfield, F. T. A. and Deneubourg, J.-L. 2011. 'Interactive robots in experimental biology.' *Trends in Ecology & Evolution* 26 (7): 369–375.
Kurland, J. A. 1980. 'Kin selection theory: A review and selective bibliography.' *Ethology and Sociobiology* 1 (4): 255–274.
Leigh, E. G. 2010. 'The group selection controversy.' *Journal of Evolutionary Biology* 23 (1): 6–19.
Lenhard, J. 2006. 'Surprised by a nanowire: Simulation, control, and understanding.' *Philosophy of Science* 73 (5): 605–616.
Lenhard, J., Küppers, G. and Shinn, T., (eds.). 2006. *Simulation: Pragmatic Constructions of Reality*. Dordrecht: Springer.
Levitis, D. A., Lidicker, W. Z. and Freund, G. 2009. 'Behavioral biologists don't agree on what constitutes behavior.' *Animal Behavior* 78 (1): 103–110.
Mitri, S., Wischmann, S., Floreano, D. and Keller, L. 2013. 'Using robots to understand social behavior.' *Biological Reviews of the Cambridge Philosophical Society* 88 (1): 31–39.
Morgan, M. S. and Morrison, M. 1999. *Models as Mediators: Perspectives on Natural and Social Science*. Cambridge: Cambridge University Press.
Nelson, R. R. and Winter, S. G. 1982. *An Evolutionary Theory of Economic Change*. Harvard, MA: Harvard University Press.
Okasha, S. 2001. 'Why won't the group selection controversy go away?' *British Journal for the Philosophy of Science* 52 (1): 25–50.
Parker, W. 2009. 'Does matter really matter? Computer simulations, experiments, and materiality.' *Synthese* 169 (3): 483–496.
Pfeifer, R., Bongard, J. and Grand, S. 2007. *How the Body Shapes the Way We Think: A New View of Intelligence*. Cambridge, MA: MIT Press.
Piliavin, J. A. and Charng, H. W. 1990. 'Altruism: A review of recent theory and research.' *Annual Review of Sociology*, 16: 27–65.
Rheinberger, H. J. 1997. *Toward a History of Epistemic Things: Synthesizing Proteins in the Test Tube*. Stanford, CA: Stanford University Press.
Riskin, J. 2007. *Genesis Redux: Essays in the History and Philosophy of Artificial Life*. Chicago, IL: University of Chicago Press.
Rohrlich, F. 1990. 'Computer simulation in the physical sciences.' *PSA: Proceedings of the Biennial Meeting of the Philosophy of Science Association*, 2: 507–518.
Schweber, S. and Wächter, M. 2000. 'Complex systems, modeling and simulation.' *Studies in History and Philosophy of Science Part B: Studies in History and Philosophy of Modern Physics* 31 (4): 583–609.
Smith, D. W. 2006. 'The concept of the simulacrum: Deleuze and the overturning of Platonism.' *Continental Philosophy Review* 38 (1–2): 89–123.

Smolin, L. 1997. *The Life of the Cosmos*. New York: Oxford University Press.
Steinle, F. 1997. 'Entering new fields: Exploratory uses of experimentation.' *Philosophy of Science* 64 (4): S65–74.
Toulmin, S. E. 1972. *Human Understanding : Vol. 1*. Princeton, NJ: Princeton University Press.
Varela, F., Thompson, E. and Rosch, E. 1991. *The Embodied Mind*. Cambridge, MA: MIT Press.
Waibel, M., Floreano, D. and Keller, L. 2011. 'A quantitative test of Hamilton's rule for the evolution of altruism.' *PLoS Biology* 9 (5): e1000615.
Winsberg, E. 2009. 'A tale of two methods.' *Synthese* 169 (3): 575–592.

4 Vanishing friction events and the inverted Platonism of technoscience

Alfred Nordmann

Molecular wires – a fitting tale

In 1997, the nanotechnologists Mark Reed and James Tour presented a remarkable accomplishment which was reported not only in *Science* but a few years later also in *Scientific American* as proof of concept for molecular electronics – that is, for building computers at a molecular scale with organic molecules serving as electric wires.

> The Yale researchers found that the molecule could sustain a current of about 0.2 microampere at 5 volts – which meant that the molecule could channel through itself roughly a million million (10^{12}) electrons per second. The number is impressive – all the more so in light of the fact that the electrons can pass through the molecule only in single file (one at a time). The magnitude of the current was far larger than would be expected from simple calculations of the power dissipated in a molecule.
>
> (Reed and Tour 2000)[1]

Reed and Tour had created a very sophisticated experimental setup, made some measurements and thereby showed that a rather large current could pass through a single organic molecule which could thus be considered a wire for the construction of electronic circuitry. That the current was impressively large valorized their research and made it look all the more important for the purposes of molecular computing. And that the current was larger than one would expect – not only impressively, but surprisingly or perhaps disturbingly large – did not bother the researchers much, even though it is easy to see why it might have. Normally, when one passes too much current through a wire, that wire tends to melt. This is because the electrons bounce around and collide with each other or against the walls of the channel and thus create heat. Reed and Tour never thought for a minute that their accomplishment violated laws of nature or upset the theories of physics in any major way. When they were asked by journal editors in another instance to provide an explanation of what they had produced, they felt a bit put upon, but didn't hesitate to "stick one in."[2]

That they can produce surprisingly large currents fits the general conception that the world is full of unexpected things at the nanoscale. After all, things get pretty complicated at this scale, because this is where classical physics and chemistry come up against quantum physics and chemistry. A theory or explanatory model of electron transport would probably be very complicated and quite local, combining mechanics and thermodynamics, classical and quantum features of the movement of electrons through this type of molecular wire. Granting this and exempting the researchers from the requirement to develop such a theory or model, what remained was to take note of their apparently naïve representation of electrons passing in single file like ducks in a row – which is not how the trajectories of electrons are typically imagined. Equally naïve appeared their visual representations that showed the wire at molecular resolution as a string of atoms, whereas the electrodes were pictured as bulk material, shiny pieces of macroscopic gold (Reed et al. 1997; Reed and Tour 2000). This naïveté appeared to be the privilege of nanotechnoscientists who do not think very deeply about theoretical matters, but manage to open a whole new world of surprising possibilities.

The story so far agrees very well with a caricature of strategic or application-oriented research and with a philosophical definition of technoscience: Science is knowledge production by *homo depictor,* possibly requiring a lot of technology to produce phenomena, theories or representations. Technoscience, in contrast, is knowledge production by *homo faber*; though it uses theories to get things to work, it does not aim to develop and test theories.[3] Instead of classifying fields of research, institutes or laboratories, individual researchers or their research programs as either scientific or technoscientific, the proposed definition is adapted to very particular contexts of research. For any given research publication, one can tell how it queries its objects of research. Does it assume the posture of *homo depictor* by ascertaining facts about these objects in order to confirm or test a hypothesis regarding their nature? Or does it assume the posture of *homo faber* by showcasing a laboratory achievement which probes the power and potential of the research objects? "Look," *homo faber* says, "we managed to synthesize this molecule, we created a deliberative setting in which citizens voiced their concerns, we passed an impressively large current through a single organic molecule."

Technoscientific theory – this is the question

Due to the criterion that theory development and hypothesis-testing comes with the representational business of science, and that the business of technoscience involves the acquisition and demonstration of capabilities of control, it is at first glance merely tautological to say that there can be no such thing as theory development in technoscience. And just as the phenomenon of dusk does not militate against the distinction between night and day, neither does the existence of apparent gray zones in the case at hand: Yes, scientific hypothesis-testing can involve experimentation in the course of which new technical capacities are developed, but this would be a case of technology being subservient to theory development. And of course, there are many theorists who model what goes on in technoscientific research. They develop new models all the time, but this usually involves the use of bits and pieces of already available theory to reproduce and literally rebuild the process at hand in the computer as a technical system. This kind of theoretical work in technoscience pursues conceptual and predictive control rather than the improvement of theories or hypotheses. One telltale sign of this is that the work of modelers rarely revolves around 'truth' or the discovery of the one best and only acceptable model. Here, the very ability to rebuild a phenomenon from elements of theory serves as an explanation that provides understanding, irrespective of the possibility of the proliferation of such 'explanations' – a term that evidently assumes a different meaning in the context of technoscientific research.

It is therefore obvious that at first glance there cannot be technoscientific theory development. But at second glance we just encountered the case of 'explanation' or 'understanding' and its different meanings in the context of science and of technoscience – in the first case having to do with veracity and intelligibility, and in the second case with the competence to build a simulacrum of a laboratory process or phenomenon. So, one might ask, if there are scientific and technoscientific kinds of explanation and understanding, might there not also be technoscientific theory – as such, a very different kind of animal than scientific theory? And what does this kind of theory take as its objects?

Lessons from nanoelectronics

In the years after Reed and Tour published their results, it became a kind of commonplace to call for better theories of electron transport. In the meantime, experiments proceeded and review articles were published that sidelined the actual need for this theory (Tao 2006). But then a theory arrived that had been there all along. As a physical theory of conductance, it took the relation between thermodynamic and mechanical aspects of the movement of electrons as its object, including experiments on entropy as it gave way at smaller scales to quantum effects. By becoming a technoscientific theory for and of nanoelectronics, it took as its objects various nanoscale devices, such as the one constructed by Tour and Reed.[4]

The theory in question has been articulated by Supriyo Datta and goes back as far as 1957 to work by Rolf Landauer. Quite in the spirit of modern empiricism, Datta's theory does not claim to rival or supersede other theories, but recommends itself primarily as a 'pedagogically attractive' way to understand resistance. Indeed, it can be viewed as a general set of instructions on how to represent phenomena of conductance by seeing them as a product of spatially separable components, namely thermodynamic and mechanical motion, entropic and ballistic behavior, the contact and the channel.[5] Ballistic motion is purely mechanical and conceives of electrons moving through a channel like ducks in a row without generating heat by way of collisions. Heat is thermal energy, associated with increase in entropy, and sets an upper bound to the amount of current that can pass through any wire: increase in current results in an increase in collisions and hence increases the amount of heat dissipated. According to Datta, one can understand phenomena of resistance and conductance in a bottom-up manner, by seeing how the mechanics and thermodynamic of electron transport come together at any given scale and for any given device.

Now, if one wants to decompose electron transport into two components for analytic reasons, a physical regime would appear as a theoretical possibility where these might actually separate out. This would result in a simple world where the thermodynamics is all in the contacts, whereas ballistic movement alone rules the passage of electrons through a channel, wire or organic molecule. This pedagogically attractive theoretical possibility is of no importance to the person who seeks to understand electron transport as a product of the two components. But when nanotechnology began to explore the physical regime where thermodynamic and mechanical action do separate out, and when Tour and Reed apparently produced an actual case of purely ballistic transport in a molecular wire, the general scientific theory of conductance became a technoscientific theory of nanoelectronic devices, telling us not so much what these are, but what they ought to be.[6]

Inertial devices – an analogy

When Supriyo Datta speaks of "Lessons from Nanoelectronics," he suggests that the nanotechnological device builders produce experimental evidence for his bottom-up approach. And when he speaks of "Lessons for Nanoelectronics," he suggests that device builders can derive design principles from his theoretical considerations. Taken together, these two suggestions indicate that the nanoelectronic devices can be classified neither as experiments nor as applications of theory. Indeed, Datta did not seek to specifically explain or model Tour and Reed's experimental system and measured values, nor did he claim that his bottom-up approach is more likely to be true thanks to molecular electronics. At the same time, Tour and Reed were not in the business of hypothesis-testing or of realizing a predicted possibility. Their molecular wire was the technical achievement of mastering a so-called break junction to obtain a measure of a current that was passing, presumably, through a single molecule. Impressive as it was, this technical achievement was to foreshadow something better, namely a perfect nanodevice that can be part of a computer that could be built from such devices – and it foreshadowed the more perfectly engineered device independently of any theory, including Datta's.

What is then the object of a technoscientific theory of nanoelectronic devices? Is it a specific phenomenon in the laboratory, or is it its perfectly engineered counterpart? Is it an idealized theoretical construct, is it the realization of a possibility, or is it a not-yet realized engineered device? A somewhat crude analogy may help us better understand what happened when Datta's scientific theory of the physics of conductance became a technoscientific theory of engineered nanoelectronic devices. It is an analogy to the most famous of physical theories, namely Newtonian mechanics. While Newton's point-masses are abstractions from or idealizations of spatially extended actual physical bodies, his first law of motion is not an idealization. For analytic purposes, it specifies one vector or component of motion, namely that – unless acted upon – bodies persist in a straight line in their state of motion. Any particular real motion is the product of several actions that each unfold in a straight line, such that to posit rectilinear uniform perpetual motion provides a backdrop to all representations of motion.

There are many ways to probe and test Newtonian theory, but there is one claim which it does not make and that is not subject to testing, namely that one could actually produce phenomena of rectilinear uniform perpetual motion. This would be to misunderstand the first law: It refers to something behind the actual phenomena of motion, to something that is underlying or prior to the phenomena. It does not to refer to a phenomenon in its own right.

But let us now imagine that scientists and engineers find a way to shield or cancel gravitational forces and all other influences. Suddenly it would appear to be not merely a theoretical but a technical possibility to produce uniform rectilinear motion, at least to approximate it. And now, Newton's theory would have become a theory of inertial devices, that doesn't take all phenomena of motion as its object, but objects at the vanishing point of engineering – that is, objects that foreshadow what might be achieved by impossibly perfect technical control. This I refer to as the inverted Platonism of technoscience.

From archetypes to prototypes

The analogy is crude, but it exhibits the Platonism of scientific theorizing with its search for structures behind the appearances, as well as the inverted Platonism of technoscientific theorizing which refers to objects just beyond the grasp of engineering. Here the term "Platonism of Science" should not be mistaken for the claim that all scientists are actually committed to Platonism, but only that they are always confronted with the question whether science reveals a true reality behind the appearances or whether it imposes a structure upon them. To be sure, empiricists of all stripes reject the metaphysical picture of a world below, behind and prior to experience, but instead endorse the view that appearances are systematically organized by scientific theories.[7] Either way, theory is the discovery of structures that intelligibly organize phenomena and all that lies within the realm of experience. And either way, the objects of knowledge are akin to *Urbilder* or archetypes that are investigated for their general features.

For a technoscientific theory of nanoelectronic or of inertial devices, the object of knowledge is the vanishing point of engineering. It is not what lurks behind the appearances or what systematically organizes experience, but it is an image of a perfection that is yet to be attained. The inverted Platonism of technoscience does not attend to immutable features of reality, but anticipates what lies just beyond reality. It is not concerned with phenomena that are instances, copies or shadows of *Urbilder* and archetypes, but is interested in proofs of concepts – that is, with exhibitions of *Vorbilder* or prototypes. Technoscientific practice and technoscientific theorizing seeks to perfectly control the phenomenal world by imagining elusive objects, as elusive as the vanishing point of engineering. All of this can be further explored by considering another example, namely technoscientific theories of friction that take so-called vanishing friction events as their objects.

Superlubricity

"Surface Grime Explains Friction": This is the headline of a news item published on February 8, 2001, on the website of *Physical Review Letters,* reporting on a paper by Martin Müser and others (Müser et al. 2001; Sincell 2001). At first sight, the headline may provoke disbelief. It conjures the image of frustrated tinkerers who were thwarted in their attempts at constructing a perpetual motion machine, utterly convinced that it was only a question of grime. If only they could get things to work properly in a pure and pristine manner, then surely there need not be a loss of energy to friction. Since, presumably, new insights about friction will not prove these crazy tinkerers right, one needs to take a closer look at the theory that explains friction by grime. The website notice helps considerably by contrasting vividly the old view and the new theory of friction.

"The standard frictional force law makes sense only if the two surfaces are identical crystals, fitting together at the atomic level like a pair of gears. New theory and computer simulations show that for mismatched surfaces, a molecular coating of grit explains static friction" (Sincell 2001). Where the standard frictional force law makes sense, grime does nothing to explain friction. But in the case of mismatched surfaces and "according to the modern microscopic theory of interacting surfaces, [friction] shouldn't exist at all" – were it not for the grime.

The old view was macroscopic, to be sure. It continues to inform the understanding of ordinary people as well as engineers, but from the microscopic point of view this old normal looks like a very special, exceptional case.

> The laws of static friction were first laid down by Guillaume Amontons, a 17th century French researcher who believed that surfaces have regular jagged edges, like the teeth in a bicycle wheel gear. When two surfaces meet, his theory said, the teeth of the upper surface settle into the grooves of the lower surface. To get the upper surface sliding, a lateral force has to lift the teeth out of the grooves.
>
> (Sincell 2001)

For the sake of illustration, the story of the interlocking teeth is sometimes told in reference to two egg cartons where the one on top is turned upside down and fails to slide across the one at the bottom because their "teeth" will soon interlock (e.g., Popov and Gray 2014, 159; Wikipedia contributors 2015). But take those same cartons and position the top one in such a way that its surface structure is askew to the one at the bottom, the two now mismatched surfaces would not interlock anymore, the one on top sliding rather effortlessly over the one at the bottom – were it not for an egg in the carton that will spoil the ride. If the two mismatched egg cartons stand for incommensurate surfaces, the egg represents grime. "Theoretical models predict zero static friction for mismatched surfaces [but] virtually every pair of surfaces lock together and resist sliding. The key is that impurities like dust, dirt, and stray molecules, coat nearly every surface and prevent smooth motion" (Sincell 2001).

Is this too good, too simple to be true – or is it a deliberately simple rendition of the world so as to make more credible that it will yield all the way down to human engineering? A more technical presentation of the theory reveals how its apparent simplicity comes at the expense of infeasibility:

> The frictional force vanishes when surfaces contact incommensurately. Since in the incommensurate contact the ratio of the primitive vector of the upper body to that of the lower body is irrational along a certain direction, neither the magnitude nor the direction of the force exerted on the upper atom coincide with each other along the direction. Therefore the frictional force, which is obtained by summing the force on each atom along the sliding direction, vanishes on incommensurate contact.
>
> (Shinjo and Hirano 1993)

We now find out just what is meant by "mismatched surfaces." On any surface there is a distribution of atoms and thus of peaks and valleys that attract or repel the peaks and valleys on another surface. If the two surfaces are incommensurate throughout, attractions and repulsions cancel each other out. When frictional forces are summed, their positive and negative magnitudes come to zero and the force vanishes. This, of course, is a statistical argument and like most such arguments, it relies on large – indeed infinitely large – numbers. Only when infinitely many magnitudes are summed will they have all cancelled out. Aside from being incommensurate throughout, the two frictionless surfaces thus need to be infinitely large, unbounded also in not having edges where atoms are configured differently.[8] So for any two infinitely large incommensurate surfaces, theory predicts zero friction – were it not for atomic grime, that is, any stray atom that might insinuate itself between these surfaces.

Idealized or simple

So far we heard of a 'theory of friction' and left open whether it is an example of scientific or technoscientific theorizing. When there is no friction whatsoever between two surfaces, their sliding motion could be perpetual. The theory thus speaks of 'superlubricity' as a state of matter akin to that of 'superconductivity'. Or is it more mundane than that, and does the theory only identify a theoretical possibility that is implied by general knowledge of attraction and repulsion between surface-atoms?[9]

Datta's theory of conductance was perfectly general when it suggested a way to analyze phenomena into their mechanical and thermodynamic components. In the context of technoscientific nanotechnology, it became a special theory of nanoelectronic devices that lie beyond the reach of engineering. The new theory of friction is perfectly general in distinguishing the cases of commensurate and incommensurate surfaces. And again in the context of nanotechnology, it zeroed in on a theoretical possibility and became a special theory of sliding devices or vanishing friction events that lie beyond the reach of engineering. And just like

Datta's, this theory upends the common view that things are highly complex at the nanoscale because classical and quantum effects intermingle there and because the most minute physical differences can make a large difference in observed effects. Instead, the nanoworld appears to be simple, if not pristine. The world gets more complex at higher scales. Here, the channel and the contact can separate out and nothing complicates the picture of ballistic electron transport. And there, frictionless sliding is possible because stray atoms can be controlled.

How is this simplicity a hallmark of technoscientific theorizing with its inverted Platonism? It would appear, after all, that all theorizing seeks simplicity, and the Platonic ideas, for example, are simple in comparison to the appearances that carry the impurities of becoming and decaying and the multitude of forces acting upon them. Plato's 'ideas' stand at the cradle of what are called 'idealizations,' whereas the simplicity of ballistic transport in a nanowire and superlubric sliding are not at all idealized, but posit real phenomena in an all but contingently impossible world of perfect technical control.[10]

Galileo's inclined plane is an idealization, if ever there was one. It discounts friction but does not posit frictionless sliding. It would be to misunderstand such idealizations if one took them as description of phenomena in their own right. Instead, idealizations provide schematic understanding or a model of the phenomena. To be sure, one can de-idealize one's models and make them more descriptive of the particulars. Such de-idealization requires empirical enrichment – one has to stick back into the model one has previously left out.

The theory of frictionless sliding does not seek to decompose all cases of static friction into the components of surface characteristic and grime. Instead, it takes the nanoworld and its complexities as simple – the infinitely large surfaces and the averaging of frictional force over all atoms is considered to be the normal case, the paradigm phenomenon in a pristine molecular world where one might get rid of atomic grime. Instead of needing to be de-idealized or enriched empirically, superlubric sliding is a theoretical possibility that has become an engineering challenge.[11]

Frictional duality

Newton's First Law is not a theory of inertial devices. And one does not test an idealized model by taking it literally, seeking to produce idealized conditions for real. But this is how the new theory was taken:

> "This is a general philosophy of friction that is very compelling," says Miquel Salmeron of the Lawrence Berkeley National Laboratory in California. "The idea that [impurities] could influence friction had been floating around for years, but people hadn't thought of it in this simple form until now." Salmeron's group is one of many that are trying to carry out the challenging experimental verification of the theory. "It is very difficult to make sure that no extra atoms are present," he says. "But the challenge is there, so people will find a way."
>
> (Sincell 2001)

The challenge is there and a group of researchers at the University of Münster took it up in 2008, receiving considerable acclaim for their effort.[12] They performed a difficult experiment and analyzed it ingeniously, thus exhibiting utmost scientific rigor while running up against the ambivalences of their impossible dream of technoscientific reason. Before the backdrop of Müser's theory, they constructed a surface and placed islands of comparatively large nanoparticles on it (up to 310,000 nm^2, that is, with a side-length of about 0.00557 mm). They gave these nanoparticles a gentle sideways push with the tip of an atomic force microscope and observed their subsequent sliding behavior. This is where 'frictional duality' came in: most of these nanoparticles displayed familiar behavior and ground to a halt; others exhibited frictionless sliding. This experimental finding was then subjected to a form of reasoning that philosophers of science call 'inference to the best explanation' or 'eliminative induction'. André Schirmeisen and his collaborators formulated four different scenarios to explain the duality. The one that survived and was adopted as the only plausible account agrees in one important respect with Müser's simple microscopic theory: "Surface grime explains duality."

> While clean interfaces may exhibit superlubric behaviour due to structural mismatch (particles may be crystalline or amorphous), even small amounts of mobile molecules (such as hydrocarbon or watermolecules) trapped between the sliding surfaces can cause a breakdown of the superlubric behaviour. Often referred to as 'dirtparticles', these molecules are able to move to positions where they simultaneously match the geometry of both top and bottom surfaces, thus augmenting the height of the bottom surface in a way that matches the (atomic-scale) undulations of the top surface.
>
> (Dietzel et al. 2008, 125505–2)

Frictional duality, however, might be referring not only to the two groups of nanoparticles. It could also refer to the various symptoms of ambivalence in a rigorously scientific paper that is dedicated to technoscientific theorizing. This ambivalence sets in on a methodological level. Hypothesis-testing takes place at the level of the scenarios that might explain the different behaviors. But how are these scenarios generated and what is the overall thrust of this research paper?

Even though there is little empirical evidence of superlubricity – and that evidence should be hard to come by according to the theory's own assumptions – the radical change of perspective from the classical 'Amontons'-paradigm is already presupposed: "If superlubricity is the 'natural' state for crystal surfaces, why is friction so common?" (Monroe 2008) Instead of being puzzled that perpetual motion should be possible at all, the question is now why not all of the nanoparticle-islands assume that state. Apparently, it does not call for explanation (e.g., in terms of measurement error) why some nanoparticles do not show signs of friction. Instead what needs to be explained is why most nanoparticles behave according to Amontons's law. But while the questioning was thus directed toward a technically possible but practically unachievable goal, there remained the worry about the very term 'superlubricitiy', which implied an absolutely frictionless state of matter. Instead, talk of 'vanishing friction events' places these on a trajectory of ever less friction, withholding judgment as to whether superlubric behavior can or will ever be manifested. But in the end, talk of superlubric behavior and vanishing friction proved interchangeable, the scruples giving way and the conflation taking hold of what has been achieved technically and what is yet to be achieved. In the spirit of inverted Platonism, what will be in a perfectly controlled world is as good as real already.

Another source of ambivalence comes from the experimental findings that appear to confirm that grime might make the difference, but in other respects do not so readily agree with the model of microscopic friction. For example, that smaller islands conformed better than large ones was not to be expected from the new theory and has since led to further work (Dietzel et al. 2013). And it would appear even stranger yet that a single stray hydrocarbon may well have been the culprit for the breakdown of superlubricity, while at the same time the phenomenon was observed nonetheless also in contaminated ambient conditions. Frictionless sliding in ambient conditions contradicts the theory of superlubricity taken literally but agrees with its orientation toward a transformation by engineering of the world as we know it: "In any case, the ambient manipulation experiments prove that the superlubric state can, if only rarely and for short distances, be observed even under strongly contaminated conditions, which gives hope for future technical applications of frictionless sliding" (Dietzel et al. 2008, 125505–4).

What could that be – the technical application under ambient conditions of frictionless sliding? It would make a contribution to that field of engineering which seeks to reduce the loss to friction in the running of machines – that is, to the field of tribology. While a general theory of friction at the macro- and microscales provides a framework, background or general terms for tribology, a theory of superlubric devices proposes to extend tribology and to continue as nanotribology where tribology leaves off. But here is, finally, another dimension of ambivalence – for what good does a nanotribology do that willfully turns on its head the knowledge and the experience of tribologists?

Tribology is the science and engineering of friction, lubrication and wear. In regard to friction and lubrication, there are three tenets of tribology or three things that everyone knows for sure. First, there is Amontons's law as the basic law of tribology: Friction increases in a linear way with contact area. Second, there is the engineering maxim that follows from it: contact area should be minimized and lubricants do much of this job because, like ball-bearings, they reduce the contact area. And third, the progress of tribology lies in the improvement of surfaces and lubricants such that there is less and less loss of energy to friction, less and less wear on machine parts, less and less use of lubricants: while frictionless or perpetual mechanical motion is impossible, the progress of tribology consists in moving toward this elusive goal. In contrast, for nanotribologists and their theory of superlubric devices, contact area might have to be maximized if the superlubric state involves two infinitely largely surfaces that are incommensurate throughout. And if not a stray atom is to be tolerated between these two surfaces, neither are lubricants, which are now just so much grime. Perpetual motion is not an absolute limit to engineering that serves simultaneously as a constraint and as a challenge to gradually approximate it, but it is the normal state which would be realized by perfect technical control in a presently imperfect world. The predicament and ambivalence of nanotribologists in the machine culture of tribology is thus revealed by the telling turn of phrase 'technical application of frictionless sliding': whereas tribology is traditional engineering that seeks to reduce friction in technical applications, technoscientific nanotribology conceives of frictionless sliding as its object at the vanishing point of engineering and as something that stands ready to be applied.[13]

The dream of technoscientific reason

The notorious perpetual motion machine is emblematic of a technoscientifically inverted Platonism, and so is the nanowire, the frictionless surface, meta-materials or graphene. The inaccessible rational objects and pure forms of Plato's world of ideas are displaced by the dreams of technoscientific reason, by inaccessible devices that can never exist and yet figure as ends of technological development – the nanoelectronic, the inertial, the superlubric devices as objects of technoscientific theorizing.

This dream calls into question the very idea of a limited world to which technology must adapt itself. It does so by showing how an in-principle possible and only practically impossible apparatus actually works. Indeed, it shows that perpetual motion is possible, but that we can never fully produce the technical conditions to actually produce it. And similarly, by offering a glimpse of 'superlubricity' or of frictionless motion, nanotribologists offer a proof of concept that is not a proof of technical feasibility. Instead of forging a link between theoretical and practical possibility (which is what proofs of concept are usually asked to do) and instead of showing what can be done from now on in the world as we know it, nanotribologists replace a theoretical notion of physical impossibility with a notion of mere practical impossibility, and they show us that they can do things that can only be done in a world that is nothing like the dirty, grimy, unclean world we know, that can only be done in another world – namely in one where engineering control is perfected and perpetual motion is possible.

Conclusion

The difference between science and technoscience is not one of pure and applied research. This has been emphasized previously, by pointing to a pure technoscience that acquires basic capabilities of manipulation, visualization or modeling. In addition, there is a pure technoscience also in the sense of developing and testing technoscientific theories. In virtue of being theories of technoscientific objects, they do not so much represent, explain or describe phenomena in order to show what these phenomena are or to elucidate their structure. The objects of technoscientific theories are proofs of concept that are oriented toward the vanishing point of engineering. These are objects of knowledge that lie not behind or below the phenomena, but just beyond the horizon of technical possibility.

Bruno Latour famously asked about science: Where were the microbes before Pasteur? His answer: they were constructed as something that always existed (Latour 1999). In contrast, there are no answers to the questions: Where is frictionless sliding, and where is the molecular wire before or after nanotribology and molecular electronics? These are constructed neither as something that always existed nor as something that will exist, but as something that could exist if we were not constrained by our conditions of life on earth, and perhaps as something that should appear on the trajectory of our engineering efforts always to overcome the constraints of our conditions of life on earth.

In this sense, somewhat like the perpetual motion machine of the past, the vanishing friction event carries the promise of enormous effort and ingenuity producing a world that is fundamentally simple, not in the sense of being mathematically analyzable and yielding to the mind, but in the sense of being rationally engineered and yielding to our (re)creations.

Notes

1 This popular presentation is based on Reed et al. (1997); compare Nordmann (2004a).
2 James Tour, personal communication referring to Chen et al. (1999).
3 Alfred Nordmann, Bernadette Bensaude Vincent, Sacha Loeve, and Astrid Schwarz, "Science vs. Technoscience – A Primer." See under *Research* at http://www.philosophie.tu-darmstadt.de/nordmann.
4 Compare this story to that of the industrial product Aerosil®, which had been manufactured for many decades when it suddenly changed its character, if not its nature, by taking on the new identity of a nanomaterial; see Nordmann (2010).
5 According to Datta, alternative top-down approaches that begin from macroscopic phenomena are "conceptually challenging" due to "the intertwining of dynamical and entropic forces. By contrast, the ballistic limit leads to a relatively simple model with dynamical and entropic processes separated spatially. Electrons zip through from one contact to the other driven purely by dynamical forces. Inside the contacts they find themselves out of equilibrium and are quickly restored to equilibrium by entropic forces, which are easily accounted for simply by legislating that electrons in the contacts are always maintained in local equilibrium. We could call this the 'Landauer model' after Rolf Landauer who had proposed it in 1957 as a conceptual tool for understanding the meaning of resistance, long before it was made experimentally relevant by the advent of nanodevices. Today there is indeed experimental evidence that ballistic resistors can withstand large currents because there is negligible Joule heating inside the channel. Instead the bulk of the heat appears in the contacts, which are large spatial regions capable of dissipating it. I consider this separation of the dynamics from the thermodynamics to be one of the primary reasons that makes a bottom-up viewpoint starting with ballistic devices pedagogically attractive" (Datta 2010).
6 For candid statements to this effect see also Datta (2004, 2012).
7 This is where the reflections in this chapter meet up with Peter Galison's contribution to this volume. Platonists and empiricists share a concern for matters of ontology, for the status of data on the one hand, the entities postulated by theory on the other hand. Similarly, not all scientists are realists, but the contest of realism and the various brands of idealism or constructivism come with the business of representation. In contrast, technoscientific research, including technoscientific 'theorizing' with its inverted Platonism, is ontologically indifferent.

8 By the same token, larger surface areas are associated with a greater probability that surfaces aren't perfectly mismatched. Accordingly, there have since been more detailed investigations of surface properties at different scales; see Dietzel et al. (2013).

9 Most researchers, including Schirmeisen and Fuchs (in conversation), are careful and a bit suspicious of the term 'superlubricity', pointing out that it concerns only the absence of static friction force which leaves the frictional effects of interactions with electrons and phonons. In Dietzel et al. (2008) the term 'superlubric' is employed, however, though the first time in quotation marks.

10 What might appear as an idealization for the purpose of scientific theorizing is a phenomenon for technoscientific theory – albeit a phenomenon that is only at the cusp of reality, on the trajectory of engineering but beyond its reach: "Vanishing friction" refers to the vanishing point of engineering. Accordingly, the presumed simplicity of the nanoworld is a model for the macroscopic world not because it provides understanding – as if macroscopic phenomena are easier to grasp when viewed from the bottom up. More likely, it is a model for the macroscopic world because it has been disclosed by feats of most advanced engineering. As opposed to the messy and impure, if not grimy, macroscopic world, the nanoworld testifies everywhere to the power and promise of technological rationality; see Nordmann (2004b).

11 Technoscience theorizes things as simple so as to render a world that is subject to technical control. This is particularly evident in the case of nanomedicine. See also the title of the paper by Müser et al. (2001) ("Simple Microscopic Theory"), which promises and delivers a disarmingly simple theory.

12 Like Müser et al. before them, their findings were promoted by *Physical Review Letters* (Monroe 2008), with reports also in several German newspapers.

13 Technoscience, so the stereotypical story goes, is what happens to the sciences once an engineering mentality gets hold of them. Technoscience transforms perfectly respectable disciplines that were once concerned with rather abstract and theoretical matters into design practices that aim to build new devices and new worlds. This is said to be driven by the ambition to take available technological and conceptual tools to control ever more complex phenomena and processes – for example, to effect things at the nanoscale where the worlds of classical and quantum physics meet, where complexity reigns supreme, and where one cannot expect new theories or laws but only new capabilities of managing complexity. According to this story, as the pure and natural sciences progress and reach their limits, they become technoscience. But we are seeing now that it can also work quite the other way around, technoscience can turn engineering on its head, adding a bewildering twist to perfectly good application-oriented research, offering new theories with attendant proofs of concept that defy common sense, that offer nothing of practical value but a way of thinking that is so unencumbered by traditional preconceptions that it appears simultaneously compelling and naïve. This is what happened when tribology 'moves forward' to ever smaller scales to become the kind of nanotribology described here.

References

Chen, J., Reed, M. A., Rawlett, A. M. and Tour, J. M. 1999. 'Large on-off ratios and negative differential resistance in a molecular electronic device.' *Science* 286: 1550–1552.

Datta, S. 2004. 'Electrical Resistance: An Atomistic View.' In *Physics of Electronic Transport in Single Atoms, Molecules, and Related Nanostructures*, edited by Friend, R. H. and Reed, M. A., special issue of *Nanotechnology* 15: S433–S451. Bristol: IOP Publishing.

———. 2010. 'Nanoelectronic Devices: A Unified View.' In *The Oxford Handbook on Nanoscience and Nanotechnology: Frontiers and Advances*, vol. 1, chapter 1, 1–21, edited by Narlikar, A. V. and Fu, Y. Y. Oxford: Oxford University Press.

———. 2012. *Lessons from Nanoelectronics: A New Perspective on Transport*. Singapore: World Scientific Publishing Company.

Dietzel, D., Feldmann, M., Schwarz, U. D., Fuchs, H. and Schirmeisen, A. 2013. 'Scaling laws of structural lubricity.' *Physical Review Letters* 111: 235502.

Dietzel, D., Ritter, C., Mönninghoff, T., Fuchs, H., Schirmeisen, A. and Schwarz, U. 2008. 'Frictional duality observed during nanparticle sliding.' *Physical Review Letters* 101: 125505.

Latour, B. 1999. 'The Historicity of Things: Where Were Microbes before Pasteur?' In *Pandora's Hope*, 164–173. Cambridge: Harvard University Press.

Monroe, D. 2008. 'To slide or not to slide.' *Physical Review Focus* 22: 10, http://physics.aps.org/story/v22/st10 (accessed April 11, 2016).

Müser, M. H., Wenning, L. and Robbins, M. O. 2001. 'Simple microscopic theory of Amontons's laws for static friction.' *Physical Review Letters* 86: 1295.

Nordmann, A. 2004a. 'Molecular Disjunctions: Staking Claims at the Nanoscale.' In *Discovering the Nanoscale*, edited by Baird, D., Nordmann, A. and Schummer, J., 51–62. Amsterdam: IOS Press.

———. 2004b. 'Nanotechnology's worldview: New space for old cosmologies.' *IEEE Technology and Society Magazine* 23 (4): 48–54.

———. 2010. 'Enhancing Material Nature.' In *Nano Meets Macro: Social Perspectives on Nanoscale Sciences and Technologies*, edited by Kjølberg, Kamilla Lein and Wickson, Fern, 283–306. Singapur: Pan Stanford.

Popov, V. L. and Gray, J. A. T. 2014. 'Prandtl-Tomlinson Model: A Simple Model Which Made History.' In *The History of Theoretical, Material and Computational Mechanics*, edited by Stein, E., 153–168. Berlin: Springer.

Reed, M. and Tour, J. M. 2000. 'Computing with Molecules.' *Scientific American* 282 (6): 86–93.

Reed, M. A., Zhou, C., Muller, C. J., Burgin, T. P. and Tour, J. M. 1997. 'Conductance of a molecular junction.' *Science* 278: 252–254.

Shinjo, K. and Hirano, M. 1993. 'Dynamics of friction: Superlubric state.' *Surface Science* 283: 473–478.

Sincell, M. 2001. 'Surface grime explains friction.' *Physical Review Focus* 7: 6, http://physics.aps.org/story/v7/st6 (accessed April 11, 2016).

Tao, N. J. 2006. 'Electron transport in molecular junctions.' *Nature Nanotechnology* 1: 173–181.

Wikipedia Contributors. 2015. 'Superlubricity,' *Wikipedia, The Free Encyclopedia*, https://en.wikipedia.org/w/index.php?title=Superlubricity&oldid=690294225 (accessed April 10, 2016).

5 From the birth of fuel cells to the utopia of the hydrogen world

Pierre Teissier

In May 2007, five members of the European Parliament proposed the Written Declaration 16/2007 to establish "a distributed green hydrogen economy" and to implement "a third industrial revolution in Europe" by 2025 (Guidoni et al. 2008). Hydrogen is expected to replace hydrocarbons and become the major source of energy of the global economy. Oil, as well as natural gas, coal and wood are carbon-based fuels, whose combustion in heat engines converts chemical energy into mechanical power (Loeve and Bensaude Vincent 2017, 194). The steam engine, improved by James Watt in the 1760s, was the first of these heat engines, fuelled by coal. This allowed the 'first industrial revolution' during the nineteenth century in Europe: the exponential growth of coal combustion sustaining the exponential growth of industrial activities, especially in textile and mining. From the 1880s onward, the electrical dynamo – a new device converting mechanical force into electrical power – allowed the building of power plants, in which heat engines were combined with dynamos to convert the chemical energy of coal into electricity. This was an important step in the 'delocalization of power'[1] through electrical networks. In the meantime, the internal combustion engine led to the massive development of the motor industry, where the multiplication of private cars in operation on the roads during the twentieth century. Electrical networks and private cars carried the 'second industrial revolution'. By the late 1960s, petroleum had become the primary source of energy in the world. In two centuries (1770–1970), heat engines and power plants replaced wind and water mills to provide energy for humanity. The thermal forces of petroleum and coal replaced the mechanical forces of water and wind. Despite a few whistle-blowers who worried about the exhaustion of oil and coal in the subsoil and the poor conversion efficiency of heat engines, the industrial carbon-based economy grew rapidly.

The 'hydrogen economy' of the third millennium

Gradually, however, experimental clues convinced more observers that an unnatural warming of the global climate was on the way (Weart 2008). The issue became a hot topic, and a large independent group of scientific experts – the Intergovernmental Panel on Climate Change (IPCC) – came together from 1988 onward to establish a worldwide analysis of the situation. All the IPCC successive reports converged to stress the influence of greenhouse gases of anthropogenic origin, especially carbon dioxide. Carbon dioxide is an atmospheric gas produced by the respiration of living organisms. It is responsible, with other gases like water and methane, for the natural greenhouse effect that sustains life on earth. It is also a product of the combustion of hydrocarbons and coal. The massive growth of heat engines in operation throughout the world had increased the level of carbon dioxide in the atmosphere by around one-third, compared with that before the industrial era. This man-made increase of carbon dioxide has "very likely" resulted in global warming during the twentieth century, as stated by the fourth (2007) and fifth (2014) IPCC reports.[2] Carbon-based fuels were made responsible of the climate change. Their relative consumption, in a given country, was proportional to the degree of industrialization and of wealth. For most rich countries, oil and coal have become 'politically incorrect' in the public opinion. Under European leadership, the 1995 Kyoto Protocol tied countries willing to reduce their domestic greenhouse gas emissions. Nevertheless, the global emission kept increasing. While the self-regulation of carbon emissions reached an impasse at the global scale, the European Union (EU) committed in 2008 to cutting its own emissions to 20 percent below 1990 levels by 2020.

The time was ripe for the EU to strike up an 'energy transition' from the carbon economy of the twentieth century to the hydrogen economy of the twenty-first century (Rifkin 2002).[3] Newspapers presented a black and blue opposition between the 'dark dirt' of carbon-based fuels and the azure cleanness of water-based fuels. High-technology multinational companies like United Technologies Corporation advertised childish visions of the hydrogen economy (Figure 5.1).

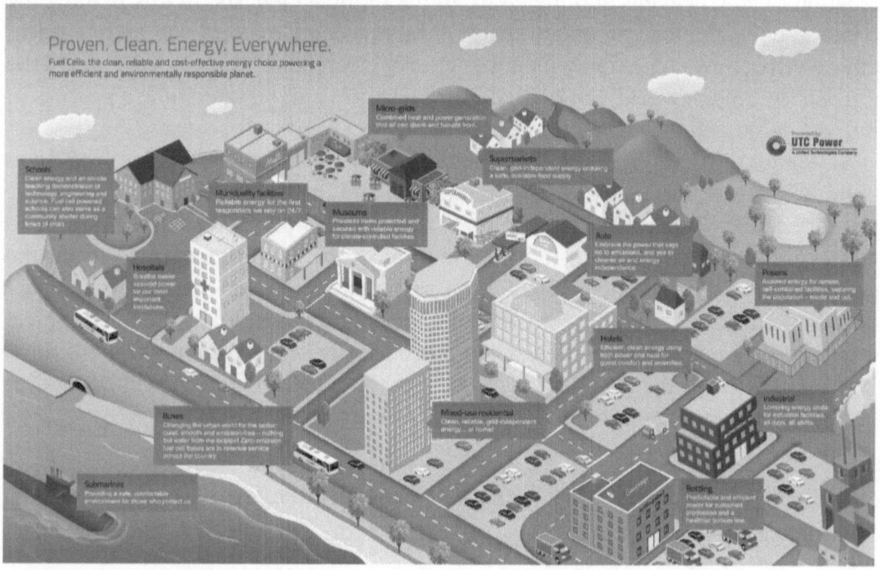

Figure 5.1 Advertisement for the hydrogen economy, UTC Power (2013).

There, fuel cells played the same role as heat engines for carbon economy by converting the chemical energy of hydrogen into electrical power. A fuel cell was an electrochemical generator that produced electrical power from the chemical reaction of two gases, hydrogen and atmospheric oxygen. There was an oxidation at the anode: hydrogen was oxidized according to the following half-reaction, $H_2 = 2H^+ + 2e^-$. There was a reduction at the cathode: oxygen was reduced according to the following half-reaction: $\frac{1}{2}O_2 + 2e^- = O^{2-}$. In between the anode and the cathode, the electrolyte allowed the electrical contact and prevented the mixing of the compounds. Instead of emitting carbon dioxide like heat engines, fuel cells produced water by the sum of the two half-reactions: $H_2 + \frac{1}{2}O_2 = H_2O$.[4] The electrons released by hydrogen gave electrical power. Heat was also released by the electrochemical reaction. In the 2000s, engineers and industrialists could design operational prototypes and commercial products with a conversion rate of energy up to 90 percent.

The fuel cell had thus become a multifunctional device providing water, heat and power with a high-energy efficiency. This aroused enthusiasm among policy makers and industrialists in the beginning of the third millennium. Fuel cells were expected to constitute the first node of the hydrogen network.

It is not a trivial matter, however, to turn expectations into reality. The hydrogen economy remains a magic formula with no impact on the world. Words are reluctant to be turned into things. It is not like in the antique tale of Genesis, where the God of the Bible once said, “Let there be light. And there was light.” The implementation of the hydrogen economy would have required a Creator to turn the wish into reality. But there is no such a thing as a demiurge in the third millennium. The fuel cell seems to resist all efforts and attempts to create a clean ‘hydrogen world’. To make this long and erratic story short, let us use a parody of the seven-day creation of ‘the heavens and the earth’ in the biblical Genesis. In this parody, the creature plays the leading part instead of the missing Creator. Each metaphorical day sets up a new context in which the ‘hydrogen creature’ is reconfigured and the desired ‘hydrogen world’ remodeled. The tale shows at the same time a process of individuation and the ‘recalcitrance’ of the fuel cell.[5]

To render the dialogue between expectations and resistances, a voice is given to both the context, through impersonal phrases, and the ‘hydrogen creature’ speaking for herself, *with ‘I’ at the end of each day*. The following narrative adopts and adapts the Genesis frame to tell the creation of the ‘hydrogen world’ from the perspective of the ‘hydrogen creature’. The dialogue between the creature and the environment aims to avoid three usual preconceptions imposed by the biographical genre: the permanence of things, the essence of beings and the continuity of time. Instead, it displays the changing patterns of the fuel cell and the correlated ‘hydrogen world’ in an attempt to characterize their ontologies.

A parody of the seven-day Genesis of the ‘hydrogen world’

Day 1 – Let there be hydrogen

Day 1 took place in 1787 when the ‘flammable air’ isolated by Henry Cavendish was renamed ‘hydrogen’ in Paris. The word ‘hydro-gen’ (*hydro-gène*) was coined in 1787 by four French chemists – Louis Bernard Guyton de Morveau, Antoine-Laurent Lavoisier, Claude-Louis Berthollet and Antoine de Fourcroy – authors of the *Méthode de nomenclature chimique.*

Hydrogen appeared among the five elements, which were 'the closest to the state of simplicity' and provided the basis for the formation of all compound substances. This first class included: light, caloric, nitrogen, oxygen and hydrogen (Guyton de Morveau et al. 1787).[6] The Book of Genesis mentioned only the first one when "God saw how good the light was" to distinguish day from night, and omitted the four other ones. Fortunately, Lavoisier came to complete God's works and all was in order! Caloric, the principle of heat took over a number of functions of phlogiston, while nitrogen (generator of nitrous compounds) and oxygen (generator of acids) replaced the usual denominations phlogisticated air (*air phlogistiqué*) and dephlogisticated air (*air déphlogistiqué* or *air vital*). The new denomination hydrogen (generator of water) disqualified the elementary nature of water inherited from the Aristotelian tradition.[7] At the same time, it opened up the possibility of obtaining this inflammable gas through a decomposition of water. The public demonstration of the composition of water orchestrated by Lavoisier and Jean-Baptiste Meusnier de la Place in February 1785 at the Paris Academy of Science resulted from the diversion of a campaign of experiments initiated two years earlier to explore means of production of inflammable air for the Montgolfier brothers'. The initial goal was to show that the production of large amounts of hydrogen to fill the balloons of aerostats with lighter-than-air gas was possible. However, the reform of chemical language downplayed this property for emphasizing the property of generating water and promoting analytical chemistry based on the concepts of simple and compound bodies.

> *In the Book of Genesis, light drew a clear boundary between Day and Night. In the first day of the hydrogen world, the Book of French chemists made the relation between Words and Bodies transparent. It also created the two airs – oxygen and hydrogen – that would allow me to breathe soon. However, I still lacked an ontological determination that would come during the second day.*

Day 2 – A missed crucial experiment

The second day opened in Paris on October 22, 1838. At a Monday session of the French Academy of Sciences, Antoine-César Becquerel, physics professor of the Musée d'histoire naturelle, read two short notes on a marginal subject of electrochemistry. The field of electrochemistry had developed since Alessandro Volta's pile invented in 1800. A voltaic pile was a stacking of two alternating metals (zinc/copper) separated by moist cardboard. It generated an electrical fluid when its two 'poles' were connected to a conducting loop of platinum wire. It was later renamed the 'voltaic battery' when it replaced the common electrical generators of the time, the Leyden jars, usually associated in battery to increase their power. In the mid-1830s, Michael Faraday established a new nomenclature: the 'poles' became the 'electrodes', one positive at high tension, called 'cathode', one negative at low tension, called 'anode', the liquid solution in which electrodes were plunged being the 'electrolyte'. The voltaic current flowing in the circuit from the cathode to the anode could decompose the chemical compounds of an electrolyte and produce simple bodies at each electrode (Pancaldi 2003, 130). This reaction known as the electrolysis of water (HO) produced hydrogen gas (H) at the cathode and oxygen gas (O) at the anode.[8] The equation was the same as for the thermal decomposition of water from the 1785 demonstration of Lavoisier: $HO \rightarrow H + O$.

Natural philosophers disagreed about the nature of the forces involved in electrolysis. For Volta, the "contact between, for example, silver and tin [gave] rise to a force, an exertion, that cause[d] the first to *give* electrical fluid, the second to *receive* it."[9] This 'contact theory' suggested an inexhaustible 'electromotive force' (EMF) generated by the contact of two different metals, which was in contradiction with the limited lifetime of a battery. It was challenged by a 'chemical theory' in the 1830s who assumed that the voltaic current resulted from a chemical transformation of the electrodes and the electrolyte (Kragh 2000, §3). It was in this controversial context about the nature of voltaic electricity that Becquerel read two short papers in 1838. The first was an extract of a letter by a university professor of physics and chemistry from Basel, Christian Friedrich Schönbein. After he had performed the electrolysis of muriatic acid (HCl) in a U-tube electrolyte, Schönbein was intrigued by the detection of weak 'secondary currents' in the opposite direction to electrolysis (Schönbein 1838). The second extract of a communication by a well-known physiologist and physicist of Forli, Carlo Matteucci noticed that two platinum plates covered by hydrogen and oxygen gases generated a current 'from hydrogen to oxygen' in an acid electrolyte (Matteucci 1838). It was the opposite direction than expected.

The two brief notes read at the French Academy of Sciences were part of an international debate about 'secondary effects' in electrochemistry. 'Secondary' referred to weak currents or polarities, which required delicate measurements and had opposite direction or polarity to usual phenomena like voltaic currents and electrolysis. For Schönbein and Becquerel, the origin of the phenomenon was chemical; for Matteucci, it was contact (Snelders 1975, 198; Kragh 2000, 138). In an article published in January 1839, Schönbein stressed the importance of the electrode materials to generate 'secondary' effects (Schönbein 1839).[10] Platinum played an active role in "the union of hydrogen and oxygen," contrary to gold and silver. The article concluded that "the secondary currents produced [. . .] by polar wires [were] due to chemical action, i.e. (in the cases mentioned) to the union of oxygen with hydrogen, or to that of chlorine with hydrogen." One month later, Grove demonstrated that when two platinum plates plunged into an acid electrolyte were put under two tubes of gas, one of oxygen, the other of hydrogen, a current was generated from the platinum in oxygen (cathode) to the platinum in hydrogen (anode).[11] As the glass tubes remained lowered, the volumes of hydrogen and oxygen decreased and the level of water rose. The interpretation was obvious for a supporter of the chemical theory like Grove: hydrogen and oxygen combined to form water ($H + O \rightarrow HO$).[12] By repeating this experiment in series, Groves envisions "to effect decomposition of water by means of its composition." Then electrochemists expected that a 'crucial experiment' would reveal the circularity of natural forces through the composition and decomposition of water. Instead, they found a 'secondary' type of electrochemical phenomenon. A distant composition of oxygen and hydrogen generated a delicate current flowing in the direction opposite to the usual one. Chemical forces could be converted into electrical forces.

The second day of the hydrogen world did not last enough time to set a crucial experiment that would have given the key to the mystery of natural forces. Natural philosophers designed, instead, an unexpected device with a tube-and-plate structure stimulated by electrical and chemical fluids. Thus my body was filled with humors and, at the twilight of the second day, I was waiting for the following dawn to breathe.

Day 3 – The first breathing battery

The third day, Grove left his research on 'secondary' effects for a couple of years when his design of a specific voltaic battery gave him access to the Royal Society in 1840. The 'Grove cell' was made of a zinc electrode in dilute sulphuric acid connected to a platinum electrode in concentrated nitric acid. It was the most powerful battery of its time. Powerful batteries, like the Daniell, Grove and Planté cells, were useful as electrical generators for the industry of electro-deposition and telegraph. When appointed professor of experimental philosophy at the Royal Institution, Grove published a full article devoted to the 'Gaseous Voltaic Battery' (Grove 1842, 417). The first name of fuel cells made them acquainted with the lineage of batteries that originated with Volta's work.

The 'gas battery' (Figure 5.2) became a 'peculiar' type of battery that, unlike others, generated electricity from gaseous fuels, namely hydrogen and oxygen. A

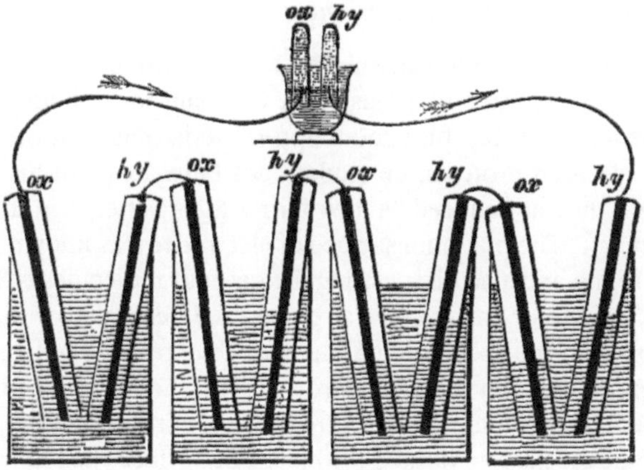

Figure 5.2 First published picture of the gaseous voltaic battery, W. Grove (1842).

Source: Wikimedia Commons.

series of at least twenty-six pairs of electrodes was able to produce gaseous oxygen ('ox') and hydrogen ('hy') in the two separate glass tubes. Water was thus electrolyzed from its elementary parts. In the electrodes, the volumes of hydrogen and oxygen decreased and the water level rose, while in the separate tubes, the volumes of hydrogen and oxygen increased and the water level was lowered. The gas battery started breathing, with the electrodes breathing in gases and the separate tubes breathing out gases. Indeed, the battery sometimes coughed and hiccupped, since the delicate anatomy was weak, brittle and leaky. But the respiration process "exhibit[ed] such a beautiful instance of the correlation of natural forces" (Grove 1846). It helped "establish that gases in combining and acquiring a liquid form evolve sufficient force to decompose a similar liquid and cause it to acquire a gaseous form." As for the partisans of the 'chemical theory', this was the proof that the contact theory was wrong: Where is the contact in this experiment, if not everywhere? This made the gas battery theoretically 'more perfect' than other batteries. It became a 'phenomeno-technical' instrument: a scientific tool used to unravel natural phenomena by producing 'purified' – theoretically constituted – technical effects.[13] Chemical forces were turned into electricity, which was used to decompose the molecule of water.

> *During the third day of the hydrogen world, I was given a Christian name and a nice portrait for Christmas 1842, and breathing capacities. Yet, as Salman Rushdie (1995) once said, "Suspiro ergo sum. I sigh, therefore I am. [. . .] We inhale the world and breathe out meaning." This ontological certificate made me so happy that I agreed to reveal the conversion of natural forces to philosophers. It was a pretty good job and I thought they would give me a break. Unfortunately, industrial entrepreneurs took over from philosophers to make me work harder the following day.*

Day 4 – A prayer for a coughing generator

The fourth day opened with the attempt by Ludwig Mond to turn the gas battery into an electrical generator for industry. A German-born chemist-industrialist, Mond, established a successful production of soda and ammonia in England. Among many research projects, he studied gas batteries with his assistant Carl Langer. In 1889, they announced "a new form of gas battery" in the *Proceedings of the Royal Society*. The platinum surface of electrodes was known to have a catalytic effect by fixing and also dissociating the gaseous molecules of hydrogen (H_2) and oxygen (O_2). Mond and Langer stressed "the necessity of maintaining the condensing power of the absorbent unimpaired" by preventing the surfaces of the electrodes from getting wet by the electrolyte. They copied "dry piles and batteries" and trapped the sulphuric acid electrolyte in a porous material.[14] The "quasi-solid form" of the electrolyte limited the electrode surfaces becoming wet without modifying its conducting properties.

Mond and Langer's program led to practical improvements: air, instead of pure oxygen, and coal gas, instead of pure hydrogen, were being tested; platinum black was preferred to platinum to increase the contact surfaces between gas and electrodes; the EMF was stabilized between 0.7 and 1V, the power between 1.5 and 2W, and the operating temperature around 40°C. Yet they failed to regulate the life time of the gas battery (Perry and Fuller 2002, S59). Impurities of the gaseous fuels poisoned the electrode surfaces and made the gas battery misfire after a while. On the scientific side, Mond and Langer pointed to significant differences between the measured EMF and the calculation based on 'Thomson's theorem' and concluded that future researchers would succeed in matching empirical and theoretical values.

In the 1880s, scholars and industrial entrepreneurs were mainly concerned with conversion efficiency. From thermodynamics, they knew that more energy was needed to prepare a given amount of hydrogen by the electrolysis of water than could be generated from the same amount of hydrogen by a gas battery. This was a major drawback for industrial applications. However, the gas battery soon became a "solid compact block with no openings except the entrances and exits for the gases" (Mond and Langer 1889, 300). The liquid electrolyte trapped in a porous solid was separated from gaseous fuels. The device was carefully sealed and partitioned to prevent leaking and the mixing of fluids. The gas battery became dry. However, in spite of endless quantitative controls, its operation remained erratic and smoky. When they understood how reluctant the gas battery was to black-boxing, the industrialists closed their programs and the scholars arranged the 'gas cell' on the shelves of science.[15]

The fourth day of the hydrogen world was not a bed of roses. Industrialists persisted in improving my delicate body to design a robust engine. But all their surgical operations failed to fix my respiration at a constant pace and intensity and to prevent me from coughing. Smoking was my pleasure, and I didn't want to become an industrial application. I was right indeed, since when I came on the job market the following day, I was left unemployed.

Day 5 – Unemployment on the market place

In the morning of the fifth day (early twentieth century), while gas batteries were considered hopeless for industrial applications, a few dreamy engineers envisaged 'gas operated vehicles'.[16] Among them, a mechanical scientist from Cambridge, Francis Thomas Bacon, contracted an obsessive interest for gas battery in the 1930s and collected information on it during his spare time. On reading a 1933 comprehensive survey on 'fuel couples' by two German chemists, Emil Baur and Jakob Tobler, he decided that basic electrolytes would give better results than acid electrolytes and chose nickel, which was much cheaper than platinum to use as electrodes (Baur and Tobler 1933).[17] As he failed to convince his employers to launch a 'hydrogen-oxygen cell' project, he resigned from his post in Newcastle[18] and followed his freelance enthusiasm for the gas battery until he was enlisted by the admiralty into the war effort. After World War II, he got financial support from the Electrical Research Association (ERA) to set up a small research group studying the gas battery at Cambridge University. His team included electrochemists, inorganic chemists, theoretical physicists and engineers. In eight years (1946–1954), they overcame a series of empirical obstacles and built "a six fuel cell battery [. . .] using 5 inch diameter electrodes and operat[ing] at 200°C and 400 p.s.i. [27 atm.] achieving individual cell EMFs of 0.8V at 230 mA.cm^{-2}." The 'Bacon cell' (Figure 5.3) was patented in February 1952.[19] It was a masterpiece of engineering work; the over-pressurized black-box held around forty-five distinct parts, including pipes, gauges and valves. Bacon's long-lasting research overcame the recalcitrance of the gas battery. This successful black-boxing made the gas battery a mobile and reliable device.

Bacon felt confident enough to publish a seminal article on "Recent Research in Great Britain on the Hydrogen-Oxygen Fuel Cell" for electrical engineers (Bacon and Forrest 1954). He renamed the 'gas battery' as a 'fuel cell' to emphasize its aptitude for refueling.[20] Contrary to other electrochemical cells whose energy was packed *inside* the box, the energy came from *outside*. This made the fuel cell highly dependent on surroundings. No private industry wanted to fund research on an electrical generator that could not operate autonomously. Therefore the ERA withdrew its support in 1956. By contrast, the (British) Welfare State funded Bacon's program through the National Research and Development Corporation (NRDC). Bacon's team could design a 6-kW operating generator with forty 150-W fuel cells exhibiting 60 percent efficiency which was ready in August 1959. Still, there was no employment for fuel cells on the marketplace, and the NRDC withdrew.

> *Tom was a stubborn chap. I couldn't help myself but trust him. He had worked hard for two decades to turn my weak body into a reliable machine. Yet, in the evening of the fifth day of the hydrogen world, I was still jobless on Earth. Waiting for the divine providence to do something for once, I was about to blaspheme against the dark Heaven when my attention was attracted by an odd light slowly crossing the dome of the sky.*

Day 6 – Space and glory

In the sky of the sixth day, flew Sputnik, an artificial satellite launched in October 1957 by the Soviet Union. This was the Cold War, and the United States counterattacked. In October 1958, the National Aeronautics and Space Administration (NASA) was created to take the lead of the space race and launch space flights. Space was a cold, airless, hostile environment where there was no room for heat engines. Electrochemical generators were the only candidates to supply onboard energy. NASA experts recognized that fuel cells presented three advantages over batteries: a higher electrical output for the expected duration of missions (one to ten days) (Bacon and Fry 1973, 439); two fuels, hydrogen and oxygen, which were also used for spaceship propulsion; and a subproduct, water, which could provide drinking beverage and

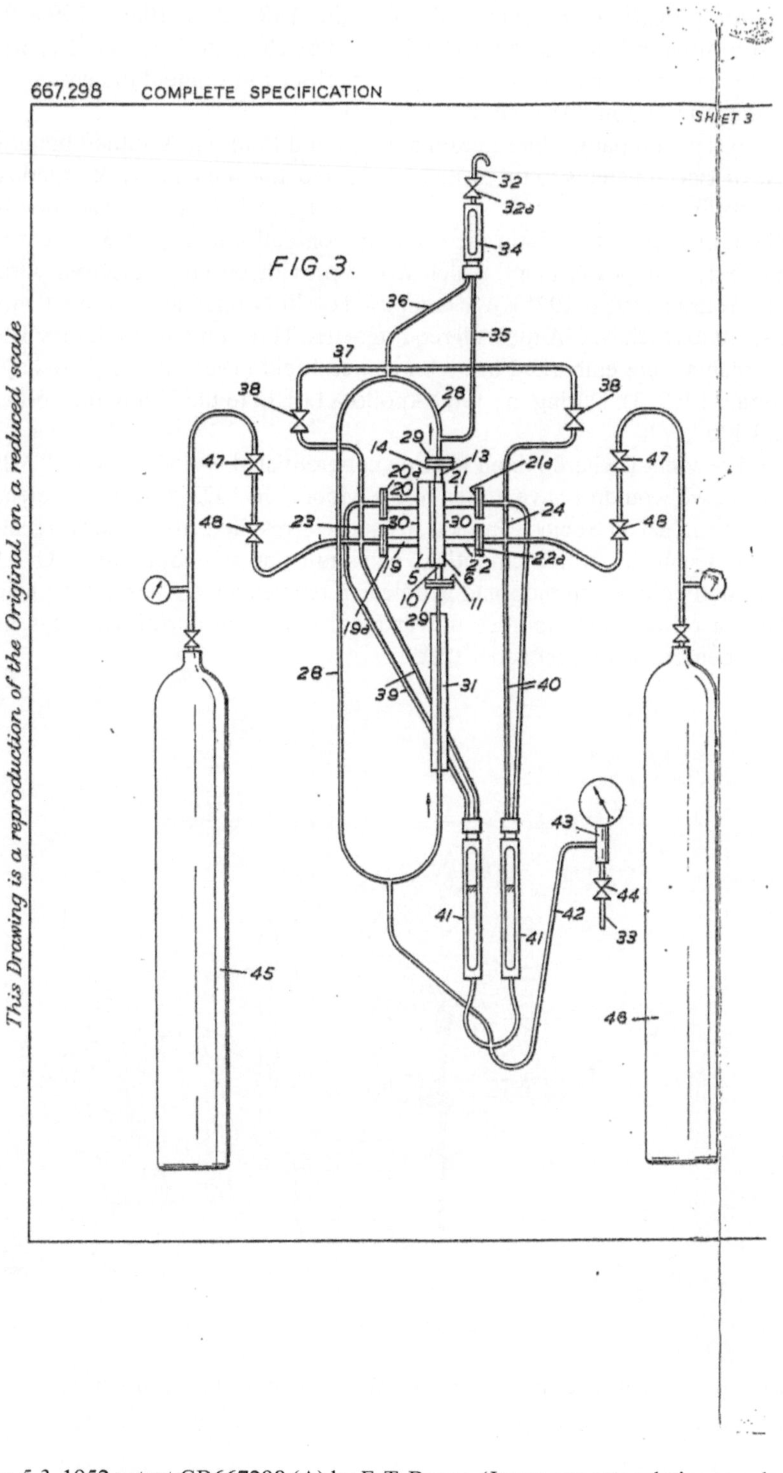

Figure 5.3 1952 patent GB667298 (A) by F. T. Bacon. 'Improvements relating to galvanic cells and batteries'.

Source: European Patent Office online database.

cabin humidification (Warshay and Prokopius 1989, 2; Williams 1994, 9). In Gilbert Simondon's terms, the fuel cell could be characterized as a 'hypertelic' machine, a machine over-adapted to one specific environment of use: interstellar space (Simondon 2012, ch. 2).

U.S.-based companies like Leesona-Moos and Pratt and Whitney bought the license of Bacon patents to the British government. A huge network of industrial and scientific institutions was set up around NASA.[21] Pratt and Whitney hired 1,000 employees to transform the 6 kW Bacon cell into a 1.5 kW power plant (Figure 5.4). The power plant, which would provide on-board electricity for the Apollo program (1961–1975), was composed of thirty-one fuel cells. $100 million were spent to reach NASA fuel cell requirements. Three Pratt and Whitney 1.5 kW power plants were embarked in each manned flight (1968–1972) (Warshay and Prokopius 1989, 3). During the 1969 Apollo XI flight to the Moon, they provided around 400 kW.h.

U.S. President Richard Nixon warmly congratulated Bacon in 1970: "Without you, Tom, we wouldn't have gotten to the Moon." In 1973, Bacon was elected a Fellow of the Royal Society. In fact, his success resulted from a convergence of scientific, technical, military, political, economic and symbolic forces. One hundred million dollars, ten thousand people and ten years were needed to make the fuel cell a reliable autonomous power plant operating during ten days in the extreme conditions of interstellar space.

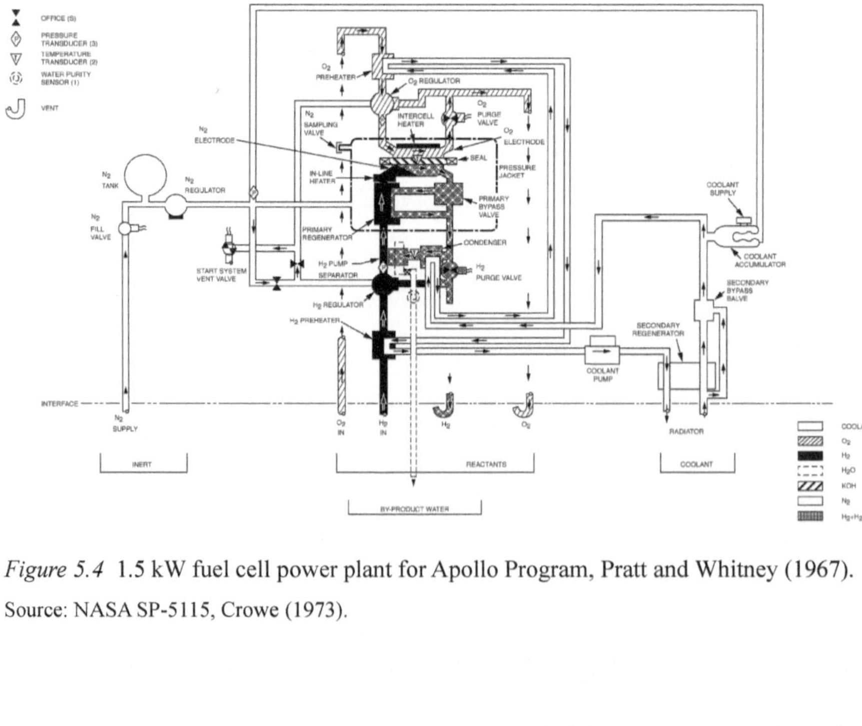

Figure 5.4 1.5 kW fuel cell power plant for Apollo Program, Pratt and Whitney (1967).

Source: NASA SP-5115, Crowe (1973).

In the sixth day of the hydrogen world, Uncle Sam and Uncle Tom on the center stage made huge efforts to overcome my recalcitrance and turn a stubborn creature into a docile robot. Their efforts and money were only rewarded by the tiny amount of power generated during a single-week manned space flight. During the space trip back from the Moon, I was already prepared for the seventh day, the Holy day for everybody to rest. However, instead of the traditional Sabbath, the hydrogen world invented the hydrogen utopia.

Day 7 – No sabbath on earth

Back to Earth the morning of the seventh day, the fuel cell was drafted for military low-energy portable units and submersible vessels (Rogers 1969; Warszawski et al. 1971). It was also recruited by the gas industry to conduct an ambitious program to generate electricity from charcoal and air. Started in 1967, the Target consortium gathered around thirty U.S. gas companies around Pratt and Whitney. Fifty million dollars were spent during six years to design a 12.5 kW fuel cell power plant fuelled with hydrocarbons.[22] In Britain, a similar project associated NRDC (renamed Energy Conversion Ltd.), British Ropes, British Petroleum, Guest Keen and Nettlefolds to develop hydrogen-air and zinc-air systems (1968–1973). The 1973 oil crisis curtailed these projects. NASA promoted automotive applications of fuel cells to justify the tremendous space budgets.[23] Yet the Lunar Rover Vehicle itself had been driven by (Planté) lead-acid batteries, not by fuel cells.[24] Gas batteries still had lower power than traditional batteries. Their second drawback for terrestrial transportation was their high price: three orders of magnitude higher than market products for the same power range (Crowe 1973, 62). This did not prevent enthusiastic projects from the mid-1960s onward, either by automobile industry (GM electrovan) or by independent scholars (Marks et al. 1967; Kordesch 1971). Fuel-cell-powered vehicles did not find their way to the car industry market that was monopolized by internal combustion engines. There were specific niches for fuel cells: space shuttles, submarine vessels and, for emergency, hospitals and airports.[25] No mass-market commercialization was expected by practitioners. In the 1990s, Japan and the United States launched the 'hydrogen race'.[26] The State of California started to build a huge hydrogen infrastructure – 'the hydrogen highway' – to equip fuel-cell-powered vehicles on the West Coast. A decade later, the EU firmly joined the race and became a staunch advocate of hydrogen economy. However, worldwide research and development efforts stumbled on ever-disappointing results. They got no significant increase of fuel cell power generation.[27]

At the sunset of the seventh day, the 'hydrogen world' had not been created yet. It remained a utopia since it strongly depended on the need to overcome three technological obstacles: (1) the limitations of the fuel cells themselves, which were highly reluctant to (co)operate; (2) reducing fuels, especially hydrogen, whose synthesis had required the combustion of hydrocarbons in the atmosphere to perform the electrolysis of water;[28] (3) hydrogen networks, like the Californian 'hydrogen highway', whose establishment were dependant on gas station networks.[29] To overcome these three obstacles would have required a deep 'cultural shift', and such radical shifts are highly improbable in history.[30] The utopia revealed more about the present than the future. Industrialists publicly displayed the 'hydrogen vehicle' of tomorrow as childish visions (Figure 5.1) to hide their current polluting business. Politicians announced the 'hydrogen economy' to make people believe they controlled energy systems. Environmentalists advertised possible 'green systems' to fight internal combustion engines and nuclear plants. For the three groups, displaying the affordances of the 'hydrogen world' mattered more than the effective results of the hydrogen creature. With all their hype, the fuel cell became the 'philosopher's stone' of today (Eisler 2009).

Epilogue – Recalcitrance and interdependence

The biblical metaphor of the seven days of creation is helpful to emphasize three major features of the genesis of technoscientific objects such as fuel cells: First, they require a gradual process of individuation. Second, it may be a long, painstaking and possibly endless process. And third, a recurrent key issue lies in conflicting interactions between the fuel cell and its milieu. If according to Simondon the 'concretization' of technical objects is characterized by a successful interaction between the machine and its associated milieu, it is clear that the fuel cell never became a concrete technical object. Each daily interaction between the creature and its environment shaped a new figure of the fuel cell: theory-testing tool, a battery kin, an unstable generator, an open black-box, a low-power stable prototype, a medium-range power plant and the dawn of a new clean industrial era. None of these figures managed to be articulated to the adequate milieu in which material and symbolic forces could have cooperated. While the relation between the creature and its environment was reconfigured at each stage, the fuel cell developed new forms of recalcitrance: the inflammable air versus hydrogen (day 1), secondary effects versus a crucial experiment (day 2), a brittle device versus a gaseous voltaic battery (day 3), a smoky gas battery versus a stabilized generator (day 4), a jobless generator versus a commercial power plant (day 5), a fragile power plant versus a reliable energy supplier (day 6) and an expensive product versus an advertising toy (day 7).

In spite of reconfigurations at each stage, the fuel cell creature kept two constant features throughout the tentative creation of a 'hydrogen world'. The first one is a strong resistance to any form of control, containment and normalization, which accounts for the remarkably slow increase of power of fuel cells (and batteries as well) over two centuries. In this respect, there is a stark contrast between energy technology and information technology. From the 1970s onward, the computer industry doubled the information storage capacity of semi-conducting materials every eighteenth months according to the famous 'Moore's law' (Brock 2006). While this self-fulfilling prophecy allowed the exponential growth of information storage capacities over four decades, the energy storage capacities of batteries barely reached a linear growth during two centuries. Matter matters less for information than for energy. Information has been quickly mastered and disciplined by electronic engineers whereas energy remains a vexing issue challenging human ingenuity and creativity. The 'information revolution' of the last decades of the twentieth century was possible because integrated circuits cooperated with engineers to turn dreams into reality and build the 'silicon economy'. On the contrary, fuel cells were recalcitrant to cooperate with engineers to turn dreams into reality. The 'third industrial revolution' that would bring about the hydrogen economy remains a utopia in the early third millennium.

A second permanent quality of the hydrogen creature is its propensity to connect things together: energies, forces, fluids, chemicals and electrons. The inhalation of vital air and inflammable air allows the fuel cell to generate water, heat and electricity. This production cannot occur without exchanges with its environment. Like living beings and technological artifacts as well, the hydrogen creature cannot live without interacting with its milieu. Even in space, the operation stopped when no more fuel was provided. The production of hydrogen requires energy and the operation of fuel cells requires adequate surroundings. Mond, Bacon and all the champions of the hydrogen economy viewed the question just the other way around. The concept of the 'hydrogen economy' is nonsense, relying on the fiction of a self-sufficient machine. Since the ontology of the fuel cell reveals the 'hydrogen world' as modern myth, it was legitimate to tell this story through the format of the antique myth of Genesis.

Notes

1 The expression *la délocalisation de la puissance* in French is borrowed from Gras (2003).
2 "Global atmospheric concentrations of carbon dioxide, methane, and nitrous oxide have increased markedly as a result of human activities since 1750 and now far exceed pre-industrial values [. . .] over the past 650,000 years. [. . .] Most of the observed increase in globally averaged temperatures since the mid-20th century is very likely due to the observed increase in anthropogenic (human) greenhouse gas concentrations" (IPCC 2007, 37 and 39). "Very likely" refers to a probability higher than 90 percent.
3 For a historical survey, see Hultman and Nordlund (2013).
4 For the detailed history of water and of the controversies regarding its composition, see Chang (2012).
5 The concept of individuation of technological objects was introduced by Gilbert Simondon (2005); the concept of *récalcitrance* by Bruno Latour (2004).
6 The French word for 'nitrogen' is *azote*, which means 'that deprives of life'.
7 "*L'hydrogène uni à l'oxigène ne forme point un acide, mais cette union constitue l'eau qui, considérée sous ce point de vue, pourroit être regardée comme un oxide d'hydrogène*" (The union of hydrogen with oxygen does not form an acid, but composes water, which, from this perspective, might be regarded as an oxide of hydrogen) (Guyton de Morveau et al. 1787, 88).
8 The Swedish chemist Jons Jacob Berzelius replaced the symbols used by John Dalton for chemical elements by the initial letters of their names.
9 Letter from Volta to Friedrich Gren (1796), cited by Kragh (2000, 134).
10 The article was dated from December 1838.
11 The postscript of the article dated from January 1839 (Grove 1839b).
12 The interpretation did not appear in the short postscript but was given two months later in Grove's correspondence to the French Academy of Sciences (Grove 1839a).
13 The concept of *phénoménotechnique* was introduced by Gaston Bachelard (2000, §6).

14 The porous solid was either "an earthenware plate [or] conducting strips [. . .] filled up with plaster of Paris" (Mond and Langer 1889, 297).
15 Early twentieth-century textbooks stressed the theoretical importance of the 'reversibility' of 'gas cells'. See Glasstone (1930).
16 The November 18, 1932 issue of *Engineering* held "two articles on pages 605–608, one headed 'Gas Operated Vehicle' and the other 'The Erren Hydrogen Engine'" (Williams 1994, 4).
17 The literal translation of *Brennstoffketten* would have been 'fuel series' but, according to personal papers of Bacon in the Churchill Archives Centre [NCUACS 68.6.97/B35], the phrase was translated for Bacon as 'fuel couples'.
18 The Churchill Archives Centre mentions a first report in December 1939 [NCUACS 68.6.97/B.36] and a second report extended version of B.36 in January 1940 [NCUACS 68.6.97/B.38].
19 The priority date of the patent, entitled 'Improvements relating to galvanic cells and batteries', was June 1949. For further information on the 1952 patent GB667298 (A), see the European Patent Office database. Bacon obtained nine patents related to fuel cells between 1952 and 1965.
20 On the Web of Science, the phrase 'fuel cells' appears in scientific journals in the early 1960s. It was coined after 'fuel' from the 1933 *Brennstoffketten* of Baur and Tobler and 'cell' from the 1939 'hydrogen-oxygen cell' of Bacon, which derived from electrochemical cells of the nineteenth century. The coining was similar in German (*Brennstoffzelle*) and slightly different in French where *pile à combustible* means 'fuel pile'.
21 From 1958 to 1963, 800 reports were written on fuel cells. NASA SP-120 report 1967, 1.
22 TARGET stood for 'Team of Advance Research for Gas Energy Transformation' (Crowe 1973).
23 More than half the 1973 NASA report on fuel cells by Crowe was devoted to non-space applications.
24 The Lunar Roving Vehicle was built by the Boeing Company for $38 million; see the 1971 Lunar Roving Vehicle *Operations Handbook*, document LS006–002–2H, Huntsville: Boeing, http://www.hq.nasa.gov/alsj/lrvhand.html (accessed July 17, 2016).
25 In 1980, the fuel cell power plant of the U.S. Orbiter mission had a 12 kW ouput according to Warshay and Prokopius (1989, 3–4).
26 For an international R&D survey, see Solomon and Banerjee (2006, 781–792).
27 A survey on Web of Science exhibited an exponential growth of scientific articles containing 'fuel cell' in their title from around 100 in 1995 to around 1,400 in 2008.
28 96 percent of the hydrogen produced in the world came from hydrocarbons and coals in 2006, and 4 percent from the electrolysis of water, according the website of IFPEN (French Institute of Petroleum and New Energies).
29 The concept of 'path-dependency' in particular makes clear why past choices matter for future choices. See Kirsch (2000).
30 The problem of energy bifurcation was analyzed by Gras (2003).

References

Austin, L. G. 1967. *Fuel Cells: A Review of Government-Sponsored Research, 1950–1964*. NASA SP-120. Washington, D.C.: NASA.

Bachelard, G. 2000. *Le matérialisme rationnel*. Paris: Presses Universitaires de France.

Bacon, F. T. 1952. 'Improvements relating to galvanic cells and batteries.' European Patent Office online database: patent GB667298 (A).

Bacon, F. T. and Forrest, J. S. 1954. 'Recent research in Great Britain on the hydrogen-oxygen fuel cell.' *British Electrical and Allied Manufacturers' Association Journal* 61.

Bacon, F. T. and Fry, T. M. 1973. 'The development and practical application of fuel cells.' *Proceedings of the Royal Society of London. Series A: Mathematical and Physical Sciences* 334 (1599): 427–452.

Baur, E. and Tobler, J. 1933. 'Brennstoffketten.' *Zeitschrift für Elektrochemie und angewandte Physikalische Chemie* 39 (3): 169–180.

Brock, D. C. (ed.). 2006. *Understanding Moore's Law: Four Decades of Innovation*. Philadelphia, PE: Chemical Heritage Press.

Chang, H. 2012. *Is Water H_2O? Evidence, Realism and Pluralism*. Heidelberg, London and New York: Springer.

Crowe, B. J. 1973. *Fuel Cells: A Survey*. NASA SP-5115, prepared under contract NASW-2173 by Computer Sciences Corporation, Falls Church, Virginia. Washington, DC: NASA.

Eisler, M. N. 2009. 'A modern "philosopher's stone": Techno-analogy and the bacon cell.' *Technology and Culture* 50 (2): 345–365.

Glasstone, S. 1930. *The Electrochemistry of Solutions*. New York: Van Nostrand.

Gras, A. 2003. *Fragilité de la puissance. Se libérer de l'emprise technologique*. Paris: Fayard.

Grove, W. R. 1839a. 'Décomposition de l'eau par le moyen de deux lames de platine en communication chacune avec l'un des éléments d'un couple voltaique, Procédé de M. Grove, communiqué par Becquerel.' *Comptes rendus de l'Académie des sciences* 1 (8): 497–498.

———. 1839b. 'On voltaic series and the combination of gases by platinum.' *The London and Edinburgh Philosophical Magazine and Journal of Science, 3rd series*, 14 (86): 127–130.

———. 1842. 'On a gaseous voltaic battery.' *The London, Edinburgh and Dublin Philosophical Magazine and Journal of Science* 21 (140): 417–420.

———. 1846. *On the Correlation of Physical Forces: Being the Substance of Course of Lectures Delivered in the London Institution, in the Year 1843*. London: London Institution.

Guidoni, U., Gurmai, Z., Prodi, V., Turmes, C. and Wijkman, A. 2008. 'Establishing a distributed green hydrogen economy and advancing a third industrial revolution in Europe trough a partnership with committed regions and cities, SMES, and civil society organisations.' Written Declaration 16/2007, *Official Journal of the European Union*, C102/E2 (24), http://www.europarl.europa.eu/sides/getDoc.do?pubRef=-//EP//NONSGML+WDECL+P6-DCL-2007-0016+0+DOC+PDF+V0//EN&language=EN.

Guyton de Morveau, L. B., Lavoisier, A.-L., Berthollet, C.-L. and Fourcroy, A. 1787. *Méthode de nomenclature chimique*. Paris: Cuchet (re-printed Paris: Seuil, 1994).

Hultman, M. and Nordlund, C. 2013. 'Energizing technology: Expectations of fuel cells and the hydrogen economy, 1990–2005.' *History and Technology* 29 (1): 33–53.

IPCC. 2007. *Climate Change 2007: Synthesis Report*. Geneva: IPCC, www.ipcc.ch/pdf/assessment-report/ar4/syr/ar4_syr_full_report.pdf (accessed July 17, 2016).

Kirsch, D. 2000. *The Electric Vehicle and the Burden of History*. Newark, NJ: Rutgers University Press.

Kordesch, K. V. 1971. 'City Car with H_2-Air Fuel Cell/Lead Battery (One Year Operating Experiences).' *Proceedings of the 1971 Intersociety Energy Conversion Engineering Conference*, 103–111. Boston, MA: SAE.

Kragh, H. 2000. 'Confusion and Controversy: Nineteenth-Century Theories of the Voltaic Pile.' In *Nuova Voltiana: Studies on Volta and his Times*, vol. 1, edited by Bevilacqua, F. and Fregonese, L., 133–157. Pavia and Milan: Hoepli.

Latour, B. 2004. *Politics of Nature: How to Bring the Sciences into Democracy*. Cambridge, MA: Harvard University Press.

Loeve, S. and Bensaude Vincent, B. 2017. 'The Multiple Signatures of Carbon.' In *Research Objects in their Technological Setting*, edited by Bensaude Vincent, B., Loeve, S., Nordmann, A. and Schwarz, A. Abingdon: Routledge.

Marks, C., Rishavy, E. A. and Wyczalek, F. A. 1967. 'Electrovan – A Fuel Cell-Powered Vehicle.' In *Society of Automotive Engineering Congress*, Paper 670176, doi:10.4271/670176. Detroit, MI: SAE Congress.

Matteucci, C. 1838. 'Conclusions d'un Mémoire de M. Matteucci sur les polarités secondaires, communiquées par M. Becquerel.' *Comptes rendus de l'Académie des sciences* 4 (2): 741.

Mond, L. and Langer, C. 1889. 'A new form of gas battery.' *Proceedings of the Royal Society* 46 (280–285): 296–304.

Pancaldi, G. 2003. *Volta: Science and Culture in the Age of Enlightenment*. Princeton, NJ: Princeton University Press.

Perry, M. L. and Fuller, T. F. 2002. 'A historical perspective of fuel cell technology in the 20th century.' *Journal of the Electrochemical Society* 149 (7): S59–67.

Rifkin, J. 2002. *The Hydrogen Economy*. New York: Tarcher/Putnam.

Rogers, L. J. 1969. 'Hydrazine-Air (60/240 watt) Manpack Fuel Cell.' *Proceedings of the 23rd Annual Power Sources Conference*. Red Bank, NJ: PSC Publications Committee.

Rushdie, S. 1995. *The Moor's Last Sigh*. London: Vintage.

Schönbein, C. F. 1838. 'Observations sur les courants secondaires.–Extrait d'une lettre de M. Schœnbein à M. Becquerel.' *Comptes rendus de l'Académie des sciences* 4 (2): 741.

———. 1839. 'On the voltaic polarization of certain solid and fluid substances.' *The London and Edinburgh Philosophical Magazine and Journal of Science, 3rd series* 14 (85): 43–45.

Simondon, G. 2005. *L'individuation à la lumière des notions de forme et d'information*. Paris: Millon.

———. 2012. *Du mode d'existence des objets techniques*. Paris: Editions Aubier.

Snelders, H. A. M. 1975. 'Schönbein, Christian Friedrich.' In *Dictionary of Scientific Biography,* vol. 12, edited by Gillispie, C. C., 196–199. New York: Charles Scribner's Sons.

Solomon, B. D. and Banerjee, A. 2006. 'A global survey of hydrogen energy research, development and policy.' *Energy Policy* 34 (7): 781–792.

Warshay, M. and Prokopius, P. R. 1989. 'The Fuel Cell in Space: Yesterday, Today and Tomorrow.' In *Grove Anniversary (1839–1989) Fuel Cell Symposium*, NASA Technical Memorandum 102366, 1–10. London: Royal Institution.

Warszawski, B., Verger, B. and Dumas, J. C. 1971. 'Alshtom fuel cell system for marine and submarine applications.' *Marine Technology Society Journal* 5 (1): 28–41.

Weart, S. 2008. *The Discovery of Global Warming*. Cambridge, MA: Harvard University Press.

Williams, K. R. 1994. 'Francis Thomas Bacon, 21 December 1904–24 May 1992.' *Biographical Memoirs of Fellows of the Royal Society* 39: 2–18.

Part II

Arenas of contestation

6 Heroin

Taming a drug and losing control

Jens Soentgen

The history of heroin is first and foremost a story of progressive scientific control over a substance. Starting with opium, then morphine, through to subcutaneous administration with a syringe, we have influenced ever more control over the drug, which with the refinement of morphine to heroin has likely reached its peak. The moment of total control of the substance seemed within reach when its negative characteristics were believed to have been eliminated through the acetylation of morphine to diacetylmorphine (i.e., heroin). At the same time, the uncontrolled proliferation of the substance was banned. In this very moment a peculiar dialectic occurred. The substance escaped the scientists and became kept under legal control and subsequently police control. Simultaneously, it escaped the narrower circle of doctors and spread – promoted by the two World Wars of the twentieth Century – throughout society as a whole. Finally, in the fight against this proliferation, scientists even lost control over the notion of the substance. From the 1950s onward, this notion was no longer determined by pharmacists or chemists, but by politicians, who saw in heroin a bacilli-like evil thing and pursued the substance with drastic measures, which only lead to its further proliferation. The prohibition and pursuit of the substance had significant side effects. In this chapter, only those historical aspects of heroin are examined that are relevant to its genesis first as a scientific object, then as a legal matter of concern and eventually as a technoscientific object being shaped by controversial debates extending into very different societal contexts. The focus of this chapter is on the German discourse of the 1920s,[1] when doctors, pharmacists and judges argued over sovereignty in matters concerning opiates.

Why couldn't become heroin a wonder drug?

Aspirin and heroin are stepsisters that were first synthesized about 120 years ago in August 1897, only a few days apart and by the same person: the chemist Felix Hoffmann, an employee at the Farbenfabriken formerly Friedrich Bayer & Co. in Elberfeld (Germany). From a medicinal standpoint, both chemicals are pharmaceuticals with notable properties and, like all effective pharmaceuticals, are associated with characteristic advantages and disadvantages. Both substances are toxic in larger doses, whereby heroin is substantially more toxic than aspirin. Both are administered today primarily to

Figure 6.1 Fake heroin in the hand of the author (replica of a Bayer-Heroin-Flask).

Photo: Author.

combat pain, among other applications. Heroin is many times more effective in this than aspirin. Compared to other opiates, heroin stands out for taking effect faster and having the same effect at lower dosages. Like other opiates, heroin, unlike aspirin, carries with it a high risk of dependence on the compound. Heroin can easily become addictive, indeed both physically and psychologically simultaneously. However, other than this, heroin has notably few side effects, even with long-term consumption.

While aspirin is considered to be a 'wonder drug', heroin is no longer recognized as a medicine at all, rather only as a forbidden addictive substance that destroys those who become dependent on it. It is considered the work of the devil, as a substance that possesses almost even magical powers. Touch it once, and you are ruined forever. It is the epitome of a 'hard drug'. Our society does not perceive heroin as a pharmaceutical, but rather as a moral and legal notion. This moral notion determines the way it is dealt with. Heroin has been, and still is in most countries, a non-trafficable, non-prescribable narcotic substance. The precursors to heroin also lie under strict control: opium poppies, from which opium, heroin and morphine can be produced, may only be grown for decorative purposes. Heroin is no longer legally produced anywhere, except in Great Britain. While aspirin is present nearly everywhere in our modern world, in billions of households and handbags, almost the exact opposite is true for heroin.

Follow the thing! – My phenomenological quest for heroin

Follow the thing! This methodological device put forward by Arjun Appadurai,[2] George E. Marcus and others is difficult to implement in the case of heroin. If you want to trace it, you have to go underground. It is present, but is also one of the best hidden substances. Even for a scientist, it is nearly impossible to get hold of true heroin.

My phenomenological quest for heroin led me to the Augsburg local police station. There I met a local official for drug crimes.[3] He did not understand exactly why I was interested in heroin. I tried to explain the GOTO project on research objects in their technological setting to him, but he was not convinced that the scientific project was the true reason for my interest in heroin. He did, however, explain the three basic pillars of the Bavarian drug policy to me: repression, persecution and precaution. He appeared to be on edge, possibly believing that I had been sent by some authority to test him. He had some dried opium poppy seeds stained with blue dye in a vessel in his cupboard. When I asked him whether there was some true heroin in his office, he was bewildered. But indeed there was – in a safe. I asked him to open the safe, but he would not allow me to take a look because "all of the items are evidence in criminal trials." I asked him what heroin smelled like. He said it smelled of vinegar, like heated water-soluble aspirin tablets. He explained to me that one gram of impure heroin was worth sixty euros on the black market in Germany. One kilogram of heroin can sell for twenty thousand euros. From an economic point of view, he explained, dealing in heroin is a quite rational thing to do: "You do not need much space to store the substance and, depending on how pure or impure you sell it, you make a big profit." I promised him to send him the results of my studies when they were published.

Some weeks after this interview I discovered that heroin was sold by our university's chemical supplier, but I was reluctant to buy it there. It was also ten times more expensive than on the black market. Finally, I decided to synthesize it at home from poppy seeds. These can be purchased in July and August for 2.50 euros each at flower shops. I blended the poppy seeds and filtered the juice. Then I added ammonia and a white substance precipitated that was soluble in acids. It was obviously an alkaloid, but the result of an opiate quick test remained ambiguous. Perhaps I had used the wrong kind of poppy.

To sum up my experiences – to get hold of heroin, you have to leave your ordinary life. Immense walls have been constructed around the substance. It seems you cannot even touch the substance without becoming a criminal. Go underground! Because an image of heroin as a highly dangerous, addictive substance has prevailed and been enforced, we can no longer accept the substance in middle-class society. We incriminate ourselves when we hold it in our hand. And what can be bought as 'heroin' on the black market is often highly contaminated and cut (mixed) with all sorts of substances. The criminalization of heroin has warped the substance itself, even if the chemical formula has of course remained the same. How could this happen?

Theriac and opium

Relatives of 'heroin' have been in use for ages and by various cultures, be it as a part of 'theriac' or of 'opium'. Opium[4] is the juice of seeds from the opium poppy, *Papaver somniferum*. It was mentioned as a remedy for ulcers in the famous *Ebers papyrus*, which was written around 1550 BC. In the *Corpus hippocraticum*, composed in the fifth to fourth century BC, poppy juice was considered a pain reliever. Dioscurides describes in detail as far back as the first century AD how one should make the cuts in the seed in order to glean the juice. Opium was widely distributed in the form of theriac, a mixture of different animal and plant products whose principal component was opium. It has been invented by King Mithridates, who thought that he could achieve a certain level of immunity against poisonous substances by ingesting this preparation. Theriac brings us also to the likely best known opium-addicted philosopher: the Emperor Marcus Antoninus, also known as Marcus Aurelius.[5] Galenius, the Emperor's court doctor, noted that the Emperor tried to get rid of the opium-component of his theriac – but then he suffered insomnia and returned to the old preparation, that included opium (Galenius 1965, 3–5). Galenius's description of Marcus Aurelius's theriac use is probably the first description of drug addiction in western history. Theriac became a cure-all remedy in Rome through the effect of the Emperor's example. It was again in use in the Middle Ages and was widespread in folk medicine until into the twentieth century. However, its use and that of other opium preparations was delicate, since the composition of the theriac and the growing conditions of the plants meant that the drug was effective to varying degrees. Underdoses, which were ineffective, as well as toxic overdoses were routine occurrences (Meyer 2004). This resulted in particular in numerous attempts to 'temper' and 'correct' opium.

Taming opium – Glauber the alchemist and Sertürner the apothecary

One example of these alchemist efforts is provided by Johann Rudolph Glauber.[6] In his Pharmacopoea Spagyricae, he describes opium as sometimes "not only making one sleepy / but also making some not want to wake up again / but to need rest up until the very Judgement Day / (now that is really a calming remedy)" (Glauber 1656, 90). Glauber then points out that many "have invested a lot of effort in correcting the opium" (Glauber 1656, 90) for this reason. Thus it was extracted with pure alcohol (*spiritus vini*) or concentrated acetic acid (*spiritus aceti*) and then crystallized, although this undoubtedly resulted in an unclean form of morphine (mixed with other alkaloids), which was then used. Glauber refined this process by treating the aqueous opium extract with sulphuric acid, filtering it and then breaking it down with an alkaline potassium carbonate solution.[7] With this method he refined the resulting opium extract, which mainly consisted of morphine. Even with today's knowledge, Glauber's was a practical method for extracting and refining morphine by first extracting it with acid and then breaking it down with bases, although today ammonium is usually chosen as a base.

Accordingly, the 'discoverer' of morphine is part of a long tradition of attempts to tame opium. In Paderborn (Germany) in 1804, the apothecary assistant Friedrich Wilhelm Sertürner isolated a gray precipitate from opium extracts that was not soluble in water, but easily soluble in acetic acid. By then adding ammonia, he got the same precipitate back, which showed that it was a stable product. Sertürner determined through an experiment he conducted on his dog that this substance caused the same sleepiness as opium and that the other components of opium were ineffectual. Sertürner concluded thus that he had found the effective component (Meyer 2004). He called the substance morphium. In today's nomenclature, the common name is morphine. Although it can be assumed that morphine had already been isolated by alchemists such as Glauber following similar procedures, Sertürner determined the basic nature of the substance, refined it carefully through recrystallization and showed that it is in fact one of the active substances in opium. His objective was to provide doctors with a pure substance of consistent quality: "He will always be able to use this substance dissolved in alcohol and acids with equal success, instead of the currently prevalent opium preparations which are not always uniform" (Sertürner 1806, 55).

Losing control – World War I and morphinism

Since taking morphine orally caused nausea, other ways to administer the drug were sought. The solution was found to be injection, a method that was developed independently in the 1850s by both the French doctor Charles-Gabriel Pravaz and the Scottish doctor Alexander Wood (Reynolds Whyte et al. 2002, 104). The first substance that Wood injected was in fact morphine. This new method of administration, with a syringe, almost made a new substance out of morphine – in any case a far more effective substance. Although he was not the first to describe the symptoms of morphine addiction (Jacob 1984, 88), it was Eduard Levinstein who coined the term 'morphinism' in a monographic publication. In his monograph, he vividly describes how the easy technique brought about a "quick, miraculous effect against pain":

> Morphine injections were rarely carried out in Germany until about a decade ago. The simple technique of the Pravaz'schen method, which afforded a quick, miraculous effect against the pain for the seriously wounded and sick of the war of 1866, and the calm which it brought, rapidly paved the way for this treatment in Germany. The range of therapeutic indications for the treatment was extended daily without distinction.
>
> (Levinstein 1877, 1)

Levinstein thus represents the point where the substance escaped the hands of professionally trained doctors and began take on a life of its own. The method and the substance spread with extraordinary rapidity, especially following the opportunity to observe the beneficial effects of morphine on the many wounded during the Austro-Prussian War of 1866.

The subcutaneous injection of morphine soon became popular and the range of therapeutic indications was expanded considerably.

> Moreover, the Great War of 1866 and 1870/71 gave those involved and their dependents plenty of cause and opportunity to use a drug that was able to eliminate sorrows and pains at a stroke: within a short time, a disease previously unknown in Germany arose – the morphine habit.
>
> (Jacob 1984, 88)

It was mainly doctors that spread the disease, as many observers noted – for example, the psychiatrist Emil Kraepelin:

> At this point serious charges must be raised against the medical profession with vehement condemnation, that it is first and foremost they whom we have to blame for the existence and the frightening spread of morphinism. If there were no doctors, there would be no morphinism.
>
> (Kraepelin 1904, 142)

The invention of heroin

In the nineteenth century, morphium was extensively used in pain therapy, particularly in the wars: the Crimean War, the Prussian-Danish War, the Austro-Prussian War as well as in the Franco-German War. With the outbreak of World War I and the many millions of injured and dead, the demand for morphine increased dramatically. In 1920, the Franz Krömeke, biographer of Friedrich Wilhelm Sertürner, wrote that "morphine has become every doctor's indispensable companion. For millions during the World War it was the comforting angel on the battlefield and in the hospitals" (Krömeke 2010, 1).

The problems with morphine, visible even before World War I, had by the late nineteenth century motivated a search for a substitute that on the one hand was able to be better tolerated and on the other unleashed no dependency-creating potential. It was in this context that Felix Hoffmann developed diacetylmorphine at the Farbenfabriken, formerly Friedrich Bayer & Co., a compound which had indeed already been synthesized by other researchers, but had nevertheless not been thoroughly tested for possible pharmaceutical effectiveness. Just as the alchemist Glauber and many others had tried to "temper the wildness" of opium (Glauber 1656, 92), morphine was to be 'tamed' by acetylization. The work at the pharmacology department at Bayer is therefore part of a centuries-old tradition; the topic was old, only the methods were new. Similar to the isolation of morphine, and then subcutaneous administration with a syringe, acetylation was meant to increase the level of control over the substance. Scientists clearly hoped to preserve the positive qualities of morphine with the new preparation, but to eliminate the negative by means of acetylation.

The plant pharmacologist Heinrich Dreser tested the compound on animals and immediately recognized its pharmaceutical potential: the substance namely slowed respiration – it made respiration more economical, as Dreser opined. After several animal trials, the preparation was put on the market as a cough medicine, and indeed at first as a respiratory sedative. The name of the new product sounded much more promising than the one that had been chosen for acetylsalicylic acid: heroin. Unfortunately the story of how the name was chosen cannot be followed as the relevant documents are no longer in the Bayer archives.

Dreser seized upon the substance despite being skeptical at first. The substance was accepted by the grateful public. Lung diseases and various types of whooping cough were widespread in the smoke-filled industrial cities of the nineteenth and early twentieth centuries. Bayer produced heroin in kilograms and many other companies, domestic and foreign, jumped on the bandwagon. The head of the board of directors at Bayer, Carl Duisberg, sent some with best wishes for a speedy recovery to a colleague plagued by coughing. Mothers gave it to their children. Only scattered voices were heard from doctors or scientists warning of dangerous side effects (De Ridder 2000, 56–61), but that had also been the case for aspirin.

In fact, the increased toxicity of the drug was well known. But the fact that the drug led to dependence in the same way as morphine was overlooked for a long time, as the experts Otto Anselmino and Adolf Hamburger describe in their commentary published in 1931 on the opium law: "The use of diacetylmorphine due to addiction was then, in 1920, not . . . known to the medical authorities of the Reich and states." (Anselmino and Hamburger 1931, 9).

Redefining opiates – politicians take control

With the onset of World War I, morphine, heroin and cocaine addiction had virtually become a national disease, at least in Germany. In 1925 the psychiatrist Walter Jacob summarized this, writing that "a large percentage have become morphine addicts as a result of wounds or psychological attrition in the field" (Jacob 1925, 94). Many of the morphine addicts were addicted to both heroin and cocaine. Psychiatrist Karl Bonhoeffer, father of Dietrich Bonhoeffer, wrote in the *Handbuch der ärztlichen Erfahrungen im Weltkriege 1914/1918* (Handbook of Medical Experiences in the Great War 1914/1918):

> There was a clear increase in morphinism. This is explained mainly by the frequent use of morphine on the wounded and by the increased number of people involved in the delivery of morphine. The increase in morphinism developed particularly recently after the end of the war, so that at some clinics the number of morphine addicts exceeds that of alcoholics.
>
> (Bonhoeffer 1922/1934, 23)

Also in America, where the preparation was usually administered intravenously, it was recognized that it had a high potential for being addictive, and in this respect presented no improvement over morphine. Heroin, like all morphine derivatives, leads to a reduction of sensitivity in the respiratory center and thereby to a depression of all respiratory activities. Like all opiates, it can trigger addiction. The dream of finding a morphine substitute that did not generate an addiction had not been realized. Attempts to employ heroin to wean morphine addicts had the same effects as similar experiments in which cocaine was used to fight drug addiction. Furthermore, heroin is more toxic than morphine. However, it is apparent that it also has advantages as a pain reliever in comparison to morphine, as the onset of the effect is significantly faster and it has fewer side effects.

The war on drugs in the United States

The battle against opiates was fostered by the U.S. government. The anti-drug movement had, according to some historians, a Christian-moralistic background (see Buxton 2006, 43). It is also argued that it served to stigmatize population groups that were perceived by American society as a threat, such as the Chinese immigrants who consumed opium, as well as blacks (Musto 1999, 4f.).[8] In any case, opiates, especially morphine but also heroin, were often administered without instructions. The result was a high number of addicts.

The United States was the driving force behind the first International Opium Conference in Shanghai. Recommendations were developed at this conference which then formed the basis of the First International Opium Convention of 23.1.1912. The agreement also regulated the use of diacetylmorphine (heroin), morphine and other alkaloids. The Opium Convention was to be ratified by the participating powers, but in the interest of the German pharmaceutical industry, the German Reich hesitated to ratify the agreement until it was forced to do so in the Treaty of Versailles of June 28, 1919 (Weber 2009, 3).

With the introduction of the prohibition, the import of opium was declared illegal in the United States in the Smoking Opium Exclusion Act of 1919 (De Ridder 2000, 111). The substance was stigmatized and excluded from normal commerce – parallel to a stigmatization and exclusion of specific population groups. The promotion of heroin to be the most dangerous of all addictive substances began in 1917, thus during the World War I. In line with the propaganda against the enemy, Germany, heroin was denounced as a sneaky and highly dangerous poison, with which it was thought Germany was weakening the resistance of the American nation (De Ridder 2000, 111). After the war was won, the heroin-critical position became the foundation for all further initiatives in the United States against drugs in general and against heroin more specifically. The very powers that were critical of regulation (i.e., the German Reich and Austria but also the Ottoman Empire) had lost the war and now had to bend to the will of the victor. The U.S. position could now prevail, and the American prohibition policy was exported internationally.

The production became internationally controlled (Buxton 2006, 39). It was limited through quotas. At the same time, the signatory states were responsible for seeing to it that the incriminated substances were used exclusively for medicinal purposes. These regulations were to be implemented in national laws. A national and international opium bureaucracy arose to control the implementation of the agreement. Even if the goal of the United States to slowly but surely stop production of the drug could not be achieved, production nevertheless sank (Buxton 2006, 57f.). The anti-drug policy of the United States was thus at first successful. In the 1920s and 1930s, the American stance on heroin and other opiates was further solidified, spurred on by this success.

In 1930, the Federal Bureau of Narcotics was established on the initiative of Steven G. Porter, a member of the American House of Representatives. Harry J. Anslinger was named as the organization's first director (Musto 1999, 206f.). He sharpened the tone concerning opiates and also led the first campaign against marijuana. He described the danger that these substances presented drastically in books, brochures and films, presenting them uniformly as types of poison. The heroin-critical stance among the American public developed into a heroin phobia that saw in the substance a highly dangerous bacillus, the very contact with which would lead to its damaging, often deadly, effects being felt.

Pharmacologists and doctors protest against the opium law

After the capitulation of the German Reich, the ratification of the Hague Convention was imposed as a condition in Article 295 of the Versailles Peace Treaty. The Weimar Republic met the demand by creating the first German Opium Act on 30.12.1920. However, the control of opiates, especially heroin, was not implemented because of the Versailles Peace Treaty alone. The regulation of opiates and cocaine was recognized as necessary by professionals such as the Berlin pharmacologist Louis Lewin, who had first described the clinical picture of morphinism.

Nevertheless, even in the first review of the law in which Lewin had been instrumental, it was noted that the "Opium Law . . . [inhibits] the duty of the doctor to help the sick – specifically with analgesics" (Lewin and Goldbaum 1928, 20). Commentators argued that addicts were sick people who "require the substance with which they have become familiar and which is dear and necessary to them as much as the stomach or the entire body requires food" (Lewin and Goldbaum 1928, 21). Withdrawal caused "alongside pain . . . the risk of acute physical and mental breakdown" (Lewin and Goldbaum 1928, 21).

The Opium Act thus confronted doctors with an unsolvable dilemma: "A doctor visited by such an addict, having been implored to help, will be [. . .] faced with the choice of violating the law or providing help" (Lewin and Goldbaum 1928, 22). Two doctors specializing in addiction medicine, Fritz Frankel and Joël Ernst, highlighted the medical dilemma even more clearly in a much-quoted article from 1927:

> When the morphinist has first truly lapsed into addiction, he gains . . . no real enjoyment from his syringe, rather only the restoration of his equilibrium. When lacking his poison, the morphinist resembles a sick person with objective symptoms. This acute morphine sickness is relieved abruptly with the

> appropriate dose of morphine and in such a way that can only be achieved with morphine (or a related opiate). [. . .] Does morphine used in cases of pronounced morphinism serve as a medicine or a stimulant? It cures the acute morphine weakness and the deprivation symptoms. It heals of course by no means the morphinism, rather it entertains it.
>
> (Ernst and Fränkel 1927, 1055)

But the two doctors drew further distinctions, in a way that has not lost its relevance:

> Any doctor who meets a morphine addict (without any other disease) will try to persuade him to follow a course of rehabilitation. . . . There are also patients that keep themselves able to work with relatively low doses and who do not improve in the slightest through treatments, rather that are damaged otherwise, and from repeated experience decline rehabilitation. Even in such cases, one would have to consider morphine prescriptions as serving a medicinal purpose.
>
> (Ernst and Fränkel 1927, 1055)

The doctors used morphine as an example in their statements. However, the statements also applied to other alkaloids, which is why Ernst and Fränkel also referred to alkaloid addiction (Ernst and Fränkel 1927, 1052). Although they clearly supported the aim of the Opium Act to curb the abuse of narcotics, it was their central wish to move the substance back into the competence area of doctors: "The decision on the existence of such a situation [namely whether prescribing opiates in individual cases fulfils a medical purpose] can, after careful examination, only be made by the doctor" (Ernst and Fränkel 1927, 1055).

The dilemma that morphinism and heroin addiction posed to doctors was thus clearly described and in a way that even today has lost none of its relevance. If the doctor helps an addict, then the addict remains addicted. If the doctor does not help him, then he leaves the addict in severe pain and with symptoms that can lead to physical damage and even to death. The distinction between medicine and stimulant, as Ernst and Fränkel stress, is not applicable in such cases. It is not a distinction that can be drawn summarily, rather only in every case individually following medical examination. From today's perspective, Ernst and Fränkel's position of easing pain by giving drugs where one cannot heal is thoroughly consistent because of the relatively common phenomenon of 'maturing out': with age, many drug addicts find the strength or the maturity to break away from the drug.

From the perspective of lawyers and politicians, however, drug addiction is in no way a disease but a bad and dangerous habit – a vice that can be wiped from the world with the threat of punishment and penalties. They also did not know how to distinguish generally and in a sufficiently convincing way between use as a medicine and use as a stimulant. Several cases brought before the Reichsgericht (high court) document the various makeshift decisions made by the judges. Time and again, the highest court of the German Reich was forced to take the role of the doctor and determine whether a medical or stimulant purpose was at hand in individual cases (see Reichsgericht 1927, 60: 365–371). Pharmacists were also urged to take on the role of medical regulators, but they protested with reference to the medicinal edict from Kaiser Frederick II of 1240, which prohibited pharmacists from practicing medicine, a norm that has existed with good reason ever since (Ries 1965, 903–905). Nevertheless, the fragile balance between lawyers and doctors continued into the 1960s. The substance of the opium law remained untouched for decades.

Battles about the cultural setting of drugs

But then the student revolt began, and many of its protagonists had a very positive relationship with all kinds of mind-altering substances, including heroin. Now the opiates, as well as other drugs, got caught in the crossfire of a cultural struggle between the younger and the established generations. The fragile balance between medical and legal perspectives on the drugs was destroyed. There was no room for subtleties in the battle of the generations. The legal side asserted itself and removed all of the conceptual dilemmas with decisions by force. The generation challenged by the student revolts responded to the large role of mind-altering substances in the student revolts of the 1960s with the execution of the United Nations' 1961 Single Convention on Narcotic Drugs. In Germany, the Opium Act of the Weimar Republic was replaced in 1972 with a new version of 22.12.1971, known as the Betäubungsmittelgesetz (narcotics law). A revised narcotics law came into force in 1982 and was barely changed until the early 1990s (Weber 2009, 4).

This law permanently passed heroin and some other substances from the hands of doctors to those of police officers and judges. In Appendix 1 of the act, heroin was declared a 'non-trafficable substance' and could therefore no longer be prescribed. Everything was now clear in legal terms. The vexatious differentiation between medicine and stimulant was gone: heroin was simply declared not to be a medicine. The dilemma mentioned by the first commentators on the Opium Act, quoted previously, was not only never resolved but negated.

This assertion succeeded because heroin had in a sense become an orphan socially. Production was stopped in Germany because after great initial success, demand had shrunk significantly and thus no longer had the support of the powerful German pharmaceutical industry. Proponents of the drugs were not yet in positions of power, but acted as 'extra-parliamentary opposition'.

The law enforcement authorities could now begin to use their resources to deter the interested and the experimenters and to push addicts into social isolation – all to the acclaim of a portion of the public that increasingly demonized heroin, developed heroin phobia, and saw in the drug the real reason for the strange behavior of the youth. Heroin was now considered evil. Drugs – heroin included – were so restricted that we can speak of an almost universal de facto ban.

This had truly cruel consequences for the addicts who, because it was incorrectly believed that they simply lacked willpower, were abandoned to impoverishment and forced into crime. The medical principle that in cases where you cannot heal you can at least offer relief was repealed for cases of addiction by the internationally coordinated opium and narcotics laws. Addicts in particular were under no circumstances to be given their substance, despite it being known since the 1920s that in many cases a complete withdrawal from opiates is not possible, but that a normal life can be obtained only by continuing to give small or medium-sized doses (Ullmann 2001, 23). But this no longer applied: "It was now the law of all or nothing: without prior abstinence, which often had to be achieved without any medical relief, there was no treatment for any serious illnesses, neither physical nor mental" (Ullmann 2001, 24).

Finally, the lawyers and politicians had withdrawn the authority over heroin and other drugs completely from doctors and assumed it for themselves. Moreover, they sought the assistance of psychologists, attempting to use their resources to establish control over the substances. This attitude was not even shaken by the HIV epidemic, which was particularly widespread among injecting drug users. Thus in 1987 a chief federal prosecutor took his own staff to court for having described supplying drug addicts with single-use syringes as legally unobjectionable. In 1995, the federal government's drug commissioner judged the establishment and operation of drug consumption rooms in Frankfurt to be a criminal offense (Körner 2001, VII). A law professor's idea of giving heroin to addicts (Adams 1994, 106–111) was immediately dismissed by drugs politicians as 'cloud cuckoo land'. It was considered more important to give the addicts the moral support to give up their addiction (Eylmann and Kusch 1994, 209–211). The concerns of doctors and pharmacologists, however, held no sway and morphinists were increasingly put under pressure (Ullmann 2001, 21).

The present situation – less control than ever

By damning heroin to being a non-trafficable and thus de facto banned substance, the Federal Republic of Germany was following the broad lines of the drug policy put forward by the United States. That is not to say that the policy had not been propagated in Germany with genuine conviction, particularly but not only among conservative politicians.

The narcotics laws, which were equivalent to prohibition, not only antagonized a deplorable situation but also produced new ones. The descent of many addicts into crime, their health and social deprivation, and their deaths were accepted as a loss and even falsely used as evidence of the danger of heroin, even though it had been known since the 1920s that addicts may well live a normal middle-class life if they are given their dose.

Heroin has, as already stated, comparatively few side effects. A medical practitioner specializing in addiction gives the following verdict:

> The substance, heroin, is . . . less toxic than often assumed. The pharmacological damaging bodily effect of heroin is comparative to that of alcohol or nicotine. The main danger of heroin consumption is attributable to the unsterile storage and application of the substance and lies in an infection with HIV or hepatitis. The often toxic cutting substances also make heroin consumption risky.
>
> (Croissant et al. 2003, 181)

The numerous deaths from hepatitis and HIV infection among drug users have their roots in the prohibition and can barely be justified ethically (Ullmann 2001, 24f.). It is undoubtedly the harsh prohibition policy, implemented by legal means enforcing international agreements since the late 1920s, that is responsible for the impoverishment and the high mortality of drug addicts, even though it is this very impoverishment that is used to justify the policy of repression. Even the supposedly healthy suffered the negative side effects of the prohibition policy. It no longer went without saying that seriously ill patients would be given opiates. Often these patients also succumbed to heroin phobia and refused to be treated with opiates.

A lesser known but equally grave side effect of the restrictive heroin policy built upon an irrational heroin phobia is the frightening undersupply of seriously ill and dying people with effective opiates. Of course, no one will share the enthusiasm of the discoverers of heroin of the early twentieth century and see heroin as an all-purpose wonder drug. But nevertheless, it is not to be overlooked that it is in fact an effective drug against strong coughing, such as the consequences of tuberculosis. But above all, it is a highly potent pain reliever. In essence it functions in the same way as morphine but takes effect faster and does not generate nausea. For terminally ill patients suffering from great pain, it can bring relief like no other medicine (Carnwath and Smith 2002, 146–153). The pain is distanced by taking heroin, without noticeably clouding the consciousness. A doctor therefore wrote in a new study on heroin: "In any fair assessment, one would have to say that even after a hundred years, heroin remains a medicine without a superior" (Carnwath and Smith 2002, 146–153).

It is not only the doctors who keep the administration of alkaloids at a low level. The patients themselves often refuse opiates for fear of becoming addicted. The concept the patients have of opiates means the bottle stays in the cabinet. The Norwegian doctor and pain expert Stein Husebö commented about terminally ill patients: "If I recommend to cancer patients that we start treatment with morphine, many say: no, I don't want that. I don't want to become a drug addict!" (Husebö 2001, 120). In Germany, on the other hand, the country in which morphine was first isolated and characterized and in which heroin was first industrially produced, the doctor and heroin historian de Ridder believes that very ill and dying patients are being undersupplied with pain relievers, especially with opiates (de Ridder 2010, 93–113). One expert estimates that the number of people suffering from tumor pain who receive insufficient support lies in the region of 190,000 (de Ridder 2010, 93–98). This may appear to be an overestimation, but it is certain that there is an undersupply of pain relievers in more than a few isolated cases, even today, despite the efforts of palliative medicine and the hospice movement to dispel the fear associated with opiates.

If we consider the enormous influence of the drug mafia and so on, then it can be seen that there are many other side effects of an economic and political nature which would not even exist were it not for the prohibition.[9] Only in the UK has the medical perspective on the subject survived. Drug addiction has always been regarded there as a health problem. Only in Great Britain and Belgium can heroin still be prescribed by a doctor.

Transferring a substance, shifting the matter of control

The chemical and medical optimization of the original substance, opium, through agent isolation, processing and the refinement of its administration, did not lead to perfect control over it. Rather, the increased potency also intensified its problematic characteristics. In the same way, the political war carried out in the name of public health against morphinism and heroin addiction has not led to an increase in general welfare, but has caused torment and misery for countless addicts. It has supported drug cartels and other underground organizations, some of which have gained so much power that they can challenge states. The historic progression from opium to morphine to heroin, with the associated extreme scientification, spun out of control. The war against drugs carried out by politicians, lawyers and the police in order to restore this control unintentionally became a war against drug cartels, but also against part of their own population: against drug addicts. The war escalated but did not end in a victory. Not only has control not been restored, but the lack of control has escalated and in some areas has shaken entire states.

Yet even in Germany a reorientation can be identified since the early 1990s. This is probably due in part to the spread of HIV, which affected addicts considerably. To control the contagion, alleviating measures such as needle exchanges were allowed; even substitution programs were made possible subject to many conditions. A further ‘pillar’ in drug policy is being given increased attention, namely that of harm reduction (Weber 2009, 5). On the other hand, the legislature is, bit by bit, facilitating the medical use of opiates, in particular thanks to political lobbying by the hospice movement and palliative medicine. Thus since 2009 it has again been possible to prescribe heroin to serious addicts in Germany, provided they meet certain conditions, thanks to a regulation amendment.

Nevertheless, in our society, handling heroin and other opiates remains bound to a restrictive regime, in part because this is additionally laid down in international treaties, and in part because a critical attitude toward drugs has now become a part of the self-definition of conservative circles. From an ethical standpoint, the statutory handling of heroin and opiates is certainly highly problematic. One is inclined to agree with the words of Hans Harald Körner, a long-time commentator on the Betäubungsmittelgesetz (narcotics act), who notes that a “comprehensive review of the BTMG is urgently required” (Körner 2001, VII).

One gets the impression that the situation in the early 1920s, when a prescription was required for opiates but the decision about their use still rested in the hands of doctors, was better than that today, since the authorities now control the permits for handling them. It seems advisable to wipe the slate and give heroin and opiates a new chance. It would be desirable that heroin as a medicine break free from its special status as a ‘prohibited substance’ step by step and again be regarded, in the same way as the other potentially addictive alkaloids, as part of the range of modern medicines – as a medicine in the hands of doctors.

Notes

1 Most of the sources are German because morphine and heroin were first presented and marketed in Germany. Some of the sources have until now not been evaluated in the broader discussion about opiates.
2 Arjun Appadurai, editor of the book collection *The social life of things* (1986), argues for an anthropology of things. George Marcus condensed the ideas of Appadurai in the famous sentence "Follow the thing!" (Marcus 1995, 106–107).
3 The meeting took place in Augsburg, May 9, 2012.
4 'Opium' comes from the Greek word 'opos' and means literally 'juice'.
5 Despite the fact that Marcus Aurelius himself does not mention theriac in his famous *Meditations*, we know through two independent sources that he regularly ingested theriac. Cassius Dio reports in his *Roman Histories* that the Emperor ate nothing at all during the day – except his daily portion of theriac (Cassius 1969, 20–23). The second source is Galenius (1965, 3–5), who also reported Marcus Aurelius' theriac consumption.
6 Glauber was a famous seventeenth century alchemist whose descriptions of the chemical processes he worked with are still comprehensible nowadays despite the use of alchemical names and symbols.
7 'Liquor Nitri' (Glauber 1656, 92) is not a nitrogen compound; see Gugel (1955, 51f.) and Link (1993, 127).
8 See also Shapiro (1995, 32–41).
9 There is a large amount of literature on this topic; see for instance Gootenberg (2005), or also Carnwath and Smith (2002, 62).

References

Adams, M. 1994. 'Heroin an Süchtige – Kollektiver Wahnsinn oder das gesuchte Konzept zur Zerstörung des Drogenmarktes?' *Zeitschrift für Rechtspolitik* 3: 106–111.

Anselmino, O. and Hamburger, A. 1931. *Kommentar zu dem Gesetz über den Verkehr mit Betäubungsmitteln (Opiumgesetz) und seinen Ausführungsbestimmungen*, edited by Anselmino, O. and Hamburger, A. Berlin: Springer.

Appadurai, A. 1986. *The Social Life of Things*, edited by Appadurai, A. Cambridge: Cambridge University Press.

Bonhoeffer, K. 1922/1934. 'Die Bedeutung der Kriegserfahrung für die allgemeine Psychopathologie und Ätiologie der Geisteskrankheiten.' In *Handbuch der Ärztlichen Erfahrungen im Weltkriege 1914/1918. Geistes- und Nervenkrankheiten*, edited by Bonhoeffer, K., vol. 4, 23. Leipzig: Johann Ambrosius Barth.

Buxton, J. 2006. *The Political Economy of Narcotics: Production, Consumption and Global Markets*. London: Zed books.

Carnwath, T. and Smith, I. 2002. *Heroin Century*. London: Routledge.

Cassius, D. 1969. *Roman History, IX*, Book LXXI, chapter 6. Cambridge, MA and London: Harvard University Press / Heinemann, Loeb Classical Library.

Croissant, B., Croissant, D. and Mann, K. 2003. 'Cannabis, Ecstasy, Heroin & Co.: Aktuelle Informationen zum Drogenkonsum.' In *Biologie in unserer Zeit*, edited by Croissant, B., Croissant, D. and Mann, K., 181. (33:3). Weinheim: Wiley.

De Ridder, M. 2000. *Heroin. Vom Arzneimittel zur Droge*. Frankfurt am Main: Campus Verlag.

———. 2010. *Wie wollen wir sterben?* München: Deutsche Verlags-Anstalt.

Ernst, J. and Fränkel, F. 1927. 'Öffentliche Maßnahmen gegen den Mißbrauch von Betäubungsmitteln.' *Klinische Wochenschrift* 6 (22): 1055.

Eylmann, H. and Kusch, R. 1994. 'Heroin an Süchtige – Wokenkuckucksheim oder professorale Genialität?' *Zeitschrift für Rechtspolitik* 27: 209–211.

Galenius, C. 1965 [1827]. *Galeni Opera Omnia*, Editionem curavit. Kühn C. G, De Antidotiis, lib. II, 20 vols, vol. 14. Hildesheim: Georg Olms Verlagsbuchhandlung.

Glauber, J. R. 1656. *Pharmacopoeae Spagyricae Ander Theil. De Vegetabilium, Animalium, & Mineralium praeparatione per Solvens Universale*. Amsterdam: Johan Janssen.

Gootenberg, P. 2005. 'Talking Like a State: Drugs, Borders, and the Language of Control.' In *Illicit Flows and Criminal Things*, edited by Schendel, W. V. and Abraham, I., 101–127. Bloomington: Indiana University Press.

Gugel, K. F. 1955. 'Johann Rudoph Glauber 1604–1670, Leben und Werk.' *Mainfränkische Hefte* 22: 51f.

Husebö, S. 2001. *Was bei Schmerzen hilft. Ein Ratgeber*. Freiburg and Basel, Wien: Herder.

Jacob, W. 1984. 'Zur Statistik des Morphinismus in der Vor- und Nachkriegszeit.' *Archiv für Psychiatrie* 76 (1925): 212–232, reprint in *Wiener Zeitschrift für Suchtforschung* 7: 1, 87–96.

Körner, H. H. (ed.). 2001. *Betäubungsmittelgesetz, Arzneimittelgesetz*. Beck'sche Kurz-Kommentare, 66 vols, vol. 37. München: Verlag C. H. Beck.

Kraepelin, E. 1904. *Psychiatrie: ein Lehrbuch für Studierende und Ärzte*. 7th ed., 2 vols, vol. 2. Leipzig: Verlag von Johann Ambrosius Barth.

Krömeke, F. 2010. *Friedrich Wilh. Sertürner, der Entdecker des Morphiums. Lebensbild und Neudruck der Original-Morphiumarbeiten*. Jena: Gustav Fischer Verlag 1925. reprint Hamburg: Severus Verlag.

Levinstein, E. 1877. *Die Morphiumsucht. Eine Monographie nach eignen Beobachtungen*. Berlin: August Hirschwald.

Lewin, L. and Goldbaum, W. 1928. *Opiumgesetz nebst Internationalem Opiumabkommen und Ausführungsbestimmungen. Kommentar*. Berlin: Verlag von Georg Stilke.

Link, A. 1993. 'Johann Rudolph Glauber 1604–1670. Leben und Werk.' Dissertation, Universität Heidelberg. Heidelberg.

Marcus, G. E. 1995. 'Ethnography in/of the World System: The Emergence of Multi-Sited Ethnography.' *Annual Review of Anthropology* 24: 106–107.

Meyer, K. 2004. 'Dem Morphin auf der Spur,' *Pharmazeutische Zeitung*, http://www.pharmazeutische-zeitung.de/index.php?id=titel_16_2004 (accessed August 12, 2010).

Musto, D. F. 1999. *The American Disease: Origins of Narcotic Control*. 3rd ed. Oxford: Oxford University Press.

Reichsgericht. 1927. *Entscheidungen des Reichsgerichts*, edited by Mitglieder des Gerichtshofes und der Reichsanwaltschaft, vol. 60. Leipzig und Berlin: de Gruyter.

Reynolds Whyte, S., van der Geest, S. and Hardon, A. 2002. *Social Lives of Medicines*. Cambridge: Cambridge University Press.

Ries, J. 1965. 'Strafrechtliche Haftung des Apothekers bei Verstößen gegen das Opiumgesetz durch Abgabe von Betäubungsmitteln.' *Pharmazeutische Zeitung* 28: 903–905.

Sertürner, F. 1806. 'Darstellung der reinen Mohnsäure (Opiumsäure) nebst einer chemischen Untersuchung des Opiums mit vorzüglicher Hinsicht auf einen darin neu entdeckten Stoff und die dahin gehörigen Bemerkungen.' *Journal der Pharmacie für Aerzte, Apotheker und Chemisten von D. Johann Barholmä Trommsdorf* 14: 55. Leipzig.

Shapiro, H. 1995. *Sky high. Droge und Musik im 20. Jahrhundert*. St. Andrä-Wördern: Hannibal.

Ullmann, R. 2001. 'Geschichte der ärztlichen Verordnung von Opioiden an Abhängige.' *Suchttherapie Sonderheft* 23, 20–27.

Weber, K. (commentator). 2009. *Verordnungen zum BtMG*. K. Beck'sche Kurz-Kommentare. München: Verlag C. H. Beck.

7 Long live play

The PlayStation Network and technogenic life

Colin Milburn

They say it's always good to start with a joke.

On May 4, 2011, the American comedian Jay Leno, during his regular opening monologue on *The Tonight Show*, reminded his audience about an ongoing current event – namely, the Sony PlayStation Network outage. The PlayStation Network, or PSN, had been offline since April 20th, and gamers everywhere were making their distress known all over the Internet. Sony officials eventually confessed that they had shut down the network as a retroactive security response to an 'external intrusion'. Apparently, a sophisticated team of hackers had managed to infiltrate the PSN databases. Sony was unaware of the hack until a couple of days after it took place. The intruders had extracted the personal information, passwords, and possibly the credit card numbers of registered PlayStation Network users – upward of seventy-seven million people. In the end, the PSN would remain down for a total of twenty-four days. Leno summed up the situation with his characteristic wit:

> Sony has apologized after the accounts of PlayStation users were hacked into. They say this could severely affect the lives of over a hundred million PlayStation users. You know something, if you're playing PlayStation all day, you don't have a life! Okay? I don't think you have to worry about your life being interrupted.[1]

Leno rehearsed versions of the same joke throughout the week, draining every last laugh out of the idea that gamers 'don't have a life'. If the joke falls a bit flat, it is not only because it disregards the risks faced by the Sony customers – identity theft and credit card fraud among them. Rather, what seems most out of touch about Leno's joke is that it overlooks the sheer scale of the affected population – millions upon millions of gamers around the world – and thus misrecognizes the nature of the risk entirely: a threat to a particular lifeworld, a technological way of life. Legions of gamers, dispersed over many different countries, had been forcibly ejected from their familiar online community and recreational space, and the Internet was now buzzing with the sound of their anxiety, their anger about the security breach mixed up with longing and adoration for the network as such. One gamer explained all the commotion with a simple assertion: "We are not nerds. We have a life."[2]

Despite all the mockery, a vibrant form of life had been dramatically interrupted by the disappearance of the PlayStation Network. Even after the network was restored, the memory of its outage would retain all the force of a *primal scene*, routinely recollected as a defining moment in the history of the network and in the personal biographies of many gamers (hence the proverbial question: "Where were you on April 20, 2011?").[3] The event powerfully illuminated the operations of contemporary technogenesis, which is to say, the mutual shaping of technics and human life in the current moment.[4] For it showed how much the individuation of PlayStation gamers as gamers (ludogenesis), together with the collective individuation of the PlayStation community as a community (sociogenesis), has involved a process of internalizing and reconstituting a particular technoscientific apparatus – the PlayStation Network itself.

Epic fail

In many ways, the PSN came to life as an object and a site of technogenesis retrospectively, reborn at the moment of its disappearance. Some gamers would later remember the network outage as a 'birthday', a genetic moment when the gaming community coalesced under conditions of shared risk and heightened emotion (Parker 2012). As one of them explained, "It made me realize what a big part of my life the PS3's online capabilities were to me, and I wasn't alone."[5] By the same token, the mainstream media only became fully cognizant of the network during its time of crisis – seventy-seven million gamers, sixty million PlayStation 3 units and hundreds of servers dispersed over more than sixty countries, suddenly disintegrated.

During the twenty-four days of the outage and for several months afterwards, gamers around the world obsessively discussed the technicalities of Sony's firewalls, server architectures and encryption standards, as well as the hardware features of the PlayStation 3 unit, its Cell processor, its various firmware upgrades and the limits of its operating system. A failure of securitization seemed to be the general consensus. At the same time, they debated the database hack itself, immersing themselves in the vocabulary of distributed denial of service attacks (DDoS attacks), SQL injections and other tools from the repertoire of hacker culture. They argued furiously about the motives of the hackers, trying to make sense of what on the one hand appeared to be nothing more than grand larceny, yet on the other hand evidently had some connection to a recent spate of cyber-protests against Sony and its corporate policy of prohibiting free and open experimentation with the PlayStation technology.

The crisis therefore exposed the technical dimensions of the PSN, its fundamental materiality and its algorithmic composition – which is to say, its radically non-human aspect – even as it brought to light the heterogeneity of human elements in the system, including ideological differences among PlayStation users about the value of understanding and having access to the technical foundations of their shared recreational activities. Faced with spectacular evidence of its *vulnerability*, gamers confronted the network as both singular – insofar as it had disappeared all at once – and inherently multiple: a modular collective of hardware components organized through an evolving set of protocols and data streams, conjoining disparate crowds of people, cultural narratives and media operations to varying degrees. The outage made clear that the PlayStation Network, like all networks, is technical as well as political, human as well as non-human (Chun 2006; Galloway and Thacker 2007).

It is, in fact, a *quasi-object*, an interface of the subjective and the objective, the social and the material. It draws diagrams of relationality, producing a certain collectivity (the PlayStation community), and likewise, in its technical individuation, apprehended as a unity or coherent system, it configures its users and opens a particular identity space (the PlayStation gamer). As Michel Serres has written,

> [The] quasi-object is not an object, but it is one nevertheless, since it is not a subject, since it is in the world; it is also a quasi-subject, since it marks or designates a subject who, without it, would not be a subject. . . . The quasi-object, in being passed [between people], makes the collective, if it stops, it makes the individual.
>
> (Serres 1980, 225)

The PlayStation Network – irreducible to its component parts yet unthinkable aside from them – moves through and among its hardware nodes as flows of digital information, always potentially connected even when disconnected. It moves through and among its users as a figure, an experience, a fiction, an embodied relation. As Bruno Latour has written, "As soon as we are on the trail of some quasi-object, it appears to us sometimes as a thing, sometimes as a narrative, sometimes as a social bond, without ever being reduced to a mere being" (Latour 1993, 89). To be sure, the PlayStation Network incarnates connectivity in and of itself, a network that makes a network – a technoscientific system that is always already political. This was never more evident than in the midst of its catastrophic failure.

Although for many gamers the political dimensions of the PSN outage were not entirely clear, and often deeply confused, there was widespread awareness that the thing at the core of all this anxiety, distress, anger, adoration and love – the network itself – had somehow become a battleground for the future of participatory science, peer-to-peer research, and do-it-yourself innovation. For some, this meant the future of democracy as such in our ever more globalized and high-tech society; for others, it represented a deplorable hijacking of private property, a cooptation of entertainment technologies for illicit purposes. If nothing else, by making visible the profound entanglement of gamers with the PSN, the intensive modes of affectivity and identification immanent to the system, the outage helped to crystallize the stakes of controlling access to the infrastructures of digital culture – one way or another, for better or worse. In this way, the network was rendered a symbolic casualty, collateral damage in a broader contest over the right to experiment with the technoscientific systems now at the heart of the world, the freedom to play with the conditions of technogenic life.

Get a life

Let's rewind a bit.

In 2006, shortly before the launch of the PlayStation 3 and the PlayStation Network, Ken Kutaragi, then CEO of Sony, declared that the new technical capabilities of the PS3 would transform and revitalize the gaming experience. Games, he said, would no longer be confined to the limits of 3D graphical representation, but would break from the screen to become 'live'. A number of gaming websites were quick to make fun of Kutaragi's hyperbolic statement, pointing out that Microsoft had already been using similar marketing language about its own gaming network – Xbox Live – since 2002 (Roper 2006; Weeks 2006). Yet Sony has continued to insist on the image of vitality, vigorously promoting its hardware devices and online network as fostering the conditions for life in the networked era – which is to say, a technoscientific form of life, sustained by fun and games. After all, according to Sony's 2007 advertising campaign, 'This is living'.

The language of technological vitalism permeates the PlayStation world. Consider the PS3's Cell processor, more formally known as the Cell Broadband Engine. Often compared to a eukaryotic cell – insofar as it features a core microprocessor supported by eight synergistic processing elements – the Cell was designed to be the 'nucleus' for multicellular networks (Standage 2005, 198–200). According to one Sony engineer, "We wanted to create a . . . processor capable of functioning as the nucleus for software interactions between networks and future computers connected to those networks" (Suzuoki 2009). Similarly, in 2011, Sony revamped its handheld PlayStation Portable (PSP) to sync with the PS3 and the PSN, renaming it the PlayStation Vita. To further flesh out the image of living through PlayStation, Sony issued a firmware upgrade for the PS3 in 2008 that enabled access to the virtual microworld of Home. Home was a 3D graphical space that served as an imaginary hub for the PlayStation Network, a place where gamers could meet each other in avatar form. Home was designed to foreground domesticity and comfort, a sense of groundedness in the expanding reaches of the global gaming system. Although Sony eventually disabled the feature in 2015, for a while Home pulsed with the lifeblood of the network. As one player put it at the time, "Home is my life."[6]

The particular form of life fostered or incubated in the PlayStation Network can also be measured by the extent to which the PlayStation system was made into a platform for biology – that is, biochemical research – through its integration with the Stanford University Folding@home project. Folding@home is an experiment in computational biochemistry that began in 2000. It uses a distributed network of PCs to simulate the mechanics of protein folding. In March 2007, only a few months after the PlayStation 3 was launched, Sony announced that it had joined forces with Stanford to save lives: "Folding@home is leveraging PS3's powerful Cell Broadband Engine™ (Cell/B.E.) – and what will be an even more powerful distributed supercomputing network of PS3 systems – to help study the causes of diseases such as Parkinson's, Alzheimer's, cystic fibrosis and many cancers."[7]

Soon thereafter, all PS3 units came pre-installed with the Folding@home software, accessible from the main navigation screen. In September 2008, Sony issued a firmware upgrade that redesigned the Folding@home portal, renaming it 'Life with PlayStation'. According to Noam Rimon, a director of software engineering at Sony, the PS3 Folding@home client and the 'Life with PlayStation' feature were designed to make the scientific experiments feel more like social games:

> [A]s video game designers we pushed hard on getting all the visualization in real-time and to allow the user to have a "virtual flight" through the field of folded proteins. We also added the globe of the world with dots for each participating machine, spreading a feeling of "togetherness", so users could see they were not alone in the folding world.[8]

In this way, the promise that PlayStation would bring games to life, providing a lifeworld for gamers around the globe, converged directly with notions of 'life itself' and the experimental systems of the life sciences (Figure 7.1).

When Sony's five-year collaboration with Stanford concluded in November 2012, Vijay Pande, a professor of chemistry at Stanford and the director of Folding@home, said, "The PS3 system was a game changer for Folding@home."[9] For Pande and his colleagues, the PlayStation system had served as a powerful research instrument, a platform for biological discoveries and pharmacological strategies, as well as an object of experimentation in its own right. Testing the capacities of the console and the network, the scientists gained fresh insights into the nature of

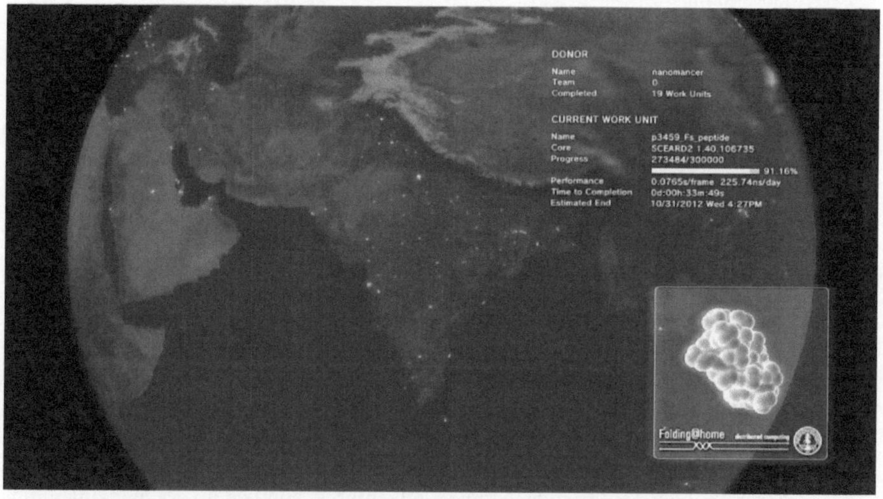

Figure 7.1 Folding@home client for the PlayStation 3. Available from 2007 to 2012, the application allowed gamers to navigate a global map of networked PS3s (represented as glowing dots) and to manipulate a simulated protein – rendering both the network and the molecule as objects of play.

distributed computing, the operations of different processors and algorithms – learning how to design better systems – in direct correlation with their studies of protein structure and potential drug candidates (Luttmann et al. 2009). Technological innovation and scientific research became inextricable.

Moreover, to the extent that the PlayStation's support of Folding@home came to symbolize a more playful form of technoscience, a research program in the mode of fun and games (significantly, Pande describes it as a 'game changer'), it also encouraged PlayStation users to think of themselves as citizen scientists, assisting the collective work of knowledge-production. As Pande said in 2007, the fact that Folding@home could be understood as "the most powerful distributed computing network ever is a reflection of the extraordinary worldwide participation by gamers. . . . Without them we would not be able to make the advancements we have made in our studies of several different diseases."[10] By 2012, more than fifteen million PlayStation users had contributed to the project, inspired to be involved in new technoscientific practices, new experimental approaches to life.

Or, as Sony's 2011 ads for the PlayStation 3 put it, "Long Live Play."

Devotion

> Hi. My name is crashsmash01 but you can call me John.b also . . . [T]he following story is the cronicles of me, my life, my friends, my family, and how Everyone at Sony and Playstation helped mold me into the fun loving but abnormnal gentelman i'm today. it all began in febuary 15th 1992. two years before playstation. back then my father and mother were divorced so i was living with my mother and stepfather most of the time . . . i was diganoised with ADHD and i also had serve depression.
>
> let us jump to march of 1994. my stepfather was on a bussiness trip . . . when he came home he gave me a huge box and it was a present, it was wrapped in red paper, when i opened it i litterly cried tears of joy and ran around the room like Crash bandicoot, it was a brand new PS1 system with a copy of the first Crash bandicoot [game], i had smothered my family in hugs and kisses and rushed straight to my room and played my heart out. a few days later my father bought me 3 games and some chocolate. it was a great day and so far a great begining to my life.
>
> as the years went by i started to grow, mature, and around this time i was in a new house, a new neighborhood, and thanks to playstation, i got my first friend ever. his name was Memo . . . if it wasn't for playstation my connection with my best friend Memo would of never been made. me and him would always hang out, playing the newest games, going outside and having fun adventures just like the characters we knew. one day for a birthday present my friend Memo and my parents got me a PS2 and some games . . . thanks to the PS2 when i traveled i started to gain new friends. had new adventures, and had a very amazing childhood . . . the other thing too was my PS2 also helped me when i was down at my lowest, from losing my friends constantly to my many family issues to money and everything. sony and playstation kept me going, and kept me running . . . many GFS [girlfriends] i had i would be playing video games with them . . . if

> anything, i'd always consider playstation and everyone who works hard on what they do, like family . . . thanks to Playstation i found my reason why i exsist after 15 years of barely any friends, a broken family, . . . My mother had died in 2006 because he[r] lungs shut down due to her ashma . . . everything began to fall apart . . . until one day. my father was uninployed but one day after school, we went out to Gamestop and he bought me a PSP system. it was my portiable gettaway to my happy place, from then i started high school . . . my PSP helped bring me together with tons of new people, new friends, and new ways to enjoy life. it was then that i figured out my reason to be here. to share and spread joy to everyone with the help of playstation. . . . things got even better when i got my first PS3 on November 11, 2006 . . . playstaion had been a part of my life sense i was almost born, it has helped me gain new friends, new loved ones, family, and much more. to this day i try my best to give back everyday to the PSN [PlayStation Network] and playstation community, wherever it be though PSN codes i find randomly, by helping someone buy a game, or jsut by being myself, playstation has helped mold myself into me. . . . thanks for reading everyone and i hope i can continue to help give back to the community and to keep this entire network strong for we shall long, live, play.[11]

In online gamer discussions, autobiographical narratives that rehearse the themes of building a life with PlayStation are common. Such testimonials of the importance of the PlayStation and the PlayStation Network for enabling meaningful friendships, a sense of camaraderie and shared culture, domestic and social belonging, comfort in times of sorrow, and personal accomplishment – indeed, all the qualities of having a life – became even more intense during the 2011 network outage. In its loss, the PSN was often rediscovered as an object of devotion, a crucial component of fulfillment, pleasure and self-completion – in the language of psychoanalysis, the *object-cause of desire*. As Jacques Lacan has written, the object of desire is fundamentally constituted by the perception of absence – "It is precisely what is subtracted from the living being" – always at risk, already lost (Lacan 1973, 198). Or, as one PlayStation gamer put it, yearning for the PSN in the midst of the crisis: "When will it be back, I feel like a part of me is down. . . lol."[12]

On the day the network disappeared, Sony issued an announcement on the official PlayStation blog: "We're aware certain functions of PlayStation Network are down. We will report back here as soon as we can with more information. Thank you for your patience" (Seybold 2011a). Two days later, Sony admitted that it had voluntarily shut down the network to address a security breach:

> An external intrusion on our system has affected our PlayStation Network and Qriocity services. In order to conduct a thorough investigation and to verify the smooth and secure operation of our network services going forward, we turned off PlayStation Network & Qriocity services on the evening of Wednesday, April 20th . . . We will continue to update you promptly as we have additional information to share.
>
> (Seybold 2011b)

Already in the earliest days of the outage, gamers began to express distress with startling rapidity. One said, "I just wish that the network was back, I never realized how dull the PS3 experience was without the network until now, after gamers around the world (including myself) have lost it. I miss talking with my friends online too."[13] Many seemed miserable: "I'm so SADDENED by this outage. GET IT UP NOW!"[14] Most were subdued in their sorrow: "miss my friends! :(."[15] Others resorted to melodrama: "going to kill myself !!!!!!!!!!!"[16] Or, "This can't be happening, wake me up from this nightmare . . . I would have thought something like this would never happen."[17] A few were more contemplative: "We have become so used to PSN being available 24/7 that withdrawal due to these events hits close to your gaming soul."[18] Yet a number of them seemed to be really suffering: "The PSN is only a part of me, right now it's the part of me that's wrenching and convulsing!"[19]

With each passing day, gamers voiced further exasperation – though some found a small degree of comfort in the fact that they could still play most single-player games or run Folding@home, whose servers were not directly linked to Sony's: "Fold your way through the PSN outage!" (Dan 2011). So while emotions were mounting, most gamers expressed confidence that Sony would restore normal operations shortly: "Hang in there everyone! I know how badly you want PSN to return, and it will. (This is bugging me too!)"[20]

On April 26th, Sony at last confessed that its databases had been hacked: "We have discovered that between April 17 and April 19, 2011, certain PlayStation Network and Qriocity service user account information was compromised in connection with an illegal and unauthorized intrusion into our network." Sony advised customers to be vigilant in protecting their online information, changing their passwords, monitoring credit card activity and so on. At this point, the emotional tide turned to outrage – much of it directed at Sony for its lax security measures, much more directed at the hackers who had perpetrated the intrusion. At the same time, several gamers admonished their agitated compatriots to remain loyal to the network itself, encouraging them to stay strong in this time of trouble. Exiled from their collective gameworlds, adrift on a flood of concentrated affect, members of the PSN community extensively debated the meanings of the network hack. Indeed, amid widespread fears of identity theft and the misery of prolonged separation from friends, many gamers also came to recognize that their online community had become a site of techno-political warfare.

Sownage

To understand the various meanings and emotions animated by the hacking incident, we need to first revisit the status of the PlayStation 3 as a technoscientific tool and a computational platform. Long before the console was released, Sony took pains to emphasize that the PS3, with its powerful Cell processor and built-in networking capabilities, would be a valuable resource for scientific researchers as well as homebrew computer hobbyists. Sony was attentive to the fact that the predecessor PlayStation 2 had often been appropriated for scientific projects

– especially those needing high-powered yet affordable computing clusters – in addition to widespread tinkering by modders, hackers, makers, and DIY science geeks. Vice President of Sony Computer Entertainment Europe, Phil Harrison, said that they were eager to support homebrew programming and software experimentation on the PS3: "the notion of game development at home using powerful tools available to anyone . . . [is] a vital, crucial aspect of the future growth of our industry."[21] Accordingly, when the console launched, its native operating system featured a function called 'OtherOS'. OtherOS enabled users to easily install a different operating system – most typically, Linux – precisely to accommodate the needs of the scientific and homebrew communities.

However, in January 2010, the young hacker George Hotz – more commonly known by his alias, GeoHot – announced that he had learned how to 'jailbreak' the PlayStation 3, gaining access to its system memory and processor. GeoHot had already attained notoriety in 2007 as the first person to jailbreak an Apple iPhone. He had now learned to do a similar thing for the PlayStation 3 – an ingenious trick that would enable new homebrew applications for the PS3, as well as potentially illicit activities, such as pirating games. GeoHot accomplished this jailbreak by exploiting the OtherOS function. To be sure, Sony had already anticipated a security risk in the OtherOS function, and despite early overtures toward the scientific and homebrew communities, the corporation removed the OtherOS feature from its 2009 PS3 'slim' model. After GeoHot published details of the jailbreak technique on his website, Sony promptly issued a mandatory firmware update for all PS3s, designed to permanently erase the OtherOS feature and simultaneously remove any dual-booting system that might have been installed.

GeoHot and other hackers – notably, the fail0verflow collective – continued working on ways to bypass the console's security. On January 2, 2011, GeoHot published the PS3 root keys on his website and announced it on *PSX-Scene* with a simple message: "keys open doors" (GeoHot 2001).

And then trouble really began. Sony filed a lawsuit against GeoHot and several other hackers, including one hundred 'John Does' (the unknown members of the fail0verflow hacking group, as well as anyone else involved in discovering and distributing information about how to jailbreak the P3S). These defendants were accused of violating the U.S. Digital Millennium Copyright Act (DMCA), the Computer Crimes Act and several other laws. Even though the U.S. Copyright Office had determined in 2010 that jailbreaking smartphones does not constitute a violation of the DMCA, apparently video game consoles present different considerations. U.S. District Judge Susan Illston ordered GeoHot to remove all information about his PlayStation jailbreak from his website, blog and YouTube account, and to relinquish his computer hardware and storage media to Sony lawyers.

As part of their pre-trial discovery efforts, Sony also demanded that GeoHot's web-provider, Bluehost, should hand over server records that could be used to identify people who may have visited GeoHot's website between 2009 and 2011. Sony simultaneously asked for data from YouTube that would reveal the identities of anyone who had looked at GeoHot's jailbreak video or posted comments about it.

Sony went further, requesting access to any Twitter accounts alleged to have discussed jailbreaking the PS3 going back to December 2010. Sony also insisted that Google should hand over all data records from GeoHot's Blogger.com site, including the IP addresses of users who had accessed the site in recent years. A U.S. federal magistrate approved all of these subpoenas in March 2011. Sony threatened to issue additional lawsuits against anyone else discovered to have participated in distributing information about the jailbreaking methods.

Thus formed the conditions for Operation Sony.

On Saturday, April 2, 2011, in retaliation for Sony's legal actions against the hacker community, the global hactivist collective known as Anonymous launched its first wave of DDoS attacks against various Sony servers.[22] Around the world, members of Anonymous focused their Low Orbit Ion Cannons (LOICs, software packages created for DDoS attacks) against the Sony empire. Between April 2nd and April 6th, Sony.com, PlayStation.com and Sony's Style.com site were all brought down, rendered completely inaccessible.

On April 4th, a small group of Anons on the OpSony IRC channel – led by a user named randomtask – suggested that the DDoS attacks were not enough. They launched a splinter operation dubbed SonyRecon: a coordinated doxing of several high-level Sony employees, the federal judge in the GeoHot case and Sony's legal representation, the Kilpatrick Townsend firm. Within a couple of days, the personal information of several Sony executives was floating freely around the Internet, the voicemail of the judge was barraged with harassing messages and the Kilpatrick Townsend website was DDoSed to oblivion.

A variety of Anonymous media releases denouncing Sony appeared in rapid succession, encouraging others to download a LOIC and join the DDoS assault. While those speaking on behalf of Anonymous were careful to point out that they were not targeting the PlayStation Network itself – "We are not after the players"[23] – at the same time as the DDoS attacks were taking place, the PlayStation Network began to exhibit signs of lag and login errors. On April 4th, Sony offered an explanation through Twitter: "PSN currently undergoing sporadic maintenance. Access to the PSN may be interrupted throughout the day. We apologize for any inconvenience."[24] Despite Sony's suggestion that the problems with the PSN were due to maintenance, many gamers speculated that it must be connected to the simultaneous DDoS attacks on other Sony servers. On April 7th, Anonymous released the following message:

> Greetings, Sony customers and PS3 users. We are Anonymous. During the last few days, Anonymous has been targeting Sony for their outrageous treatment of not only PS3 users and jailbreakers, but also of the general public. Their propaganda regarding jailbreaking implies that it encourages piracy and thereby makes people lose their jobs, whereas jailbreaking actually just means you are making YOUR device do what it should do. . . . The fact that their litigation demanded information on everyone who had viewed the material was completely unacceptable. This is a threat not only to the gaming community, but to freedom of information in general. . . . Anonymous decided it could not allow this to stand. . . . This attack is aimed solely at Sony, and we will try our best to not affect the gamers . . .

> We are Anonymous,
> We are legion,
> We never forgive,
> We never forget,
> Expect us.[25]

On April 11th, Sony announced that it had reached a settlement with GeoHot. Hotz had consented to a permanent injunction, stating in the Sony press release: "It was never my intention to cause any users trouble or to make piracy easier. I'm happy to have the litigation behind me."[26] He was forbidden from discussing the nature of the settlement, but on the same day as the settlement was announced, he posted on the *GeoHot Got Sued* blog that he supported the general Sony boycott that Anonymous had called for:

> As of 4/11/11, I am joining the SONY boycott. I will never purchase another SONY product. I encourage you to do the same. And if you bought something SONY recently, return it. Why would you not boycott a company who feels this way about you?
>
> (Hotz 2011)

While GeoHot may have settled, he did not appear to have settled down.

The following day, Anonymous released another news update, stating their intention to cease the DDoS attacks, shifting instead to other tactics. A video featuring a modified version of the press release quickly went viral (Figure 7.2). The video

Figure 7.2 Anonymous message to Sony, April 12, 2011. It depicts a scene of experimentation – the foremost PS3 is running homebrew software – ripped from an earlier video where GeoHot had demonstrated his jailbreak method.

shows a row of PS3s; as one fires up, its launch screen dissolves in static, replaced by the titular character from *V for Vendetta* in his familiar Guy Fawkes mask:

> GeoHot has taken a settlement with Sony. . . . In the eyes of the law, the case is closed. For Anonymous, it is just beginning. By forcing social networking sites such as YouTube and Facebook to hand over IP addresses of those who have viewed GeoHot's videos, they have performed an act of privacy invasion. We, Anonymous, will not allow this to happen. The attacks on the websites of Sony have been ceased. Sony's poor attempts to explain the system outages through maintenance amuse us. Therefore, we are finding other ways to get Sony's attention. This April 16th, grab your mask, a few friends, and get to a local Sony store by you. Use the IRC and the official [Anonymous] Facebook page to organize a protest in your area. Make sure the people know the injustices performed by this corrupt company. Boycott all Sony products, and if you have recently purchased any, return them. It is time to show large corporations and governments that the people, as a collective whole, can and will change injustice in society, and we will make a great example out of Sony. Sony, prepare for the biggest attack you have ever witnessed – Anonymous style.[27]

Nevertheless, when the PSN went down on April 20th, Anonymous was quick to deny responsibility. The IRC #OpSony headline was changed to read: "#OpSony is over, if you are here to baww about PSN, it wasn't us." AnonOps also posted a press release on the *AnonNews* site, claiming, "For Once We Didn't Do It": "While it could be the case that other Anons have acted by themselves, AnonOps was not related to this incident and does not take responsibility for whatever has happened."[28]

Some Anons also insisted that they were gamers, too, and that the various Sony outages over the past weeks should be understood as important achievements of OpSony, carried out in the name of gamers, hackers and DIY scientists everywhere. To whatever degree OpSony might have been related to the PSN outage, the Anons wanted even this to be understood as drawing attention to issues of technological justice:

> We have attacked Sony in order to send a message that gamers worldwide have certain rights, and are not merely sources of income. . . . Anonymous are gamers too. And we support the rights of people worldwide, and will stand up for the right of having access to the device that you BOUGHT . . . Sony has decided not to sue Geohot; that is a victory. And that's one of our purposes in OpSony. Thus, we have achieved what we wanted. Mission accomplished.[29]

No one, it seemed, anticipated how long the PSN would be down – and of course, Sony did not admit the extensive data intrusion until a full week later. In response to this massive data theft, the U.S. House of Representatives held a hearing on May 4, 2011, to which they invited Sony to explain themselves. Sony declined to send representatives but instead sent a letter detailing the circumstances of the data theft.

Sony claimed that the DDoS attacks by Anonymous provided cover for the PlayStation Network intrusion, because Sony's cybersecurity agents were so busy dealing with the DDoSing that they did not detect the PlayStation attack when it was happening. When the security admins later confirmed the data theft (first suspected when some of the PSN servers rebooted themselves unexpectedly), they apparently

> discovered that the intruders had planted a file on one of those servers named 'Anonymous' with the words 'We are Legion'. Just weeks before, several Sony companies had been the target of a large-scale, coordinated denial of service attack by the group called Anonymous. The attacks were coordinated as a protest against Sony for exercising its rights in a civil action in the United States District Court in San Francisco against a hacker.[30]

Having thus implicated Anonymous in the criminal intrusion, Sony called upon the U.S. Congress and other legislative bodies around the world to combat all forms of hacking and hactivism with "strong criminal laws and sanctions . . . Worldwide, countries and businesses will have to come together to ensure the safety of commerce over the Internet."[31]

Anonymous immediately proclaimed innocence, suggesting that they were being framed. In the meanwhile, the hacker community was not backing down. Even days after the network was restored to normal operations in North America and Europe (slightly later in Japan), the hackers struck again. And again. And again. LulzSec, a splinter group from Anonymous formed earlier in 2011, announced on its Twitter account that it was launching a new campaign: "Sownage (Sony + Ownage) Phase 1 will begin within the next day. We may have a pre-game show for you folks though. Stay tuned."[32] Days earlier, LulzSec had already broken into the servers of Sony Music in Japan, and the following week, this group successfully hacked into several secured Sony databases, including Sony Pictures, Sony BMG Belgium, Sony BMG Netherlands and others, swiping millions of user records, passwords and other information. LulzSec taunted Sony on Twitter during the attacks: "Hey @Sony, you know we're making off with a bunch of your internal stuff right now and you haven't even noticed? Slow and steady, guys."[33] A lot of the stolen information was posted to Pastebin or torrented through Pirate Bay. Between April and October, Sony's global computational infrastructure was struck dozens of times by a number of different groups. 'Sownage' became a popular term to mean complete computational smackdown: pwned like Sony.

Whereas the majority of AnonOps' anti-Sony activities were focused on disrupting the corporation itself, claiming to protect customers from harm, LulzSec and other 'black hat' hackers gleefully purloined user data from Sony's servers and threw information up on the Internet for all to see. As a result, a number of PlayStation gamers came to believe that LulzSec must have also been behind the April 18th intrusion into the PSN. But LulzSec likewise disclaimed involvement in that particular operation, tweeting on May 31st: "You Sony morons realize we've never attacked any of your precious gaming, right?"[34]

Within a year, international law enforcement agencies arrested a number of Anons who had participated in the DDoS attacks against the Sony websites. More devastatingly, the FBI, working together with British and Irish intelligence officials, later identified and eventually arrested most of the members of LulzSec between July 2011 and March 2012, all of whom were charged with criminal activities related to hacking the databases of Sony Pictures and other media organizations, as well as government agencies such as the CIA and SOCA. No one has yet been directly connected to the intrusion that prompted Sony to close down the PlayStation Network itself for so long. It remains something of a mystery.

At his first court appearance in London on August 1, 2011, eighteen-year-old LulzSec member Jake Davis (alias Topiary) carried a copy of Michael Brooks's *Free Radicals: The Secret Anarchy of Science.* Photographs of Topiary brandishing the book became iconic, seemingly confirming a number of popular intuitions about the hackers' motives (and, incidentally, boosting the book onto the bestseller lists). Brooks's account of the history of science, much like Paul Feyerabend's *Against Method,* argues that scientific innovation is rarely accomplished by adhering to method, rules or proper decorum, but is more frequently driven by reckless experimentation, playful silliness and even illicit behavior. The image of Topiary on trial holding onto Brooks's defense of scientific anarchy, as if advocating creative misconduct in the name of science and knowledge, provided an instantly legible rationale for the spate of hacking incidents. The image implied that, as much as the cyber-attacks against Sony and other organizations were unlawful hijinks, they could also be understood as experiments in techno-politics: high-tech interventions against the suppression of information, the corporate domination of thought and the lock-down of potential futures. Indeed, shortly before he was taken into custody, Topiary had posted an evocative message through Twitter: "You cannot arrest an idea."[35]

The Sownage saga exposed the degree to which online lulz, fun and games, can no longer be considered separately from issues of technical innovation and governance – for our computational networks are contested territories, pervious to corporate control and state securitization as much as revolutionary insurgency. On the one hand, gaming platforms like the PlayStation Network might afford ways of democratizing the technoscientific imagination, for example, in the success of projects like Folding@home, but on the other hand, they are embedded in intellectual property regimes that often foreclose the legitimacy of DIY experimentation in advance. For some science and technology enthusiasts, this internal contradiction appears intolerable. Hence, although the motives of Anonymous, LulzSec and other groups who participated in the pwning of Sony were heterogeneous, exhibiting as much buffoonery as activism, they converged in a common desire to liberate the technoscientific objects of everyday life from those who would restrict access, those would lay down the law to prevent us from playing with the root keys of our technogenic lifeworlds.

Resurrection

For many devotees of the PlayStation Network, the politics of the outage were both highly visible and thoroughly mystifying. Some were quite skeptical about Sony's version of events, claiming the whole thing was likely fabricated to create a political interest in regulating the Internet. Others were sympathetic with the hactivists, understanding Sony's legal maneuvers in the context of other security actions against media piracy (e.g., Pirate Bay), whistle-blowing (e.g. WikiLeaks), and the freedom of information in general. Most PSN users, however, seem to have preferred to be left alone to play games. One player, summing up a common sentiment, said:

> This Is Ridiculous It Seems Like Nothing is good enough for hackers . . . Just leave the DAMN networks alone stop tryna make a big statement go to the corporate offices and make a fuss and get locked up that way stop dragging everyone else in this who DON'T GIVE TWO PENNIES AND A NICKEL about your cause i just wanna play my games and now i cant do that because you wanna make statement think about how other people feel about this.[36]

Yet even if they did not care to think about the techno-political implications, for many of the distraught gamers who suffered eviction from their digital homeland, their preferred way of life, the stakes were made quite palpable.

In time, normal network services were restored. Sony offered an apologetic 'Welcome Back' program to its customers, including a couple of free games and a month-long subscription to the PlayStation Plus service. Gradually, gamers began to recover from the trauma of the PSN outage – what some were calling the 'ApocalyPSN'. Trying to make sense of the whole thing, a number of gamers relied on the tropes of speculative fiction to explain their post-apocalyptic condition. For example, "Since the ApocalyPSN, many of us were disconnected from GD [Sony PlayStation General Discussion Forum], our true home. We suffered Forum Deaths. . . . But alas, some have started to return from the dead, like Zombies!"[37]

Imagining themselves as zombies, internalizing the resurrected PlayStation Network as a way of conceptualizing the undead self, these gamers crafted an ironic narrative framework for living with the instabilities of digital culture, the risks of life with PlayStation. In playfully rediscovering their virtual communities through zombie imagery, mixing the language of resurrection with the language of cyberspace, these survivors of the PlayStation apocalypse once again adapted themselves to the phantasmatic quasi-object: the post-vital gaming platform, a digital warzone where the boundaries between corporate interests and technoscientific experiments, serious business and silly games are still being hotly negotiated. To be a zombie in this high-tech zone is to accept the risks, the indistinctions. In other words, it is to embrace the circumstances of having a life that some say is not a life – to have a life and to make a life in a world of endless speculation, precarity and uncertainty. After all, that's what it's like to live in the world today.

Long live play.

Notes

1 Jay Leno, 'Opening Monologue,' *The Tonight Show*, 4 May 2011, NBC.

2 Matt1246, response to Cybernetic56, in response to Joe Wilcox, 'Happy Day! PlayStation Network Is Back up – Well, Almost,' *BetaNews*, 14 May 2011, http://betanews.com.

3 ted2112, 'Remembering the Great PSN Outage of 2011,' *Homestation Magazine*, 16 April 2012, http://www.hsmagazine.net. The date has become commemorative: "April 20 will live in infamy among the PlayStation Faithful" (Moriarty 2012).
4 On the processes of technogenesis as the individuation of technical objects, subjects and collectives altogether, see Simondon (1958) and Hayles (2012).
5 ted2112, 'Remembering the Great PSN Outage of 2011.'
6 ADAMPwns, message #31, 07–21–2009, response to gts1234567890, 'Getting a Life on Home,' PlayStation®Home – PlayStation Community Forums, 20 July 2009, http://community.us.playstation.com.
7 Sony Computer Entertainment, 'Sony Computer Entertainment Joins Stanford University Folding@Home Program to Further Medical Research,' 15 March 2007, http://us.playstation.com/corporate/about/press-release/386.html.
8 N. Rimon quoted in Tach (2012).
9 V. Pande quoted in 'Termination of Life with PlayStation,' *Life with PlayStation*, 6 November 2012, http://www.playstation.com/life/en/index.html.
10 Pande quoted in Dutka (2007).
11 crashmash01, 'A "Regular" Story about Me and My Life with PlayStation,' *PlayStation® Community Forums*, 7 April 2012, http://community.us.playstation.com.
12 c_m_f_g, comment #89, 04–26–2011, response to Dobra (2011).
13 sephron9, comment #55, 04–24–2011, response to Dobra (2011).
14 RyuuSkyez, response to Nygård (2011).
15 Sebastian Nygård, 'PSN Network Is Down!,' YouTube, 23 April 2011, http://www.youtube.com/watch?v=__yLhnw0vSc.
16 erick_34, comment #50, response to Seybold (2011c).
17 Ambr0ster, comment #464, 04–24, 2011, response to Seybold (2011c).
18 Jay, response to Robinson, 'Why Is the PSN Network Down?'
19 stevev363, response to Gideon (2011).
20 Sophronia, response to Seybold (2011c).
21 Harrison quoted in Melanson (2007).
22 On the history and culture of Anonymous, see Olson (2012), Ravetto (2013) and Coleman (2014).
23 'A Statement of Purpose from Your Anonymous Friends' reposted by Anonymous Is Beast, 'Anonymous: Sony Website Hacked (April 1st 2011),' YouTube, 5 April 2011, http://www.youtube.com/watch?v=zMSVTLMyEqI.
24 Ask PlayStation, 'PSN Currently Undergoing Sporadic Maintenance. (9:35 AM),' Twitter, 4 April 2011, https://twitter.com/AskPlayStation/status/54945167689515008.
25 Anonymous, 'OpSony Update, to All,' *AnonNews*, 7 April 2011, http://www.anonnews.org/?p=press&a=item&i=797.
26 Hotz quoted in Seybold (2011d).
27 Anon19861, 'Anonymous_Message to Sony WE RUN THIS. . . ,' YouTube, 12 April 2011, http://www.youtube.com/watch?v=0GzPAa9YrTo. The video was widely reposted. For the full press release, see Anonymous, '#OpSony Update, Geohot Settlement, and April 16th IRL Protest,' *AnonNews*, 12 April 2011, http://anonnews.org/?p=press&a=item&i=809.
28 AnonOps, 'For Once We Didn't Do It,' *AnonNews*, 22 April 2011, http://anonnews.org/?p=press&a=item&i=848.
29 Anonymous, statement provided to *VGN365*, quoted in Jim, 'Anonymous Responds to Halting Attacks on Sony; Releases Statement,' *VGN365*, 22 April 2011, http://vgn365.com.
30 Kazuo Hirai, Chairman of the Board of Directors, Sony Computer Entertainment America LLC, letter to Mary Bono Mack, Chairman, and G. K. Butterfield, Ranking Member, Subcommittee on Commerce, Manufacturing, and Trade, United States Congress, 3 May 2011. Posted to the PlayStation.Blog's Flickr photostream on May 4, 2011, http://www.flickr.com/photos/playstationblog/5686965323/in/set-72157626521862165/.

31 Ibid.
32 The Lulz Boat, '#Sownage (Sony + Ownage),' Twitter, 29 May 2011, 1:16 p.m., http://twitter.com/LulzSec/status/74932233550569472.
33 The Lulz Boat, 'Hey @Sony,' Twitter, 31 May 2011, 2:09 a.m., https://twitter.com/LulzSec/status/75489095371079680.
34 The Lulz Boat, 'You Sony Morons,' Twitter, 31 May 2011, 10:12 a.m., https://twitter.com/LulzSec/status/75610558178668544.
35 Topiary, 'You Cannot Arrest an Idea,' Twitter, 21 July 2011, 7:02 p.m., https://twitter.com/atopiary/status/94225773896015872.
36 Donne, response to Robinson, 'Why Is the PSN Network Down?'
37 Wraith07, 'Zombies of the ApocalyPSN,' *PlayStation Forum*, 30 April 2011, http://www.community.eu.playstation.com/t5/General-Discussion.

References

Chun, W. 2006. *Control and Freedom: Power and Paranoia in the Age of Fiber Optics*. Cambridge, MA: MIT Press.

Coleman, G. 2014. *Hacker, Hoaxer, Whistleblower, Spy: The Many Faces of Anonymous*. New York: Verso.

Dan. 2011. 'Fold your way through the PSN outage,' *One of Swords*, 28 April, http://oneofswords.com.

Dobra, A. 2011. 'PlayStation Network still down, might not be fixed for another two days (updated),' *Softpedia*, 22 April, http://news.softpedia.com/news.

Dutka, B. 2007. 'Guiness recognizes power of Folding@home,' *PSX Extreme*, 31 October, http://www.psxextreme.com/ps3-news/2080.html.

Galloway, A. and Thacker, E. 2007. *The Exploit: A Theory of Networks*. Minneapolis: University of Minnesota Press.

GeoHot. 2011. 'Geohot: Here is your PS3 root key!,' *PSX-Scene*, 2 January, http://psx-scene.com/forums/f6/geohot-here-your-ps3-root-key-now-hello-world-proof-74255/.

Gideon. 2011. 'PSN outage: An opportunity,' *Homestation Magazine*, 24 April, http://www.hsmagazine.net.

Hayles, N. K. 2012. *How We Think: Digital Media and Contemporary Technogenesis*. Chicago: University of Chicago Press.

Hotz, G. 2011. 'Joining the Sony Boycott,' *GeoHot Got Sued*, 11 April, http://geohotgotsued.blogspot.com.

Lacan, J. 1973. *The Four Fundamental Concepts of Psycho-Analysis: The Seminar of Jacques Lacan, Book XI*, translated by A. Sheridan. New York: Norton, 1998.

Latour, B. 1993. *We Have Never Been Modern*, translated by C. Porter. Cambridge, MA: Harvard University Press.

Luttmann, E., Ensign, D. L., Vaidyanathan,V., Houston, M., Rimon, N., Øland, J., Jayachandran, G., Friedrichs, M. and Pande, V. S. 2009. 'Accelerating molecular dynamic simulation on the cell processor and PlayStation 3,' *Journal of Computational Chemistry* 30 (2): 268–274.

Melanson, D. 2007. 'Sony's Phil Harrison talks PS3 homebrew possibilities,' *Engadget*, 23 April, http://www.engadget.com.

Moriarty, C. 2012. 'One year later: Reflecting on the great PSN outage,' *IGN*, 20 April, http://ps3.ign.com.

Nygård, S. 2011. 'PSN Network Is Down!,' YouTube, 23 April, http://www.youtube.com/watch?v=__yLhnw0vSc.

Olson, P. 2012. *We Are Anonymous: Inside the Hacker World of LulzSec, Anonymous, and the Global Cyber Insurgency*. New York: Little, Brown.

Parker, M. 2012. 'Happy Birthday – A Year on from the PSN Hack,' *RinseWashRepeat*, 2 May, http://rinsewashrepeat.co.uk.

Ravetto, K. 2013. 'Anonymous: Social as political,' *Leonardo Electronic Almanac*, 19, http://www.leoalmanac.org.

Roper, C. 2006. 'PS biz brief 06: PS3 is "live",' *IGN*, 14 March, http://www.ign.com.

Serres, M. 1980. *The Parasite*, translated by L. R. Schehr. Minneapolis: University of Minnesota Press, 2007.

Seybold, P. 2011a. 'Update on PSN Service Outages,' *PlayStation.Blog*, 20 April, http://blog.us.playstation.com.

———. 2011b. 'Updates on PlayStation Network / Qriocity Services,' *PlayStation.Blog*, 22 April, http://blog.us.playstation.com.

———. 2011c. 'Latest Update on PSN Outage,' *PlayStation.Blog*, 21 April, http://blog.us.playstation.com.

———. 2011d. 'Settlement in George Hotz Case: Joint Statement,' *PlayStation.Blog*, 11 April, http://blog.us.playstation.com.

Simondon, G. 1958. *Du mode d'existence des objets techniques*. Paris: Aubier, 1989.

Standage, T. 2005. *The Future of Technology*. London: Profile Books.

Suzuoki, M. 2009. 'Cell – The Dream Processor,' *Sony Global*, 'Interviews with Engineers, vol. 4,' http://www.sony.net.

Tach, D. 2012. 'After contributing to the fight against Alzheimer's, Sony's Folding@home lead reflects on "overwhelming" user support,' *Polygon*, 16 November, http://www.polygon.com.

Weeks, K. 2006. '4D Kutaragi "live" and unhinged,' *Joystiq*, 15 March, http://www.joystiq.com.

8 A biography of a disorder that didn't want to be diagnosed

Simone van den Burg

Biographies are usually written out of a desire to know a specific person, whose life attracts attention because it is more eventful, glamorous, tragic, adventurous or notorious than that of most other people. A lot of biographies are written – and probably also read – out of a hunger for examples of great people to look up to, and a desire to be entertained. While many biographies will try to satisfy this curiosity, and form storylines that reveal secret 'truths' that throw a specific light on the person who is described, more interesting biographies also reflect on their own truth-telling aspirations and will reveal how difficult it is to know a specific person, whether historic or alive, with the characteristic ways he or she thinks, feels, acts and relates to other people.

The difficulty to 'know' the object of a biography is dealt with in a particularly interesting way in the biography *Wittgenstein's Poker* (Edmonds and Eidinow 2002). This biography focuses not on the life of the philosopher Ludwig Wittgenstein, but on a ten minute lasting argument that he had with the philosopher of science Karl Popper on October 25, 1946. The authors of the biography, David Edmonds and John Eidinow, approached this argument by means of interviews with the people who witnessed it. While the report of these witnesses' memories could invite questions about 'what *really* happened', the acts of remembering eventually also tend to eclipse the event itself: what deserves attention seems to be not the Wittgenstein-Popper argument, but what it means to people.

When a biography focuses on activities – telling, remembering and interacting – rather than an object, it abandons the presupposition that a person or event has a nature in itself which can be adequately described by an individual who experiences it. This fits with the idea of Wittgenstein's famous private language argument which states that it is impossible that an aspect of the world would cause an experience in an individual spectator, because experience of the world depends on the language that spectators use to talk and think about it (Wittgenstein 1951, §243–§271). As language is acquired in histories of social interaction, or 'language games', it would be impossible to experience the world without this social and interactive history, which may differ in various contexts in which individuals may take part.

The brief biography provided in this chapter will take this Wittgensteinian approach to experience. Its object was – at the time when the first tentative draft of this chapter was written (winter 2012) – most appropriately described as a 'disorder' that was talked about as a collection of anomalies in the morphology and development of a four year old boy, which could be observed, measured and cared for, but for which the physicians could provide no adequate diagnosis. While the disorder was part of the world of many people – parents, other siblings, pediatricians, nurses, physical therapists – who gave meaning to it in the course of their daily discourses and interactions, it lacked a 'name' and therefore it was difficult to refer to it as 'one' instead of a collection of complaints.

In the summer of 2013, however, a diagnosis was produced by whole exome sequencing (WES), the sequencing of the coding part of the genome (the exome). This new methodology succeeded in linking the complaints of the child to Kabuki syndrome, which was already described in the biomedical literature. While Kabuki syndrome is described by the geneticists and neurologists as a once and for all representation of the reality of the child's condition, this chapter will attempt to reveal how it is accommodated in a variety of meaning giving activities – biomedical, caring, social and spiritual. It will describe that the diagnosis offers new possibilities for stakeholders to interact with it, but also demands them to close off other ways to talk and think about it, and deal with it. An innovative technological interaction with the disease which offers a biomedical diagnosis also opens up a new chapter in the biography of the disorder which allows rethinking its history and expectations for the future.

The background story

The project from which the following account derives was a response to current demands for the translation of scientific findings to the clinic. These last years there have been pleas by policy makers as well as funding organizations all over the world to make more effort – and spend more money – to ensure a more effective movement ('translation') of basic scientific findings to relevant and useful clinical applications (from 'bench to bedside').[1] While there are different ways to define what translational medicine is, the most common understanding is that this type of research should make scientific findings available and useful for patients and clinicians (Van der Laan and Boenink 2012). The call to translate scientific findings to the clinic has been particularly forceful in areas like genetic (or broader, molecular) medicine, because the wealth of scientific knowledge in this field rarely succeeded to benefit patients, nor did it lead to a concurrent number of innovations to improve health care (Khoury et al. 2007; Collins 2011).

The translational research project on which this chapter focuses involves partners from different disciplines, including geneticists, pediatric neurologists, health economists and moral philosophers in the Radboud University Medical Center (Nijmegen, the Netherlands). Geneticists involved in this project are already renowned for their successful use of whole exome sequencing (WES) for diagnostics of children with rare neurological disorders (Vissers et al. 2010, 2011; de Ligt et al. 2012). The project was formulated in order to compare the value of WES as a novel diagnostic tool with the standard diagnostic trajectory that is currently used to diagnose children who are presented to pediatric neurologists with developmental delay. The standard trajectory is composed of many tests, which may (amongst others) include EEG, MRI, Genomic microarray, blood tests (for biochemical components of the blood), a muscle and/or liquor biopsy and Sanger sequencing of 1 to 5 genes. The comparison focuses on (1) the amount of diagnoses that WES and the standard diagnostic trajectory are able to produce, (2) the time it takes to get to a diagnosis, (3) the costs, and (4) the experience of patients and their parents. Together with Lotte Krabbenborg, I made the fourth comparison. We aimed to interview patients. Patients, however, were most often not able to respond to questions in an interview, so in practice we interviewed parents in 98 percent of the cases.

All of these patients suffer physical and/or psychomental developmental delay. Complaints are various, and may include impaired movement, neuromuscular abnormalities, epileptic seizures, metabolic or mitochondrial dysfunction, blindness, hearing problems and combinations thereof. Currently, 50 percent of the patients stay without a diagnosis after the standard diagnostic trajectory is completed. This high percentage of undiagnosed patients can be explained by the rarity of the conditions and the variety of complaints: on the basis of a clinical examination of a patient, it is difficult for a neurologist to select the appropriate gene to test with Sanger sequencing. Usually Sanger sequencing is repeated on one to seven genes for a patient. WES, however, allows looking at the entire coding part of the genome with a single test, which decreases the chance that a mutation is missed.

The examination of fifty patients included in our trial eventually lead to eighteen clear diagnoses. The disorder described in this chapter is one of them. There are different ways to describe how this previously undiagnosed disorder receives a diagnosis. The story that would probably fit best in current pleads for translational medicine would sketch the history of the discovery of Kabuki syndrome worldwide and then move on to the diagnosis of this particular patient, which is the moment when the linear translation from science to benefits for patients has been successfully completed. But the story could also start with a condition of a patient and describe how it is understood by the people around it, and what role various technologies – including WES – play in those understandings. This second way is most appropriate to the ambitions of this chapter. But since it is not feasible to tell the full story (as researchers we are not present all the time), the description will be limited to a series of snapshots that attempt to provide insight into shifts in meaning giving activities of parents, neurologists and geneticists in response to information provided by different diagnostic technologies.

A brief biography

Snapshot 1 – A first encounter in the doctor's cabinet

A Thursday morning in February 2012, the pediatric neurologist receives me in her office in order to make me more acquainted with her work. She sees a series of children, with their parents, and I watch her do tests with these children and talk kindly to the parents. One of her patients this morning is a four-year-old child who is brought into her cabinet by his parents, and sits lazily in his wheelchair, as if without force. His large brown eyes are turned upward. While he seems unaware of his environment, paying no attention whatsoever to things in the room or to us, he does press buttons on a toy and listens to the peeping sounds that it produces. During the fifteen-minute visit, the toy now and then falls on his lap; he sits still and his eyes move quickly from left to right.

In the minutes after the visit, the pediatric neurologist explains to me that different tests have been performed on this patient, which allows some explanation of his complaints:

- The EEG, for example, pointed out that he has epileptic seizures.
- Chemical analysis of the blood revealed an immunodeficiency, which leads to frequent infections.
- Analysis of a muscle biopsy indicated a low level of adenosine triphosphate (ATP), which plays a pivotal role in the metabolism of cells for the transport of energy. Low ATP indicates that the mitochondria in the cells don't succeed to extract sufficient energy from food, which produces weak muscles and a low level of energy to move and to develop.
- A genetic test was carried out to test whether the patient's complaints were features of Kabuki syndrome, but this test was negative.

During the conversation with the neurologist, the results of these tests figure as explanations of characteristic features of the child that I witnessed. The quick movement of the eyes is explained by the neurologist as a sign of epileptic seizures taking place; low ATP functions as an explanation of the weakness of his appearance and forcelessness of his posture. But while each of the tests offers a way to interpret part of his appearance, none of them explains all of them at once. Low ATP, for example, suggests that his complaints are an effect of mitochondrial disease, but the neurologist explains that further tests failed to justify this diagnosis.

> Neurologist: "What you see in him is that the ATP production is lower, so the amount of energy that the muscle can make is diminished. But we saw no complex deficiency. And that is difficult. Mitochondria are the energy producers of our body, and energy production is a procedure that follows several steps, starting from complex 1 to 5. You have to go through all those steps in order to produce energy. So we saw that the end product, the energy, is lower, but it is possible to measure all separate complexes and we did not find any anomalies there. So I always told mother: it is a secondary mitochondrial problem, meaning that low ATP is not explanatory for his other complaints, but that the mitochondrial dysfunction is more likely an effect of something else."
>
> *(Extract from interview with the author, February 9, 2012)*

While the tests failed to provide a diagnosis, some of them are the basis for treatment decisions. For example, the child was given medication to diminish the amount of infections and to reduce the number of epileptic seizures, and the test that reveals low ATP motivated changes in his feeding regime: at first, more regular 'feeding moments' were introduced (day and night) in order to avoid hypos (energyless moments), and when that did not sufficiently improve his condition, it was decided to tube-feed him directly into the stomach in order to keep the energy level more even throughout the day.

Snapshot 2 – A first interview with the parents

At their kitchen table in an old house in a small village in the south of the Netherlands, two tired and sorrowful parents consented to answer my questions. The child is strapped securely into a chair (to prevent his forceless body from falling) near to us and presses buttons on his noisy toys. The parents tell me about their history with him. What is most revealing in their story is that even though a diagnosis was never produced, they still use the tests that have been performed on him to interpret his needs. The discovery of the mitochondrial dysfunction (low ATP) plays a particularly large role in their history with the child, because it provided them an explanation for a lot of their difficulties: since he was born, the child did not drink his milk, couldn't swallow it, or threw up, and taking care of him meant trying to feed him, day and night, trying different types of bottles, different types of milk. But it all didn't help. The test which revealed low ATP offered them a way to understand their child's needs, and now that the child is tube-fed it liberated the parents from their daily struggle to feed him, which was a relief. Other complaints – like his frequent infections (in ears, eyes, skin, bladder, etc.) and epileptic seizures – also demand attention, but managing the child's lack of energy figures most prominently in the parent's story about how he should be cared for.

> Mother: "Depending on his energy we compose the day of periods of rest and periods of activity. Anything takes energy: sitting, bathing, even looking at things. When he sat up for a while you have to let him rest afterwards. He likes to be taken for walks in his wheelchair, but when the weather is too hot or too cold it tires him. Regulating his temperature takes energy. On energetic days we put him in a walking chair for about ten minutes, to train his legs. But after that he sometimes sleeps for hours."
>
> *(Extract from interview with the author, April 6, 2012)*

Next to an interpretation of the present condition of the child, the establishment of his low ATP also informs the parent's evaluation of the professional care offered to their child right after he was born. At that time he was already in a bad condition, and a test revealed a duplication on chromosome 1627. While the doctors who tested him admitted having little knowledge about this duplication, the parents felt that they used this anomaly to explain all of their child's complaints. Even though they went to see doctors frequently to ask for help, they never felt that they were being taken seriously: the child's complaints (including his eating troubles and his infections) were all ascribed to this chromosomal anomaly. Based on their experience, the parents were thus quite critical toward the first doctors they consulted.

> Mother: "We had a child who was very ill. The duplication in his chromosome was irrelevant to the problems he was experiencing. This puts all that happened in a different perspective, because you are thinking: something could have been done, if it had been clear what was wrong. Maybe he could not have been cured, but maybe his development would have been better."
>
> Simone van der Burg (SvdB): "What kind of help are you thinking of?"
>
> Mother: "Food. When he was eight months and [. . .] he received the proper tests, he was malnourished. So much went wrong. He lost so many of his capacities in those first eight months. He was just malnourished. For a child with his disease that is very bad."
>
> *(Extract from interview with the author, April 6, 2012)*

The parent's critical attitude toward the doctor's negligence to really look at the condition of the child is informed by their knowledge about mitochondrial diseases. Children with mitochondrial diseases need to be fed regularly in order to keep their general condition as good as possible, for energy is a precondition for their growth and development. If the energy level is low, their development will slow down, or even stop, and most often they will not retrieve the functions that they lost. Based on this information, the parents started to develop an understanding of what went wrong in the beginning of the child's life: according to them, doctors mistakenly blamed his complaints on a chromosomal anomaly, while they think the doctors should have seen that he could not take the energy out of his food and needed to be fed more regularly. They blame the doctors for being negligent: the doctors failed to really look at the origin of his complaints and take appropriate measures.

The interpretative framework within which the parents understand the needs of their child and seek the most appropriate response to it is shaped by a combination of the discovery of low ATP, the parents' troubles feeding the child and the information provided by *Energy4All,* the patient organization for children with mitochondrial diseases. Furthermore, the parents also value the patient organization as a social community in which they feel at home, and experience support from other parents with children with comparable problems. Finding such a supportive environment is a relief to them, for they feel unable to share their sorrows with anybody prior to their participation in the organization.

> Father: "You know that these parents have similar problems, so you know they do not think you are whining when you talk about your sorrows [. . .] We have come to grow attached to these parents and their children."
>
> *(Extract from interview with the author, April 6, 2012)*

Snapshot 3 – A diagnosis for the disorder

In August 2013 the results of whole exome sequencing (WES) are communicated to the pediatric neurologists. Kabuki syndrome is a rare pediatric disorder which produces congenital and intellectual disabilities, which may include heart defects, hearing loss, reduced muscle strength, epileptic seizures, frequent infections during childhood and mild intellectual disability. Furthermore, patients with Kabuki syndrome have a characteristic appearance such as long eyelashes, long eyelids and a lower eyelid that curls upward, a flat nose, large ears and bluish corneas. While the morphological characteristics are specific for Kabuki children, a lot of the other complaints are not: many syndromes include reduced muscle strength, epilepsy, developmental delay, feeding problems and problems talking.

In the original diagnostic trajectory which took place prior to the study which uses WES, the child was already tested for Kabuki syndrome. But at that time the test was negative, so the neurologists concluded that this could not explain his characteristic complaints. In the meantime, scientific research on Kabuki syndrome had continued, and a different genetic mutation had been identified in patients with a Kabuki phenotype.

> Neurologist: "We did look for a mistake in the genetic material that indicates Kabuki syndrome, but he did not have a mutation in that gene. [. . .] At the time we said, 'OK, we did not find this mistake, so we cannot justify the diagnosis.' Afterwards, I did not think of testing him again for this syndrome on a new gene that had been found for Kabuki. That is daily practice. You cannot keep up with all the most recent literature on all the syndromes. Now this result has been produced with whole exome sequencing."
>
> *(Extract from interview with the author, August 29, 2013)*

The geneticist who analyzed the exome of the child explains this as an exemplary advantage of the use of WES in diagnostics for this group of patients. Sitting in front of two large computer screens, she shows how she analyses the genome, scrolling through different 'packages' of genes that relate to various diseases and syndromes. For this research she looked at the package that is associated with developmental delay and sought *de novo* mutations that appear for the first time in the child but are absent in the genes of the parents. When she found a mutation in KDM6A, she checked a general database that contains all (recent) scientific literature on rare neurological disorders and discovered that this mutation had been identified in patients with Kabuki syndrome in the months preceding his diagnosis (Miyake et al. 2013a, 2013b).

> Geneticist: "It is protocol in Dutch hospitals to test MLL2, which was first identified as the Kabuki-gene in September 2010. [. . .] But studies on this gene pointed out that not all patients with a Kabuki phenotype had a mutation on MLL2. In January 2012 there is a first publication about a different mutation in KDM6A which is found in patients who were clinically identified as patients with Kabuki syndrome, but who tested negatively on MLL2. After that initial discovery, more patients were found worldwide with this mutation in KDM6A and a clinical diagnosis of Kabuki. In September 2013 a first publication reveals an overview of all patients with Kabuki syndrome with a mutation on KDM6A. The patient of this case study would also belong to this group, since he received a negative test result on MLL2 in 2011, and a positive one on KDM6A in 2013."
>
> *(Extract from interview with the author, September 4, 2013)*

The use of WES in diagnostics for children with rare neurological disorders is very valuable, according to the geneticist, because it allows using all the recent scientific findings about rare diseases for patients. While Kabuki syndrome is a syndrome that could be tested with a targeted test (Sanger sequencing) on the two genes that are related to Kabuki syndrome, it is far from certain that this would have happened in the present diagnostic practice. Usually not all recent scientific knowledge is used in diagnostics because clinicians are far too busy and cannot keep up with all recent knowledge about all rare diseases. This means that a child may test negatively on a disease or syndrome, while he does actually have it, simply because the pediatric neurology fails to test him on newly discovered genetic anomalies associated with this particular disease or syndrome.

> Geneticist: "In this case [scientific discovery] went very fast: in July 2011 this patient was tested on MLL2, in January 2012 KDM6A was discovered. After that it would have been possible to test on this gene too. This did not happen. This does not mean that it would never have happened, maybe it would have. But [. . .] that [. . .] depends on the physician. [. . .] I think a lot of the times clinical neurologists will feel responsible to keep up to date with knowledge about their own favorite syndrome, and will know everything about that. But if the patient has something else, or if the clinical neurologist with this interest is at a conference on the day that the patient with this syndrome comes in, than there is a chance that you miss it. So I think it is worthwhile to work together with geneticists who use whole exome sequencing, and who keep up to date with all the recent scientific discoveries about rare disorders producing developmental delay."
>
> *(Extract from interview with the author, September 4, 2013)*

It is satisfying to the geneticist to be able to deliver such a clear and informative diagnosis to this patient. While she is one of the most renowned geneticists in her field of science, she asserts finding it her most important task to make knowledge available to patients. Although she never met the parents or saw the patient, she is confident that this will help them. But, while it was possible to set a clear diagnosis in this particular case, she adds that it still has to be investigated whether the clinical characteristics of this patient fit with this diagnosis. Until now it is unclear whether there is a difference between Kabuki patients with a mutation in MLL2 and patients with a mutation in KDM6A. Now that two mutations have been identified, it is possible to compare the phenotype of both groups of patients and see if they differ. These genotype-phenotype studies have not yet been done at the time when this chapter is written in the fall of 2014.

One of the surprising aspects of this diagnosis is that mitochondrial dysfunction is not part of biomedical descriptions of Kabuki syndrome. This aspect was also not mentioned in the information that the geneticist received from the neurologist about the phenotype of this patient. The neurologist thereafter explains that the discovery of low ATP in this patient did not make her conclude that he had mitochondrial disease at all. She suspects that low ATP is an effect of this syndrome in this particular patient. Many diseases, including a common flu, cause mitochondria to function less well. As Kabuki syndrome weakens the condition of a person, it is unsurprising that it also leads to mitochondrial dysfunction.

Snapshot 4 – A second interview with the parents

The second interview with the parents takes place a few weeks after they received the diagnosis. When I arrive, the now six-year-old child sits actively in front of the television, trying to touch the colors and shapes on the screen. He looks better than before.

When asked what they think of the diagnosis, especially the mother articulates appreciation. Knowing about Kabuki syndrome, according to her, provides insight into the complaints that many children with this syndrome have. As children with Kabuki syndrome often get scoliosis and hearing problems, she aims to keep check of that in the future and maybe prevent some of these complaints. At the same time, however, she also expresses doubts about the accuracy of the diagnosis, because she notices differences between descriptions of children with Kabuki syndrome and her own child.

> Mother: "He does not have the curled lower eyelid, nor a flat nose or large ears, such as Kabuki children have. He has long eyelashes, but a lot of children have that. And the mitochondrial dysfunction is not included in descriptions of Kabuki syndrome. [. . .] And I read that children with Kabuki syndrome have mild intellectual disability, well . . . [laughing] . . . please give me that! [. . .] Mitochondrial dysfunction is also not mentioned in descriptions of Kabuki children. So I have a 1000 questions to ask [the doctor] when we see her next week, and for now I conclude that he has both mitochondrial dysfunction and Kabuki."
>
> *(Extract from interview with the author, October 4, 2013)*

At the time of the interview, the father had not yet felt ready to look at the information about Kabuki syndrome and compare it to his son. He avoided looking at it. But now that he is engaged in an interview, he feels especially worried that it will change the relation that he and his wife developed with the patient organization *Energy4All*, which has meant a lot to him. He grew attached to other parents that he encountered in that organization and felt that engagement in fund-raising activities helped give some meaning to an otherwise sad and irreparable situation. Now that his child does not appear to have mitochondrial disease, membership of this patient organization may no longer be self-evident and taking part in fund-raising activities may lose its value as a purpose that he and his wife should be engaged in. Furthermore, with respect to their fund-raising activities, it is significant for both parents that their child's disorder is a syndrome and not a disease. For a disease, it is in principle imaginable that adequate treatment will be developed in the future, which cures the disease, or at least improves the prognosis. But a syndrome can never be treated. Having a syndrome is in their opinion just a matter of 'bad luck'; you cannot do anything about it.

Father: "[Raising money] gives a very good feeling. We are powerless, for there is no medication for mitochondrial diseases . . . If we can at least bring money together to pay for research, this may not benefit our own child, but it felt good that we could at least do something for other children who are born in the future with a mitochondrial disease, and that they would be able to lead more normal lives. That really gave me a positive feeling. Like you can give a positive turn to something really bad. But with this diagnosis I feel put on another track."

SvdB: "But if you knew earlier that it was Kabuki syndrome, you would have joined a Kabuki patient organization."

Father: "Yes, but what can you do there? Because then we would have known. . . . He has a syndrome, but you cannot do anything about a syndrome."

(Extract from interview with the author, October 4, 2013)

Discussion

The snapshots of which this brief biography is composed offer a fragmented insight into the different meanings and values that the disorder – or syndrome – has for various stakeholders, depending on their own (history of) actions and responsibilities. The pediatric neurologist appreciates WES for offering an accurate diagnosis more effectively, which contributes positively to her role as a doctor. It enables her to provide more adequate information about the prognosis of the child, which helps parents to prepare for the future. The geneticist is also satisfied. Apart from the fact that WES diagnostics responds well to the societal demand to translate new genetic knowledge to the clinic where it can be used to benefit patients. For parents, however, the diagnosis has more ambiguous value. While obtaining a diagnosis is the end of a long and demanding search, it also unsettles previous understandings of the disorder (based on the muscle biopsy) that used to inform their selection of caring activities, and allowed them to join a patient organization that offered a supportive social surrounding.

A translation of scientific knowledge about new diagnoses into the clinic seems to be well served by a consideration of biographies, such as the one offered in this chapter. Pleads to translate genetic knowledge into the clinic often privilege a scientific approach to 'disorder' or 'disease', about which patients and parents just need to be informed appropriately.[2] But the snapshots that compose the biography of the disorder in this chapter reveal that the scientific terminology that is used to talk about the disease – such as 'causality' and 'genes' – can only be mastered if all stakeholders, including parents, are initiated into this new language and learn how to use its words appropriately. What such a transition involves is illustrated in the discourse of the parents in this chapter, who in the past years appropriated words like 'mitochondria', 'low ATP' and 'energy management' and at the same time started to 'see' powerlessness in their child as demanding rest, and to consider doctors who ignore this as 'bad doctors'. Learning how to use scientific words correctly, for them, means learning also how to evaluate and choose appropriate action plans, selecting appropriate health care providers, and connecting with a community (the patient organization), which enhances these understandings with information and emotional support.

Understanding just what learning to speak a different scientific language involves means looking also at the related evaluative (and even moral) discriminations. While science is often understood as a neutral discourse about facts, the accounts of the different stakeholders – especially the parents – shows how scientific accounts also inform evaluations in a context of action.[3] The badness and undesirability of the child's condition is the reason why the parents are interested in what science has to offer: it helps them to interpret their child's needs and select the most appropriate response. For them, the production of a diagnosis demands integrating it into their daily interactions with the condition, and into their lives. As the needs of parents are not the primary driver of the scientific search for a diagnosis – for it is the availability of WES that motivates this research – the transition that parents have to make deserves more attention.[4] This means that new scientific understandings of disease should not only be made available to patients and their parents, but they should be helped to integrate these understandings into their perception of their child, their interpretation, their caring response and their evaluations, which may also involve a need to find a different community of allies.

Notes

1 See, e.g., Academy of Medical Sciences (2003), Sung et al. (2003), Zerhouni (2005), European Commission (2007), Raad voor Gezondheidsonderzoek (2007) and Medical Research Council (2008).

2 In philosophy of medicine such scientific understandings of disease have been defended by authors such as Schramme, who thinks that descriptions of disease are value-free just like other scientific concepts such as molecule, gravity or H_2O (Schramme 2007).

3 Care contexts always presuppose an evaluation of the physical and mental condition of a patient, because it "has its basis in the existence of a perceived *problem*" (Nordenfelt 2007, 7).
4 Authors such as Hofmann, Stempsey and Boenink preserve an evaluative understanding of disease, but reveal technology as a driving force of medical development (Hofmann 2001a, 2001b, 2008; Stempsey 2006, 2008; Boenink 2010).

References

Academy of Medical Sciences. 2003. Strengthening Clinical Research, http://www.acmedsci.ac.uk/index.php?pid=48&prid=18#terms.

Boenink, M. 2010. 'Molecular medicine and concepts of disease: The ethical value of a conceptual analysis of emerging biomedical technologies.' *Medicine, Health Care and Philosophy* 13 (1): 11–23.

Collins, F. S. 2011. 'Reengineering translational science: The time is right.' *Science Translational Medicine* 3 (90): 1–6.

de Ligt, J., Willemsen, M. H., van Bon, B. W., Kleefstra, T., Yntema, H. G., Kroes, T. and Vissers, L. E. 2012. 'Diagnostic exome sequencing in persons with severe intellectual disability.' *New England Journal of Medicine* 367 (20): 1921–1929.

Edmonds, D. and Eidinow, J. 2002. *Wittgenstein's Poker: The Story of a Ten-Minute Argument between Two Great Philosophers*. New York: HarperCollins Publishers.

European Commission. 2007. Translating Research for Human Health: Integrating Biological Data and Processes: Large Scala Gathering and Systems Biology, http://www.irc.ee/7rp/valdkonnad/tervis/materjalid_2007/health_integrating_en.pdf.

Hofmann, B. M. 2001a. 'Complexity of the concept of disease as shown through rival theoretical frameworks.' *Theoretical Medicine and Bioethics* 22 (3): 211–236.

———. 2001b. 'The technological invention of disease.' *Journal of Medical Ethics: Medical Humanities* 27: 10–19.

———. 2008. 'Why ethics should be part of health technology assessment.' *International Journal of Technology Assessment in Health Care* 24 (4): 423–429.

Khoury, M. J., Gwinn, M., Yoon, P. W., Paula, W., Dowling, N., Moore, C. and Bradley, L. 2007. 'The continuum of translation research in genomics medicine: How can we accelerate the appropriate integration of human genome discoveries into health care and disease prevention?' *Genetics in Medicine* 9 (10): 665–674.

Medical Research Council. 2008. MRC Translational Research Strategy – A Summary, http://www.mrc.ac.uk/consumption/groups/public/documents/content/mrc004551.pdf.

Miyake, N., Koshimizu, E., Okamoto, N., Mizuno, S., Ogata, T., Nagai, T., . . . and Niikawa, N. 2013b. 'MLL2 and KDM6A mutations in patients with Kabuki syndrome.' *American Journal of Medical Genetics Part A* 161 (9): 2234–2243.

Miyake, N., Mizuno, S., Okamoto, N., Ohashi, H., Shiina, M., Ogata, K., Tsurusaki, Y., Nakashima, M., Saitsu, H., Niikawa, N. and Matsumoto, N. 2013a. 'KDM6A point mutations cause Kabuki syndrome.' *Human Mutation* 34 (1): 108–110.

Nordenfelt, L. 2007. 'The concepts of health and illness revisited.' *Medicine, Health Care and Philosophy* 10: 5–10.

Raad voor Gezondheidsonderzoek. 2007. *Translationeel onderzoek in Nederland. Van kennis naar kliniek*. Den Haag: Raad voor Gezondheidsonderzoek; publicatienr. 55.

Schramme, T. 2007. 'Lennart Nordenfelt's theory of health: Introduction to the theme.' *Medicine, Health Care and Philosophy* 10: 3–4.

Stempsey, W. E. 2006. 'Emerging medical technologies and emerging conceptions of health.' *Theoretical Medicine and Bioethics* 27 (3): 227–243.

———. 2008. 'The geneticization of diagnostics.' *Medicine, Health Care and Philosophy* 9 (2): 193–200.

Sung, N. S., Crowley, W. F. and Genel, M. 2003. 'Central challenges facing the National Clinical Research Enterprise.' *JAMA* 289 (10): 1278–1287.

Van der Laan, A. L. and Boenink, M. 2012. 'Beyond bench and bedside: Disentangling the concept of translational research.' *Health Care Analysis* 23 (1): 1–18.

Vissers, L. E., de Ligt, J., Gilissen, C., Janssen, I., Steehouwer, M., de Vries, P., Arts, P., Wieskamp, N., del Rosario, M., van Bon, B. W., Hoischen, A., de Vries, B. B., Brunner, H. G. and Veltman, J. A. 2010. 'A de novo paradigm for mental retardation.' *Nature Genetics* 42 (12): 1109–1112.

Vissers, L. E., de Ligt, J., Gilissen, C., Janssen, I., Steehouwer, M., de Vries, P., van Lier, B., Arts, P., Wieskamp, N., Hoischen, A., Brunner, H. G. and Veltman, J. A. 2011. 'Unlocking Mendelian disease using exome sequencing.' *Genome Biology* 12 (228): 64–75.

Wittgenstein, L. 1951. *Philosophical Investigations*. Oxford: Basil Blackwell.

Zerhouni, E. 2005. 'Translational and clinical science – time for a new vision.' *The New England Journal of Medicine* 353 (15): 1621–1623.

9 The plasticity and recalcitrance of wetlands

Kevin C. Elliott

Introduction

Wetlands have received a great deal of scientific and regulatory attention in recent years. This is partly because they are being degraded and destroyed more rapidly than other ecosystems (Millennium Ecosystem Report 2000). Since the founding of the United States, more than 50 percent of its original wetlands have been destroyed (Dahl 1990). This loss is particularly problematic, because wetlands are exceedingly important from both an ecological and a social perspective. They have been called the "kidneys of the landscape" because they play a central role in purifying water, recharging aquifers and contributing to chemical cycles (Mitsch and Gosselink 2007, 4). They also play valuable roles in alleviating floods, protecting against storms, and providing habitat for numerous species, including commercially valuable fish and waterfowl species.

As society has come to appreciate the value of wetlands, policy toward them has shifted 180 degrees. One hundred years ago, it was considered environmentally optimal to drain wetlands, whereas considerable efforts are now taken to preserve them and even to restore degraded wetlands or create new ones (Meyer 2004, 84–100). In the United States, this policy has been enshrined under the slogan of 'no net loss'; when damages to existing wetlands cannot be avoided, they are to be replaced by new or restored wetlands. This chapter argues that, because of the ways in which wetlands have been conceptualized, categorized and measured in an effort to preserve them, they constitute fascinating examples of technoscientific objects.

As explained in the introduction to this volume, the concept of technoscience has been used as a way of describing the complex relationships between contemporary science and technology (Haraway 1997). According to proponents of this concept, it has become increasingly difficult to distinguish the scientific activity of representing phenomena from the technical goal of manipulating phenomena. As a result, the seemingly distinct categories of natural objects (studied by science) and human cultural creations (generated by technology) have become increasingly blurred (Nordmann 2011, 19–30). 'Artificial' wetlands that have been created or restored in an effort to generate important ecosystem services seem to be a perfect example of technoscientific objects that blur the lines between the natural and the cultural (Kivaisi 2001, 545–560). But this chapter explores a somewhat more subtle sort of technoscientific object. Rather than focusing on 'artificial' wetlands, it explores the ways in which even seemingly 'natural' wetlands constitute technoscientific objects.[1]

The next section of the chapter provides a brief history of wetland science and policy (focusing, for the sake of simplicity, on wetlands in the United States). This paves the way for two sections that analyze the technoscientific character of natural wetlands. First, we will see that both the term 'wetlands' and definitions for the term were generated in large part as a result of practical efforts to preserve and regulate them. Thus, the conceptualization of wetlands as a particular entity or object is a function of 'technological' as well as 'natural' factors. Second, wetlands have now been commodified for the purposes of exchanging them in market transactions. As a result, processes for categorizing and measuring them have become a sort of hybrid practice that is guided both by scientific and technical considerations. This analysis of wetlands illustrates an important feature of technoscientific objects; namely, they can display a fascinating interplay of plasticity and recalcitrance in response to scientific and technical practices. Partly as a result of this plasticity and recalcitrance, they interact with these practices in an iterative fashion that alters science and culture as well as the technoscientific objects themselves.

A brief history of wetlands in the United States

The following is a very selective history of wetlands in the United States for the purposes of contextualizing the subsequent sections of the paper. Perspectives on wetlands in the United States and around the world are particularly intriguing insofar as they have changed so dramatically over the past four hundred years (Meyer 2004, 84–100). Although many early European settlements in the United States were established near freshwater or saltwater marshes in order to take advantage of the abundant food and fiber that they supplied, wetlands tended to have a decidedly negative valence for the early colonists. As historian Ann Vileisis notes, when Governor John Winthrop declared that the Puritans' new community in Massachusetts should be a "city on a hill," he "projected a moral landscape onto the physical landscape of the New World" (Vileisis 1997, 30). As she puts it, "Before they even set foot in America, the Puritan colonists understood the topographic tension between pious and pure hilltops and the dark, dismal lowlands" (Vileisis 1997, 30). For centuries, wetlands had been depicted as corrupt and dangerous (Mitsch and Gosselink 2007, 16). In the *Divine Comedy*, Dante described a marsh in Upper Hell as the destination for the wrathful, while Grendel (the monster of *Beowulf*) dwelled in marshes and fens, and the swamp served as a symbol for sinful barriers to spiritual redemption in John Bunyan's *Pilgrim's Progress*. These negative attitudes toward wetlands became even stronger as it became clear that the Native Americans could use them to their advantage during military campaigns.

These sentiments contributed to popular enthusiasm for draining wetlands. Besides the negative cultural and religious connotations of swamps, they were regarded as unhealthy sources of disease. Moreover, there were often financial motivations for converting marshland into cities and swamps into farmland (Vileisis 1997, 47). In the south, and especially in South Carolina, many low-lying wetlands were converted into rice fields during the late eighteenth century. Economic pressures to manipulate wetlands for the purposes of improving farmland and preventing flooding continued across large portions of the country, and especially in the Midwest and southeast, throughout the nineteenth century.

These ongoing efforts to eliminate wetlands were limited in their success until the period before and after World War II (Vileisis 1997, 47). During the Great Depression, government programs provided new money and expertise for drainage efforts. The boom in economic growth after the war created even more pressure on wetlands. Besides the ongoing pressure to improve farmlands, wetlands were severely damaged because of expanding suburbs, new airports, oil wells, interstate highways, dams, power plants and the widespread application of pesticides (Vileisis 1997, 205–206).

During this same period of time, however, pressure to protect wetlands began to develop. As early as the middle of the nineteenth century, Henry David Thoreau had expressed enthusiasm for the beauty and spiritual significance of swamps. But this sentiment was fairly unique during that time period, although a few other figures began to note the unique and intriguing features of southern Cypress swamps (Vileisis 1997, 106–109). This nascent interest in wetlands expanded dramatically at the turn of the twentieth century when local chapters of the Audubon Society, along with a number of hunting societies, began to realize that the destruction of wetlands contributed to declines in many bird populations. By the 1950s and 1960s, it was becoming clear that wetlands provided a host of other important services, including flood control, recharge of groundwater supplies, storm protection and habitat for important seafood species.

Given the economic incentives to proceed with development projects, however, it proved very difficult to stem the tide of wetland destruction. Early efforts at protecting wetlands focused especially on slowing federal projects that had previously been designed to drain wetlands (Vileisis 1997, 229ff.). By the 1970s, legislation such as the Clean Water Act, the National Environmental Policy Act, the Coastal Zone Management Act and the Endangered Species Act created regulatory policies that limited wetland destruction even on privately owned lands. Perhaps the most important piece of legislation was Section 404 of the Clean Water Act, which required the Army Corps of Engineers (in consultation with the Environmental Protection Agency) to provide a permit to individuals who wish to dredge or fill wetlands on their property (Hough and Robertson 2009, 15–33). At the present time, federal policy aims for 'no net loss' of wetlands; ideally, they are to be destroyed only if these negative impacts are mitigated by seeking to avoid impacts where possible, attempting to minimize unavoidable impacts and working to create or restore wetlands to make up for those that have to be damaged. Nevertheless, the sober fact remains that the United States has lost more than 50 percent of its original wetlands, at an average rate of 60 acres lost each hour from the 1780s to the 1980s (Dahl 1990).

The genesis of wetlands as technoscientific objects – terms and definitions

AGENT MULDER: "Are there any swamps around here?"
SHERIFF: "We used to have swamps – till the EPA made us take to callin' 'em wetlands."

(Robertson 2000, 469).

This chapter explores the technoscientific character of natural wetlands in two ways. The first part of the analysis, developed in this section, reveals that the very concept of wetlands – both the term and its definition – developed as a result of practical interests in controlling these entities. The next section explores the second aspect of wetlands' technoscientific character; namely, they are categorized and measured in a hybrid manner that combines both scientific efforts at characterizing them and technical efforts at trading them.[2] In keeping with the goals of this book project, these two sections analyze both the *genesis* and the current *ontology* of wetlands as technoscientific objects.

Turning first to the term 'wetlands', it is striking that it did not appear until the 1950s and was not in widespread use until ten or twenty years later. In the 1948 edition of Mencken's *American Language*, the term 'wetlands' did not appear, but there were thirty meanings for the word 'swamp' (Walker 1976, 75–101). A host of other terms were also employed for describing various wet ecosystems: marsh, fen, moor, bog, wet heath, carr, salt meadow, morass, mire, lowland, hayland and overflowed land (Mitsch and Gosselink 2007, 215–225; Wheeler and Proctor 2000, 187–203). In the United States, the most common terms were 'swamp', which commonly referred to land that was to be drained for agricultural purposes, and 'marsh', which tended to have more positive connotations as land that could be used for recreation or animal habitat (Moss 1980, 219).

Although there were some previous references to 'wet lands', the first influential use of the single word 'wetlands' as an all-encompassing term appears to be in a 1953 document of the U.S. Fish and Wildlife Service (FWS) (Shaw and Fredine 1956).[3] It is not accidental that a government agency – in one of the first major documents designed to identify lowlands that merited conservation – introduced the term. Subsequent authors consistently identify two motivations that played a crucial role in the introduction and widespread use of the 'wetlands' concept.[4] First, it covered many different lands with a unified term. Richard Walker has argued that all these lands have relatively little in common from a natural science perspective but are unified from a social perspective insofar as people would like to preserve them (Walker 1976, 76). Second, in contrast to terms like 'swamp', the new 'wetlands' terminology was non-pejorative. David Moss (Moss 1980, 215–225) argues that adoption of this term strengthened pro-conservation attitudes. Walker claims that it had positive symbolic value, and the National Research Council acknowledges that it "appears to have been adopted as a euphemistic substitute for the term 'swamp'" (Walker 1976, 75–101; National Research Council 1995, 43).

Subsequent trends in the use of wetlands terminology strengthen the case that, whether deliberately or not, this terminology rose to prominence because of the conservationist agenda. Moss (1980, 215–225) emphasizes that, until the early 1960s, the word 'wetlands' primarily appeared in state and federal wildlife agency publications. Starting in 1961, it was used in federal legislation designed to protect and acquire these areas. And beginning in 1963, the term began to be used by members of the public involved in conservation efforts (Moss 1980, 220).

But the introduction of the term 'wetlands' is not the only indication that this concept is a technoscientific one introduced to advance the preservation and regulation of a particular group of entities. Perhaps even more striking is the way that efforts to define it have been governed by the same technoscientific goals. Moss's commentary on the state of wetland definitions in 1980 is notable:

> The term "wetlands" was defined differently within each state regulatory law and reflected ". . . different perceptions of what is important about the [area] being categorized". . . . Definitions and terms for wetlands . . . have not been settled by extensive adjudication or scientific discussions, but have evolved from a synthesis of the definitions found in federal publications, state statutes, and from the public's attitudes toward their natural, recreational, and esthetic importance in different time periods.
>
> (Moss 1980, 221–222)

This perspective is echoed in a very influential FWS report from the 1970s, which could hardly express the technoscientific character of the wetlands concept any more explicitly:

> Effective management requires legislation; out of such legislation, legal definitions are born. . . . [B]ecause the reasons for defining wetland vary, a great proliferation of definitions has arisen. Our primary task here is to impose arbitrary boundaries on natural ecosystems for the purposes of inventory, evaluation, and management.
>
> (National Research Council 1995, 50)

In their magisterial textbook on wetlands, William Mitsch and James Gosselink (2007) contextualize these statements with a brief history of efforts at defining wetlands in the United States. They note that it was unimportant to define wetlands in the nineteenth century, when the main policy imperative was to drain them.[5] Wetland definitions became more important when these ecosystems were recognized as valuable and when legislation was passed to protect them. And the drive for highly precise definitions was almost entirely a function of social goals:

> Even as the value of wetlands was being recognized in the early 1970s, there was little interest in precise definitions until it was realized that a better accounting of the remaining wetland resources in this country was needed and

> definitions were necessary to achieve that inventory. When national and international laws and regulations pertaining to wetland preservation began to be written in the late 1970s and afterward, the need for precision became even greater as individuals recognized that definitions were having an impact on what they could or could not do with their land.
>
> (Mitsch and Gosselink 2007, 22)

Thus, while the scientific community was not initially motivated to provide precise definitions of the wetlands concept for their scientific purposes, they were enlisted in efforts to characterize wetlands for social purposes. As the National Research Council put it, "Scientists have not agreed on a single commonly used definition of wetland in the past because they have had no scientific motivation to do so. Now, however, they are being asked to help interpret regulatory definitions of wetlands" (National Research Council 1995, 43).

Perhaps the best illustration of how social goals influenced wetland definitions occurred in the late 1980s and early 1990s. The National Wetlands Policy Forum proposed in 1987 that there should be a national policy of 'no net loss' of wetlands in the United States. In an effort to improve his record on environmental issues while he was running for president in 1988, Vice-President George Bush adopted this 'no net loss' policy as part of his campaign. As a result, the four major federal agencies involved in wetland policies attempted to develop a unified definition of wetlands for the first time in 1989 in order to facilitate unified government action (Vileisis 1997, 318). However, once Bush was elected president, his administration came under attack from development interests that were hampered by the 'no net loss' policy. Moreover, the new 1989 definition received intense criticism because it included some agricultural wetlands that had been excluded under previous Army Corps of Engineer guidelines (Vileisis 1997, 319). The Bush administration responded to this criticism in 1991 by proposing a new definition that would have eliminated as much as a third of the wetlands that would have been protected under the previous definition (Schiappa 1996, 218). This proposed redefinition was met with such intense opposition, however, that it had to be abandoned (Schiappa 1996, 219; Vileisis 1997, 321–322).

The ontology of wetlands as commodities

The previous section argued that the technoscientific character of natural wetlands can be clearly seen in the genesis of the wetlands concept during the latter half of the twentieth century. This section turns to the current ontology of wetlands and argues that one can learn much about their technoscientific character by studying how they have been commodified in wetland mitigation banks. Recent work by geographer Morgan Robertson (Robertson 2000, 2004, 2006) is particularly valuable for clarifying the process of mitigation banking, its significance and its limits.

The pressure to develop a mitigation banking system has stemmed from the permitting process for dredging and filling wetlands under Section 404 of the

Clean Water Act (CWA). The notion of 'mitigation' was not an important issue in early discussions of the CWA, perhaps because it was assumed that serious damage to wetlands would not be permitted (Hough and Robertson 2009, 17). However, as it became clear that neither the Environmental Protection Agency (EPA) nor the Army Corps of Engineers (Corps) were very inclined to deny permits, the FWS and the National Marine Fisheries Service (NMFS) began to request that mitigation measures be included with some permits.[6] Palmer Hough and Morgan Robertson emphasize that environmental mitigation has typically been conceptualized by the federal government as a three-part concept, consisting of impact avoidance, impact minimization and impact compensation (Hough and Robertson 2009, 18). This three-part mitigation sequence was ultimately enshrined in a crucial 1990 Memorandum of Agreement between the EPA and the Corps that discussed how they would implement Section 404 permits (Hough and Robertson 2009, 27). However, Hough and Robertson lament that despite its significant limitations, the third option, compensation, "is so central to discussions of mitigation that 'compensation' is often mistakenly held to be synonymous with 'mitigation'" (Hough and Robertson 2009, 23). The Corps currently allows roughly 22,000 acres of wetlands per year to be impacted by development activities under its permitting program, and it requires 40,000 to 60,000 acres of compensatory mitigation in return (Hough and Robertson 2009, 23–24).

Traditionally, this mitigation was carried out by the contractors who were engaged in wetland destruction. They were required to restore, enhance, or preserve other wetland areas in order to compensate for the wetlands that they were destroying (Hough and Robertson 2009, 24). Unfortunately, there was often little effort to ensure that these compensatory mitigation efforts were satisfactory; some research suggested that these projects were rarely fully successful and that large numbers of them (in some cases up to 80 percent) were never even begun (National Research Council 2001; Froelich 2003, 130). Partially because of these problems and partially because of a desire to streamline the permitting process, some contractors pioneered wetland banking as an alternative approach. A mitigation bank consists of a wetland area that is preserved or restored to compensate for a variety of wetland areas that are subsequently destroyed. Banks of this sort were initially developed for internal use by state departments of transportation and a few other large developers in the 1980s. Beginning in the early 1990s, entrepreneurs began to develop wetland banks so that they could sell mitigation 'credits' to developers who needed them. By 2005, there were over 350 active banks, 75 sold-out banks and over 150 banks under review (Hough and Robertson 2009, 25). Annually, almost $3 billion are spent on compensatory mitigation, primarily for either contractor-operated replacement wetlands or for banked wetland credits (Hough and Robertson 2009, 24).

This banking process is socially significant, because it constitutes the most highly developed contemporary market in ecosystem services. Neoliberal, market-based approaches to environmental protection have become popular in recent years, but they have previously focused on relatively easy-to-measure phenomena such as the emission of specific pollutants. Developing a market for ecosystem services such as flood protection is a significant new step (Salzman and Ruhl 2000, 607–694; Robertson 2004, 361–373). The banking process also illustrates a second 'level' or 'register' in which wetlands are technoscientific. Not only was the concept of wetlands initially conceptualized and defined based on a mixture of scientific and technical goals, but these objects are now measured and categorized based on a hybrid system created with this same mixture of goals in mind. (Jennifer Gabrys's chapter on plastic garbage patches in this volume provides a similar identification of two levels or registers for studying technoscientific objects.) Rather than focusing solely on developing *scientific* representations of wetlands, Morgan Robertson claims that those involved in wetland banking are representing them as "reservoirs of 'capital'" (Robertson 2000, 465). He even cites the philosopher of technoscience Donna Haraway, who claims that this sort of capitalist interpretation of nature involves an implosion of the artifactual and the natural, such that "nature itself . . . has been patently reconstructed" (Haraway 1997, 245).[7]

Let us scrutinize this process of commodification in a bit more detail. Robertson suggests that it involves at least four 'moments': (1) scientific abstraction of natural entities such as wetlands into functional categories; (2) monetary valuation of those abstracted categories; (3) spatial abstraction of those categories; and (4) establishment of the exchange process (Robertson 2000, 464). For the purposes of understanding how natural wetlands serve as technoscientific objects, the first moment in this process is particularly important. As with any commodification process, the messy materiality of both natural and artificial wetlands is reduced to an abstract bundle of features (namely, particular functions or ecosystem services) that are regarded as relevant for exchange. This abstraction is performed via 'rapid assessment methods' (RAMs), which consist of algorithms that convert a variety of data about a wetland into a numerical score that represents its functional value for providing services such as species habitat or water quality or recreation (Robertson 2000, 367). Natural and artificial wetlands are made commensurable through the use of these RAMs "to ensure equivalence on both sides of the transaction" (Robertson 2004, 367).

A variety of social pressures have influenced the process of assessing wetlands using RAMs. For example, the banking process is more straightforward when a wetland is represented by one main score rather than a variety of different functional scores. Moreover, wetland bankers need assessments to be authoritative, cheap and quick (Salzman and Ruhl 2000, 665). Therefore, because plants are relatively easy to look at, and because botanists have developed some relatively high-quality analyses of the plants in various sorts of wetlands, the function of 'floristic biodiversity' has become a rough indicator for most other wetland functions. For example, the Chicago wetlands banking market, which has become a leader in this industry, uses a RAM called the floristic quality assessment (FQA) to evaluate wetlands based on a specific list of plants that are present or absent. As Robertson notes, "wetland banking markets have largely followed the Chicago consensus and based the value of their traded credits on metrics generated from a list of plants present at the bank site" (Robertson 2006, 374). Thus, for the purposes of regulatory policy under the Clean Water Act, natural wetlands have been rendered equivalent with artificial wetlands by essentially reducing them to simple scores generated from a select body of data via RAMs. In actual practice, however, even this much analysis is often regarded as too onerous, so natural and artificial wetlands end up being compared solely in terms of the number of acres present (Salzman and Ruhl 2000, 661; Ruhl and Gregg 2001, 381).

Robertson himself contends that "rapid ecological assessment" is essentially a "science-policy hybrid field" that has developed in an effort to create techniques for measuring "units of ecosystem function" that are important for developing a market in wetlands but that do not have a straightforward meaning from an ecological perspective (Robertson 2006, 368). He recounts a variety of research activities that are highly strained from a scientific perspective but that are necessary for the purposes of these ecological assessments. For example, the floristic quality assessment used by the Corps to assess wetlands requires identifying plants at the species level. However, performing a scientifically reliable identification of this sort is not feasible because the guidelines also require that the identification be performed in May or June, when most plants have not yet flowered. Therefore, assessors end up depending on what Robertson calls "shared myths" about plant features that may indicate the identity of particular species but that are not compelling by rigorous botanical standards. He clearly contrasts this technoscientific hybrid research practice with more traditional scientific work:

> [T]he primary directive for monitoring technicians is *not* to produce falsifiable results that can circulate within a hypothetico-deductive paradigm. . . . It is instead to produce data that successfully circulate in the networks of law and economics. As workers in a forum of articulation between science and capital, we made use of scientific codings and principles, but, ultimately, scientific operational logic was rejected in favor of the ad hoc logics (our 'shared myths') that worked better to bridge the two systems.
>
> (Robertson 2006, 377)[8]

The plasticity and recalcitrance of technoscientific objects

The analysis of the technoscientific character of wetlands in the preceding two sections highlights an important feature of technoscientific objects. Namely, they can display a fascinating interplay of plasticity and recalcitrance in response to scientific and cultural practices. In part as a result of this plasticity and recalcitrance, they can interact with these practices in an iterative fashion that alters science and culture as well as the technoscientific objects themselves.[9]

Consider, for example, efforts to define wetlands. The Fish and Wildlife Service has clearly acknowledged the plasticity of the wetlands concept, claiming, "There is no single, correct, indisputable, ecologically sound definition for wetland because the gradation between totally dry and totally wet environments is continuous" (National Research Council 1995, 50). Other experts make similar claims. They widely acknowledge that wetlands are distinguished by three features:

(1) the presence of water, (2) hydric soils, and (3) vegetation adapted to wet conditions (Mitsch and Gosselink 2007, 209–230). However, they insist that these features are not sufficient for generating a single wetland concept that all actors can agree on. According to Mitsch and Gosselink, various groups (e.g., ecologists, biologists, geologists, economists, and lawyers) arrive at different concepts partly because they deal with wetlands differently and have varying objectives (Mitsch and Gosselink 2007, 40). In general, Mitsch and Gosselink note that the definitions of wetlands employed by scientists tend to be more inclusive than definitions developed in the legal context, because for regulatory purposes it is important to identify wetlands in a relatively simple fashion. Thus, regulatory definitions tend to focus on the presence or absence of specific features of vegetation or soil that are fairly easy to identify but that somewhat narrow the scope of the concept (Mitsch and Gosselink 2007, 40).

Nevertheless, there are limits to the plasticity of the wetlands concept, and this has contributed to important political tensions over regulating wetlands. Edward Schiappa documents the fact that many opponents of Bush's 1991 wetland redefinition accused the Bush administration of abandoning good science for the sake of politics in formulating a definition (Schiappa 1996, 220). Schiappa is somewhat critical of this perspective, because he emphasizes that wetland definitions invariably reflect particular social and political interests that need to be evaluated in a democratic fashion (Schiappa 1996, 220–221). But we may develop a more sympathetic understanding of this debate by keeping in mind that scientific information about wetlands generates an element of 'recalcitrance' that limits which definitions can be reasonably formulated. As Ann Vileisis notes, the process in this case was taken over by White House officials who had so little knowledge of wetlands that "they had to be given a special wetlands glossary at their decisive meeting" (Vileisis 1997, 321). Therefore, these officials appear to have gone 'beyond the pale' by ignoring fairly basic scientific information about wetlands.

Similar elements of plasticity and recalcitrance are apparent when one considers the practices of measuring and categorizing wetlands for banking purposes. Robertson highlights these contrasting features of wetlands by describing what he calls the conflicting "logics" of capital, law and science (Robertson 2004, 361–373). From an economic and legal perspective, regulators would like to treat natural and artificial wetlands as easily interchangeable, and they would like to think that these wetlands can be successfully characterized using simple measurement techniques. To some extent, wetlands have indeed been 'plastic' enough in these regards to allow regulators to develop a successful mitigation banking market. Nevertheless, some scientists have also highlighted ways in which wetlands are recalcitrant and resist these efforts to treat them as interchangeable. For example, many ecologists have argued that artificial wetlands have failed to fully replace the ecosystem services provided by those that were destroyed (Schiappa 1996, 209–230; Froelich 2003, 130). A perfect example is a study from the 1990s, indicating that while the state of Maryland claimed to have gained 122 acres of wetlands between 1991 and 1996, it actually lost 51 acres of wetland functions (Salzman and Ruhl 2000, 662).

Thus, using Robertson's terminology, the logics of capital and law encourage regulators to assume that wetlands are plastic enough so that one wetland can easily be replaced by another and so that the essential features of these ecosystems can be easily characterized using simple tools. In contrast, the logic of science pushes toward acknowledging the recalcitrance of wetlands – namely, the ways in which more detailed ecological studies uncover important differences between wetlands that are destroyed and those that are created. This point is expressed beautifully by an EPA official quoted by Robertson, who joked that efforts to become more and more ecologically precise could lead to markets in

> habitat for middle-aged great blue herons who don't like shrimp, or something. Obviously, I can't imagine even trying to do that. . . .you can define a unit so that you're going to have flourishing mitigation banking. You can also define a unit so that, should there ever be one exchanged, it would be environmentally precise. And those are at potentially different extremes.
>
> (Robertson 2004, 368)

Robertson further emphasizes that the more one pursues an ecologically precise representation of a wetland, the more its functions appear to be the result of its relationships with the surrounding landscape. But if the functions of a wetland are spatially embedded, it becomes dubious whether a natural wetland in a particular setting is actually commensurable with an artificial wetland located elsewhere (Robertson 2004, 368–369). Thus, these ecological insights again serve as a source of recalcitrance, limiting the liberties that regulators can legitimately take in abstracting from wetlands' spatial locations and surroundings for the sake of commodifying them. As Robertson puts it, "there remains some element of our scientific understanding of nature which is uncaptured by state and capital, and which may provide leverage for reshaping and resistance in these other social spheres" (Robertson 2006, 367–387).

In part because of this mixture of plasticity and recalcitrance that wetlands have displayed, one can observe an iterative series of interactions between wetlands and various scientific and cultural practices, altering them all in the process.[10] As scientists began to generate information about the important services that wetlands provide, concerned citizens and legislators recognized the need to develop new regulatory policies so that these valuable ecosystems could be protected. But as policy makers sought to formulate new regulations, they created new terminology and definitions for wetlands that contributed to their social goals. The development of these regulations also altered the practice of ecology, both by forcing scientists to develop clearer definitions for wetlands and by generating simplified methods for assessing them. These efforts at defining and regulating wetlands have in turn had a profound effect on the status of wetlands throughout the United States, both by limiting damage to some ecosystems and by allowing the destruction of others in exchange for the creation and restoration of other wetlands. Finally, ongoing ecological studies of these wetlands are now uncovering ways in which current regulations may need to be changed because of their failure to do justice to the complexity of these ecosystems.

Conclusion

This chapter has argued that natural wetlands constitute particularly fascinating examples of technoscientific objects, in part because their technoscientific character is not immediately obvious. While they initially appear to be straightforwardly natural in character, both their genesis and their current ontology reveal intriguing ways in which they turn out to be technoscientific. First, the concept of 'wetlands' as a distinct set of entities was generated as a consequence of practical efforts at preserving particular ecosystems. Second, natural wetlands have now been turned into commodities that can be exchanged in market transactions with artificial wetlands. This process of commodification requires that wetlands be categorized and measured at least partly in ways that fit legal and economic exigencies rather than scientific ones. These observations are not intended to challenge the possibility of studying wetlands from the perspective of scientific disciplines. For example, ecologists have now developed their own definitions for wetlands, and they have developed ways of measuring wetlands that challenge the methods often used for the purposes of commodifying them. The point of this chapter is that the economic and legal context gave birth to these entities and continues to provide conceptualizations and ways of studying them that can clash with the approaches taken by particular scientific disciplines. Moreover, the interest that these disciplines have taken in wetlands has itself been generated at least in part by the social context in which they operate.

This clash between different ways of conceptualizing and studying wetlands highlights one of the central lessons of this chapter. Namely, technoscientific objects like wetlands can display a striking combination of plasticity and recalcitrance. On one hand, they are 'flexible' enough that a variety of regulatory and technical goals can influence how they are defined, measured and categorized. On the other hand, they exert limits on how far these technical efforts can go and how well they can succeed. For example, while scientists acknowledge that the wetlands concept is not precise, the Bush administration discovered that they could not just define wetlands however they liked. Similarly, while regulators have taken a number of liberties in the ways that they have measured and compared wetlands for mitigation banking purposes, they are discovering that their simplified assessment methods sometimes fail to capture all the ecologically relevant features of the wetlands that they are trying to protect.

Besides advancing the literature on technoscience, this analysis of wetlands can assist in developing a more sophisticated understanding of how politics and science interact in the regulatory arena. When ecologists call for more 'science-based' approaches to regulating wetlands, they offer a helpful corrective against those who think that wetlands are so flexible that regulators can evaluate and compare them without paying attention to detailed ecological information. But it is also important to recognize that, in order to make progress in the regulatory context, policy makers sometimes need to alter standard scientific approaches to make them quicker and easier to apply. The development of rapid assessment methods provides an excellent example of this 'hybrid' science. To the extent that citizens are interested not only in developing accurate descriptions of wetlands but also in developing a successful regulatory system for preserving them, these efforts appear to be legitimate. The key is for policy makers and scientists to determine how much they can bend their definitions, measurements and categorizations of wetlands to their practical regulatory purposes without becoming so unrealistic that they fail to achieve their ultimate goals.

Notes

1 Throughout the remainder of this chapter, the words 'natural' and 'artificial' will not be placed in quotation marks when they are used to describe wetlands. The quotation marks were included at the beginning of the chapter to emphasize that these terms should not be taken at face value, but for the sake of simplicity, they will be eliminated throughout the rest of the chapter because it should be clear that these terms are not straightforward.

2 A third way of arguing for the technoscientific character of natural wetlands would be to point out that humans have so altered the face of the globe (and are doing so to an even greater extent via climate change) that any seemingly natural wetland is bound to reflect subtle human influences. This is an important point that can at times be relevant for restorationists who insist on returning ecosystems to their 'natural' state (see Throop 2000); nevertheless, the two arguments provided in this chapter are particularly significant because they highlight ways in which the technoscientific character of wetlands goes beyond that of other ecosystems.

3 'Wetlands' did not appear in the 1948 edition of Mencken's *American Language*, but there were thirty meanings for the word 'swamp'; see Walker (1976, 75–101).

4 See, for example, Walker (1976, 75–101), Moss (1980, 215–225) and National Research Council (1995).

5 This may not be altogether correct, insofar as historian Ann Vileisis notes that there were important debates about how to define 'swamps' in the nineteenth century because of legislation like the Swamp Land Act of 1850; see Vileisis (1997, 73–90).

6 In 2004 and 2005, only 0.25 percent of permit applications were denied; see Hough and Robertson (2009, 17).

7 Cited in Robertson (2000, 464).

8 Emphasis in original.

9 In this volume, Pierre Tessier's chapter on fuel cells provides another excellent example of the plasticity and recalcitrance of technoscientific objects. Scientists have gone to great length to manipulate these research objects, but they have still stubbornly resisted many efforts at increasing their power.

10 For an analysis of different forms of iteration in scientific practice, see Elliott (2012, 376–382).

References

Dahl, T. 1990. *Wetland Losses in the United States 1780s to 1980s*. Washington, DC: U.S. Department of the Interior, Fish and Wildlife Service, http://www.npwrc.usgs.gov/resource/wetlands/wetloss/index.htm (accessed July 18, 2012).

Elliott, K. 2012. 'Epistemic and methodological iteration in scientific research.' *Studies in History and Philosophy of Science* 43: 376–382.

Froelich, A. 2003. 'Army corps: Retreating or issuing a new assault on wetlands.' *BioScience* 53: 130.

Haraway, D. 1997. *Modest_Witness@Second_Millennium*. New York: Routledge.

Hough, P. and Robertson, M. 2009. 'Mitigation under Section 404 of the Clean Water Act: Where it comes from, what it means.' *Wetlands Ecology and Management* 17: 15–33.

Kivaisi, A. 2001. 'The potential for constructed wetlands for wastewater treatment and reuse in developing countries: A review.' *Ecological Engineering* 16: 545–560.

Meyer, W. 2004. 'From Past to Present: A Historical Perspective on Wetlands.' In *Wetlands*, edited by Spray, S. and McGlothlin, K., 84–100. Lanham, MD: Rowman and Littlefield.

Millennium Ecosystem Report. 2000. *Ecosystems and Human Well-Being: Wetlands and Water in II Millennium Ecosystem Assessment*. Washington, DC: World Resources Institute, http://www.maweb.org/documents/document.358.aspx.pdf (accessed July 15, 2012).

Mitsch, W. and Gosselink, J. 2007. *Wetlands*. 4th ed. Hoboken, NJ: John Wiley and Sons.

Moss, D. 1980. 'Historic changes in terminology for wetlands.' *Coastal Zone and Management Journal* 8: 215–225.

National Research Council. 1995. *Wetlands: Characteristics and Boundaries*. Washington, DC: National Academy Press.

———. 2001. *Compensating for Wetland Losses under the Clean Water Act*. Washington, DC: National Academy Press.

Nordmann, A. 2011. 'The Age of Technoscience." In *Science Transformed? Debating Claims of an Epochal Break*, edited by Nordmann, A., Radder, H. and Schiemann, G., 19–30. Pittsburgh: University of Pittsburgh Press.

Robertson, M. 2000. 'No net loss: Wetland restoration and the incomplete capitalization of nature.' *Antipode* 32: 469.

———. 2004. 'The neoliberalization of ecosystem services: Wetland mitigation banking and problems in environmental governance.' *Geoforum* 35: 361–373.

———. 2006. 'The nature that capital can see: Science, state, and market in the commodification of ecosystem services.' *Environment and Planning D: Society and Space* 24: 367–387.

Ruhl, J. and Gregg, R. 2001. 'Integrating ecosystem services into environmental law: A case study of wetlands mitigation banking.' *Stanford Environmental Law Journal* 20: 381.

Salzman, J. and Ruhl, J. 2000. 'Currencies and the commodification of environmental law.' *Stanford Law Review* 53: 607–694.

Schiappa, E. 1996. 'Towards a Pragmatic Approach to Definition: "Wetlands" and the Politics of Meaning.' In *Environmental Pragmatism*, edited by Light, A. and Katz, E., 218. London: Routledge.

Shaw, S. and Fredine, C. 1956. *Wetlands of the United States, Their Extent, and Their Value for Waterfowl and Other Wildlife*. Washington, DC: U.S. Department of Interior, Fish and Wildlife Service, Circular 39.

Throop, W. 2000. *Environmental Restoration: Ethics, Theory, and Practice*, edited by Throop, W. Amherst, NY: Humanity Books.

Vileisis, A. 1997. *Discovering the Unknown Landscape: A History of America's Wetlands*. Washington, DC: Island Press.

Walker, R. 1976. 'Wetlands preservation and management on Chesapeake Bay: The role of science in natural resource policy.' *Coastal Zone Management Journal* 1: 75–101.

Wheeler, B. and Proctor, M. 2000. 'Ecological gradients, subdivisions and terminology of North-West European mires.' *Journal of Ecology* 88: 187–203.

10 The life and times of transgenics

Hugh Lacey

The transgenics (TGs) – or genetically modified organisms (GMOs) – that have been planted and harvested in agricultural practices are offspring of the liaison of technoscience and multinational agribusiness. Ambiguities linked with this liaison – having to do with harms, risks, benefits and alternatives, and how to investigate them scientifically – underlie competing narratives of the life and times of these TGs and the turmoil that has marked it. A legitimating narrative emphasizes their technoscientific parentage and alleged scientific support for claims concerning the benefits and risks of their rapidly spreading use and the alleged absence of alternatives. A critical narrative contests that there is such scientific support, highlights agroecological alternatives, and emphasizes that the socioeconomic environment into which TGs were born, in which great stock is placed on technoscientific innovation that contributes to economic growth, has been a fertile one for their rapid spread into agricultural practices in many parts of the world. By clarifying what is at stake in these competing narratives, we are better able to understand the ontology and distinctiveness of TGs.

Figure 10.1 Golden Rice refers to varieties of TG rice that have been engineered to contain beta carotene, a source of vitamin A. Since 2000 it has been undergoing field trials and is not yet available for agricultural use. The legitimating narrative holds that, especially in impoverished regions of the world, it will be able to deaf effectively with problems, such as blindness, caused by deficiency of vitamin A. The critical narrative disputes this, and regards golden rice as a simplistic, techno-fix solution to such problems. Thanks to the International Rice Research Institute (IRRI) for giving permission to use this image.

Transgenics – technoscientific objects

TGs owe their existence to technoscientific research, development and innovation (R&D&I) – to technologically accelerated science (molecular biology and genetics) and scientifically informed technology (biotechnology). Their origins lie in the celebrated discoveries that genomes of organisms contain sequences of DNA and that separating and recombining them is the major mechanism underlying such biological processes as sexual reproduction, and in the development of engineering techniques that have enabled an expanded range of these sequences to be recombined. These techniques involve inserting into the genome of a seed or plant tissue-culture DNA sequences, typically taken from organisms of unrelated species, which bring it about that the mature or growing plants acquire designated properties. TGs are modifications, induced by these (and potentially other) techniques,[1] of plants that exist in agricultural fields or natural ecosystems. Yet, although they are biological organisms, they could not have arisen by means of the mechanisms of natural selection or traditional cross-breeding used by farmers and conventional plant breeders. TGs are technoscientific objects.

It is a matter for theoretically informed experimental investigation and further development of the techniques of genetic engineering to discover what properties plants could be engineered to have. They certainly include herbicide/pesticide resistance and toxicity to certain pests, and if ongoing R&D is successful, they will also come to include such properties as providing sources of nutrition, producing higher yields and ability to grow in salt depleted, mineral deficient, dry or waterlogged soils. This R&D is motivated by the convictions – none of which gain credibility simply from the fact that TGs are efficacious products of well conducted technoscientific research – that it is advantageous for crop plants to have properties like these regardless of the mechanisms of their origins, that these mechanisms are irrelevant for assessing the risks and benefits of growing them and consuming their products, and that TGs are needed (i.e., they can provide benefits that overall surpass those of other available agricultural options).

The space of agricultural options

Scientific research relevant to determining whether TGs are needed would first have to address questions about the space of agricultural options (see Lacey 2005, ch. 10, 2015a, 2016, forthcoming). If available agricultural approaches – 'conventional', organic, biodynamic, agroecological, indigenous and others – were appropriately combined, used with locally specific adaptations, accompanied by viable distribution methods and informed by appropriate scientific research, would they be sufficiently productive to meet the food and nutrition needs of the whole world's population in the foreseeable future, sustainably so and relatively free from serious risks?

If the evidence supports 'yes', then TGs are not needed for these agricultural ends, Even so, it would be pertinent to ask: What would be the advantages, if any, and for whom and under what conditions, of granting a place (and how significant a place?) to TG-oriented approaches among the approaches in use?

If 'no', then a third question becomes pertinent: What are the limitations of currently available approaches, and does scientific research support that introducing TG-oriented farming (utilizing TGs of what plants and with what properties?) could overcome them?

R&D&I of TGs was not a sequel to research that addressed questions like these. It went ahead without much input from farmers, their organizations, agricultural scientists and others working to address the food and nutritional needs of poor people. Research dealing with the space of agricultural options was not conducted or even contemplated. Thus, TGs were not introduced in response to scientific evidence supporting, e.g., that they have a role to play in overcoming limitations of prevailing farming practices. Nevertheless, if TGs were to be engineered successfully to have such properties as containing sources of nutrition, ability to produce higher yields or to grow in salt depleted or mineral deficient soils, then it might be plausible to consider it a matter of priority to investigate their effectiveness (and the side effects of their use) in comparison with those of plants that can be produced by, for example, agro-ecological methods. The TGs actually being used in agriculture today, however, do not have these properties.

The interests of multinational agribusiness

Commercial interests have dominated the R&D that produced the TGs currently being planted and harvested for major crops. Early on, for technical, economic and strategic reasons, attention was given to developing and marketing TGs with the properties of toxicity to certain classes of insects or resistance to herbicides whose active ingredient is glyphosate. (Except in passing, I will explicitly discuss only herbicide resistant TGs.) Technically, it is simpler (and less costly) to engineer TGs of these kinds, and they can be developed more quickly than those (e.g., higher yielding plants) that, if they can actually be produced, will require more complex engineering techniques. Economically, they could be put to use quickly to recoup the costs of the R&D&I and begin to produce profits. Strategically, agribusiness corporations, as a consequence of gaining intellectual property rights (IPR) to genetic engineering techniques and newly developed transgenic varieties, have been able to use those protections to control most of the research that is conducted on TGs.

TGs (including varieties of corn, soy, canola, rice and cotton) of these two kinds and their products have been successfully marketed in many countries, especially to large-scale producers, manufacturers of processed food and companies that market food for growing livestock. In addition, raw and processed products of TGs have been successfully marketed to consumers in supermarkets throughout the world, but in the United States, for example, they are not labeled as TG products and marketed as such. Successful marketing to farmers is connected with the claims that glyphosate may be used efficaciously for weed control without harming the growing TGs; that using it with TG crops requires less use and fewer applications of pesticides – factors said to be conducive to higher yields or reduced crop losses, higher profits and a more congenial work environment. It is also said to be less toxic to human beings than most other available herbicides and more friendly to the environment since it dissipates rapidly in soils. Many farmers (not all), especially large-scale producers, based on their experience of using TGs, testify to benefits like these and continue to engage in TG-oriented farming. Others engage in it because (linked with the merging of seed companies with those that produce TGs) they find it difficult to access non-transgenic seeds and because the conditions needed for engaging in other forms of farming have been weakened.

TGs with herbicide resistance and toxicity properties exist, and are now widely used, because using them is deemed advantageous by agribusiness and its clients. Technoscientific R&D&I brought them into existence, and it provides evidence for the efficacy of their use under specified conditions. Nevertheless, this does not suffice to provide scientific backing for the social value or legitimacy of these uses of TGs, for the questions about the space of agricultural options, which involve ecological and social factors, fall outside of its compass. This kind of R&D&I lacks the methodological resources, for example, to appraise the potential impact of forms of farming like agroecology.

Agroecology admits no place for using TGs; it is aimed at satisfying simultaneously, and in a balance determined by farmers and their communities, a variety of objectives, including productivity, sustainability of agroecosystems and protection of biodiversity, health of members of the farming communities and their surroundings, and strengthening of their culture and agency (Altieri 1995). In many ways, it is a development of traditional agricultural approaches that is informed by scientific knowledge and the living record of its practitioners.[2] It exemplifies practices in the space of agricultural options with proven success and whose potential needs to be more fully investigated empirically, and it provides a point of contrast that enables us to discern more clearly what TGs are.

Commodities or renewable, regenerative resources[3]

Traditionally, crop seeds have been (for the most part) renewable regenerative resources, sources and parts of farmers' harvested crops, reproduced in and selected from crops using methods that have been adapted and improved by numerous farmers, informed by local, traditional knowledge that has been accumulating over the centuries. Crop plants, grown from seeds selected in traditional ways (and contemporary agroecological extensions of them), tend to be integral parts of sustainable ecosystems that generate products that meet local needs, and cultivating them is compatible with local cultural values and social organization. Crop seeds, which embody traditional knowledge (that can be consolidated, corrected and expanded in the light of practical experience and scientific knowledge), have been considered to belong to the common patrimony of humankind, available to be shared as resources for replenishing and improving the seeds of fellow farmers.

In recent times, ever intensifying efforts have been made to replace crop seeds that are predominantly renewable regenerative resources by seeds that, like currently used TGs, are for the most part commodities. As commodities, they have features and uses that are integrally connected with the availability of other commodities (e.g., chemical inputs and machinery for cultivation and harvesting); increasingly they are grown for uses other than human food production (e.g., biofuels), often implicated in regimes of intellectual property rights (IPR), developed by professional breeders and scientists and produced largely by capital-intensive corporations. They are not reproduced in and selected from farmer-harvested crops; they are not components of stable ecosystems, and they are not available to be shared freely with fellow farmers. The commoditization of the seed is an integral part of the transformation of the social relations of farming in the direction of the dominance of agribusiness and large-scale farming. It depends on breaking the traditional unity whereby seeds (renewable regenerative resources) are simultaneously sources and parts of crops and harvested grain is both source of food and seed for new plantings. This transformation predates the discoveries that enabled the development of TGs. It was initiated with the introduction of 'high-intensity' models into agriculture, models based on growing monocultures with consequent weakening of ecological sustainability and biodiversity, mechanization, the extensive use of chemical inputs and agrotoxics, and further developed by planting hybrids that do not reproduce reliably, so that new seeds must be bought regularly from seed companies. The introduction of TGs, and their being protected by IPR, have reinforced and intensified pressures for the breakdown of the traditional unity.[4] Furthermore, the IPR protections underlie a socioeconomic mechanism that, if needed, would function to block decisively even the remote possibility that TGs might themselves become new renewable regenerative resources; their manufacturers make use of the IPR to prohibit farmers selecting seeds from their crops and saving them for subsequent plantings.

Transgenic plants share many biological features with all plants, and unlike hybrids, some of them may reproduce reliably for some generations. However, there are biological mechanisms that virtually ensure that TGs cannot themselves become new renewable regenerative resources. First, it is unlikely that TGs would reproduce reliably beyond a few crop generations, for they are complex biological, ecological and social objects in a complex environment, open to multi-causal analysis implicated in various levels of organization, higher level properties, and feedback loops between higher and lower levels of organization. This kind of complexity involves deep uncertainty, for example, about possible switching of the location (or breaking up) of the inserted genetic materials across generations and about its consequences (Mitchell 2009, ch. 5). Second, any variety of TGs is likely to be usable for only relatively few generations. Consider, for example, crops that are resistant to glyphosate. Research supports that using glyphosate for spraying these crops is efficacious (it kills targeted weeds leaving the crop unharmed) for some generations. When glyphosate-resistant TGs were introduced, however, it was anticipated that in accordance with well known evolutionary mechanisms, after an unpredictable number of generations (that could be extended by using glyphosate in accordance with appropriate regulations and guidelines), weeds would appear ('superweeds') that themselves would be resistant to it.[5] Then, using these TGs increasingly becomes obsolescent. Consequently, if herbicide-resistant TGs are to be used efficaciously over the long haul, new varieties must be introduced regularly that are resistant to herbicides (successors to glyphosate, e.g., one known as '2,4-D') (Pollack 2014) required to deal with new generations of superweeds. In other words, a regular sequence of new {TG variety, herbicide} pairs is required, more generally of {TG variety, herbicide, fertiliser, . . .} 'n-tuple packages'.

TGs are typically components of n-tuple packages. What the components of the packages are, and the mechanisms leading to obsolescence of earlier varieties, varies with the kind of TGs involved. TG-oriented farming depends on regular innovations of new varieties of TGs and generally of n-tuple packages that contain externally provided inputs required for using them efficaciously. This exacerbates the breakdown of the traditional unity (that nurtures seeds as renewable regenerative resources), and furthers the dependence of agriculture on agribusiness, with questionable ecological and social consequences.

Risk assessments and the legitimacy of using transgenics

I indicated previously why priority was given to R&D&I of varieties of TGs that have herbicide resistant and toxicity properties. Once developed and their use confirmed to be efficacious, and – as normally expected of any biotechnological innovation – they had passed standard risk assessments [SRAs], they were rapidly introduced into agricultural practices. SRAs are empirical (laboratory or small-scale field) studies concerning the effects – described as potentially harmful for human health or the environment and occasioned by physical, chemical, or biological mechanisms – of implementing innovations in socio-economic practices and of their seriousness, probability of occurrence and capacity for being effectively regulated and thereby contained (Lacey 2005, ch. 9, 2015b).

Agribusiness and public regulatory bodies usually agree that, when the use of a variety of TGs has been confirmed to be efficacious, the legitimacy of using it and introducing it widely into agricultural practices depends, in addition, only on it passing an adequate array of SRAs and then being used according to regulations informed by the SRAs. Moreover, they generally maintain that the TGs actually in use have met this condition. In accordance with this, they do not engage in or take into account outcomes of research pertaining to the space of agricultural options. It is as if the quick, thinly mediated movement from efficacy to legitimacy gains its rationale from a covert, unarticulated, unchallenged ethical principle – the Principle of the Presumed Legitimacy of Technoscientific Innovations [PLT] (Lacey 2016) – that applies generally to technoscientific innovations:

> Ceteris paribus, it is legitimate to implement efficacious technoscientific innovations in social practices without delay, and even to tolerate a measure of social and environmental disruption in doing so, provided that, after adequate and sufficient research has been conducted, compelling evidence has not been obtained to demonstrate that the implementation would occasion serious harm or risk of it – and normally this condition is satisfied if the array of SRAs performed is judged to be satisfactory by technical experts in risk assessment.

The tumultuous times of transgenics

The 'common sense' of our times tends to take for granted that technoscientific innovations are indispensable for solving the big problems facing the world today, that valuable practical uses will soon be found for virtually any technoscientific innovation and that questions of the legitimacy of using innovations rest only upon their efficacy and the quality and sufficiency of the SRAs conducted on them – and so lie within the authority of technical experts in risk assessments. Concerning TGs, most of the experts, who consult with agribusiness and regulatory bodies, endorse the legitimacy and social value of using them. From this perspective, then, the R&D&I of TGs should have been uncontroversial.

Nevertheless, it has occasioned a lot of turmoil. TGs have experienced tumultuous times. The necessity, social value and legitimacy of using them has been questioned by an assortment of groups for a variety of reasons, and the resulting controversies tend to be marked by confrontational tactics and discourse often marked by rhetorical excesses (Lacey 2005, ch. 6). This should not obscure, and the critical and legitimating narratives that I will sketch should also not obscure, that the turmoil is principally about the social value of R&D&I of TGs – about the legitimacy of their immediate implementation, intensive utilization and widespread diffusion throughout the world in the agricultural practices that produce major crops, and about the place that should be accorded them (in relation to other forms of agriculture) in national and international agricultural policies (Lacey 2005, ch. 6).

The critical narrative

In the critical narrative that has emerged, the social value of using TGs is challenged, as also is the claim that there is sound scientific backing for the legitimacy of using them (Lacey 2005, forthcoming). The most fundamental proposal of this narrative, as I interpret it, is that better ways to farm (using agroecological methods), which could reap more significant benefits with less risk for most people, can be identified in the space of agricultural options (Lacey 2005, 2015a). It also cites empirical evidence that supports that using TGs (1) is irrelevant for meeting the needs of vast numbers of smallholder farmers especially in poor regions and for contributing to worldwide food security, and (2) is actually causing serious harm for agroecosystems and many poor rural communities. It occasions risks (i.e., potentially harmful consequences [Lacey 2015b]), concerning human health, the environment, social arrangements and worldwide food security, that lie outside of the purview of SRAs. Hence, these two conclusions are taken to indicate that SRAs are insufficient for adequate risk assessment and for informing regulations;[6] and, in view of the role played by PLT, that judgments made by regulatory bodies about risks (that only consider the results of SRAs) are not 'ethically neutral' technical judgments. Furthermore, the critical narrative maintains the central role accorded to TGs in many national and international policies does not take into account empirical investigation concerning risks (that cannot be dealt with in SRAs) and agroecological alternatives, or the interests of those who may be affected negatively by the policies – largely because political and economic power is exercised on behalf of interests of capital and the market in ways that (among other things) make it difficult to conduct independent research on TGs, and keep results of research that might run counter to the interests of the producers of TGs out of policy and regulatory deliberations.

According to the critical narrative, the deep roots of the turmoil lie in the fact that the introduction of TGs has not been informed by research dealing with the space of agricultural options, and that due attention has not been paid to the kinds of scientific methodologies needed to conduct such research and adequate risk assessments. This will be elaborated in my critical comments on the legitimating narrative in the section, 'The two narratives in critical interaction'.

The legitimating narrative

For the most part, agribusiness and regulatory/policy bodies simply dismiss the claims made in the critical narrative; they tend to take PLT for granted, and so they boil issues of legitimacy down to matters of efficacy and the technical adequacy of the SRSs. They appeal to specialized science to provide a cover of legitimacy for the offspring of the liaison of technoscience and agribusiness. Moreover, they do not concede the ethical high ground to the critics. Deeply rooted in the 'common sense' of our times, in which PLT is secreted, they consider it virtually an ethical imperative to prioritize technoscientific 'solutions' for the big problems of the world (e.g., hunger in poor countries), as well as for any harmful effects that may be caused by technoscientific innovations themselves (e.g., environmental damage), and an ethical failing to cast doubt on the potential or the legitimacy of R&D&I that may lead to such 'solutions' (Lacey 2016).

This 'common sense' provides the background and key categories for the unfolding of the narrative in which the social value of TGs (GMOs)[7] is articulated (in the press, at public hearings, on websites of agribusiness corporations, etc.). In addition to claims about the benefits of using TGs (see the "The interests of multinational agribusiness" section), the legitimating narrative incorporates claims like the following:[8]

1 GMOs are the latest development in a long line of genetically modified organisms going back to the beginnings of agriculture. They are biological organisms just like – 'substantially equivalent to' – their predecessors, and they need no more scrutiny and regulation than new varieties introduced by older methods of selection and cross-breeding.
2 GMOs are novel technoscientific inventions – produced in the first instance, not by biological mechanisms, but by techniques of genetic engineering informed by knowledge and skills gained in molecular biology and biotechnology – and so, unlike their predecessors, they and the techniques that produce them can be incorporated into regimes of intellectual rights (IPR).
3 R&D is well advanced on GMOs with a range of 'highly desirable' properties (e.g., providing sources of nutrition and ability to grow in mineral depleted soils), so that potentially using GMOs could provide benefits for all farmers, as well as solutions to hitherto intractable problems of hunger and malnutrition and of dealing with certain kinds of pests. Herbicide resistance and insect toxicity are prioritized in the 'first generation' of GMOs; these have provided a kind of testing-ground for the kinds of GMOs that will follow, which require more complex engineering techniques, and so take more time and investment of resources to develop.
4 Agribusiness corporations are contributing to deal with problems of hunger and malnutrition, by licensing (frequently free of charge) the use of patented materials for developing crops with some of the properties (mentioned in item 3) to research institutions like CGIAR that aim "to reduce poverty and hunger, improve human health and nutrition, and enhance ecosystem resilience through high-quality international agricultural research, partnership and leadership."[9] The GMOs that these institutions develop have nothing to do with profits for agribusiness; they may be given as 'gifts' to the farmers; and (in some cases) farmers may be permitted to save seeds from the harvest for planting for future harvests.
5 GMOs occasion no serious risks to health or environment that cannot be managed under scientifically informed regulations, and empirical studies confirm that those already released for commercial use, having passed an appropriate array of SRAs cause no significant harm (cf. the 'Risk assessments and the legitimacy of using transgenics' section).
6 Unless GMO-oriented farming becomes widespread, there is no way to provide for the food and nutrition needs of the world's population over coming decades – there is no alternative to continued and prioritized R&D&I of GMOs that does not occasion the risk of massive hunger throughout the world.[10]
7 Items (1), (3), (5) and (6) have the backing of science.

The two narratives in critical interaction

In this section, principally by exploring some of the implications of the legitimating narrative, I will identify key points at which the legitimating and critical narratives come into sharp conflict.

Dualist ontology

TGs are, among other things, biological organisms that under appropriate conditions will grow into mature plants from which, for example, grain will be harvested, and products of genetic engineering, not of mechanisms of natural or farmer-aided selection. The legitimating narrative maintains that, as biological organisms, TGs are just like any other crop plants (item 1); but, as technoscientific objects, they are unlike them (item 2). It incorporates a kind of dualist ontology: in the context of risk assessment, TGs are said to be just like any other crop plants; in commercial contexts they are hailed as novel technoscientific inventions, for which (unlike for other types of crop plants) intellectual property rights (IPR) may be claimed. What they are in the commercial context (products of genetic engineering) is considered irrelevant to what they are in the context of risk assessment, as if their different kinds of origins must make no difference to their functioning as plants, and to their (possible) effects on human beings and the social and ecological environment. Hence, according to the narrative, risk assessment (item 5) does not involve investigating the effects (on people, social arrangements and ecologies) that using TGs might occasion in virtue of mechanisms that derive from their being commercial objects and property. Just as Descartes's dualist account of human beings required keeping the mind out of accounts of the causal order of the material world, so does the narrative's dualist ontology of TGs serve to keep their commercial and property aspects out of causal analyses pertinent to harms, risks and alternatives. It also contributes to isolate the alleged role of TGs in solving problems connected with hunger and malnutrition (item 6) from investigations of the social causes of the problems and their persistence, and thus to undercut investigations concerning whether or not using them may strengthen corporations that are themselves integral components of the very socioeconomic system that maintains and deepens these problems (Lacey 2005, ch. 8).

Are transgenic seeds just like seeds used in agroecology?

TGs are typically components of n-tuple packages (see the "Commodities or renewable regenerative resources" section). The efficacy of using them depends on planting them, not in potentially sustainable ecosystems, but in ecosystems that receive and continue to receive the required industrially produced inputs. This cannot be ignored when they are used by farmers, and it is important for commercial considerations. Based on appeal to item 1, however, it too is ignored in risk assessments, as is the fact that the spread of TG-oriented farming must undermine the conditions needed for preserving the traditional unity of seeds as both sources and parts of crops. Contrary to item 1, TGs are not just like seeds and plants used in farming practices (like agroecology) that aim to preserve seeds as renewable regenerative resources. The two kinds of farming are incompatible. With the deployment of TGs on an ever larger scale over an extended period of time, the conditions for practicing, for example, agroecology would be further eroded. Then, not only would it remain the case that the productivity (as well as sustainability, capacity to provide food security for everyone, etc.) of TGs has not been empirically compared with that of farming practices like agroecology, but also it would become less and less possible to engage in empirical research that could make such comparisons.

The safety of transgenics – scientifically based or convenience for agribusiness?

TGs are outcomes of genetic engineering (item 2), not of natural or farmer-aided selection. Appeal to item 1 serves to ground ignoring the different mechanisms of the origins of TGs and the seeds used, for example, in agroecology when discussing items 5 and 6, although research that could test the 'substantial equivalence' of their respective products (item 1) would have to take into account the considerations about complexity mentioned previously (see the "Commodities or renewable regenerative resources" section). Because of the biological complexity of TGs and their being components of n-tuple packages, SRAs need to be conducted one-by-one for each {TG-variety, environment} of use, and to be accompanied by ongoing monitoring of their actual uses, partly to pick up potential oversights of the SRAs that have been conducted and new risks that might arise, and partly because environments change in response to agricultural use. Although these features of well-conducted SRAs are generally acknowledged by regulatory bodies, they are downplayed in practice – and also in the legitimating narrative, which claims that using TGs is safe (item 5) and will remain so for the new varieties that will be regularly introduced in the future. Since the efficacy of using particular varieties of TGs is likely to be short lived ("Commodities or renewable regenerative resources" section), claiming (without significant qualifications) that TGs do not occasion serious unmanageable risks serves to deflect concerns about the safety of future varieties. This claim is also part of the argument that developments of TGs with herbicide and toxicity properties are just initial steps (item 3) that demonstrate that using TGs can be efficacious and free from serious risks, and that provide the technical context for testing and perfecting more complex engineering techniques that are fundamental for realizing the promised potential of R&D&I of TGs (item 6). SRAs need to be conducted one by one for each {TG variety, environment} of use; however, those conducted on currently used TGs[11] cannot provide evidence that the TGs of the future will be safe. Asserting item 5 reflects commercial, economic and political policies; it is not backed by the results of well-conducted scientific research (contra item 7).

Objects that embody the values of capital and the market

The TGs actually in use are commodities and/or objects for which IPR may be held (item 2). IPR serve as instruments for ensuring profits. . . . The legitimating narrative maintains emphatically that holding IPR to TGs is indispensable for their development; without them, agribusiness corporations would not provide funds for R&D&I of TGs, for they would not readily be able to protect their investments and rapidly recoup the costs of the R&D&I.

While not highlighted in the narrative, IPR are also deployed to gain competitive advantage, to gain control both of R&D&I of TGs and of as many aspects of the agricultural economy as possible, and to prevent unauthorized (i.e., independent) research on the risks of using TGs. These deployments are crucial for understanding the social environment that has nourished the development of TGs. Agribusiness corporations – by developing the 'first generation' of TGs rapidly (item 3) and (in the process) gaining IPR protections to numerous varieties and genetic engineering techniques – have been able to shape and largely control the research agenda of TGs. Furthermore, in the name of item 6, they have used their power to push for control of the whole agricultural research agenda, so that TG-oriented farming is prioritized and pressures are exerted (not without resistance) to sideline other forms of farming. It is of the nature of these TGs to spread widely throughout the world by way of mechanisms of the market and intellectual property.

Item 4 maintains that TGs may be used in service to a variety of values, not only increasing profits for agribusiness and its clients, but also values like those pursued by CGIAR (item 4). To date, few of the kinds of TGs promised for dealing directly with the needs and problems of impoverished areas have actually been introduced commercially. Leaving that aside, however, the more fundamental point is that the range of values that TGs currently in use, and those anticipated, may come to serve is inherently limited because these TGs embody values of capital and the market. This is reinforced by the fact that TGs cannot be used without inputs that usually are only available commercially, and that introducing the TG-oriented farming in poor countries requires also some degree of penetration of the institutions of capital and market – furthering the breakdown of the traditional unity (that nurtures seeds as renewable regenerative resources), and undermining the social relations that go hand in hand with practices that incorporate it. TGs, developed to serve values of poverty reduction and the like, are not exceptions, for the research that produces them cannot proceed without licensing agreements (which are vulnerable since they may be revoked) with the corporations.

Organizations like CGIAR often criticize agribusiness for excessive concern with profits and for not giving enough attention to the needs of the poor, and they do not prioritize research on TGs to such an extent that all alternatives in the space of agricultural options are completely sidelined. Nevertheless, like those engaged in directly commercially motivated research, they tend to presuppose a negative answer to the first of the questions about the space of agricultural options (the "The space of agricultural options" section), and to presuppose that TGs are not only indispensable for the agriculture of the future (item 6), but also that many immediate problems and needs cannot be addressed without developing and using appropriate TGs. These presuppositions lead to prioritizing research related to TGs and leave few resources for investigating other alternatives in the space of agricultural options. Consequently, risks that cannot be assessed in SRAs –that using TGs threatens to undermine the conditions (e.g., the traditional unity) for engaging in forms of farming, such as agroecology, especially suited for smallholder farmers in poor countries – will not be adequately investigated. Then, relevant kinds of evidence will not be brought to bear on item 5, so that it will remain inadequately tested empirically. And item 6 is simply presupposed (perhaps because it seems to fit so well with the 'common sense of our times'). Certainly, like item 5, it is not supported by scientific evidence; and it could not be, unless R&D&I of TGs were to become embedded in research that addresses the space of agricultural options. The seeds developed in organizations like CGIAR may be used outside of the dominant market mechanisms, but their existence and the forms they have depend on the commercially instigated research that has been conducted, and on how they fit into national and international policies tied to economic growth – and they remain subject (to greater or lesser degrees) to claims of IPR. They embody the same kind of knowledge as commercially exploited TGs, and like them embody (albeit to a lesser degree) values of capital and the market (Lacey 2015a).

Must TGs embody values of capital and the market? If research on the space of agricultural options had shown that some varieties of TGs were needed in some environments, and if the resulting R&D were not dominated by institutions linked to policies that seek innovations that contribute toward economic growth, the life and times of TGs might have been different. Even so, it would remain that TGs are components of n-tuple packages and that agricultural approaches like agroecology have enormous unfulfilled promise. Nevertheless, it cannot be precluded a priori that some types of TGs (perhaps not yet anticipated) might have a substantial place in the agriculture of the future.

Does the use of TGs have the backing of science?

The legitimating narrative puts the authority of science behind the R&D&I of TGs, appealing not only to the technoscientific parentage of TGs, but also to science for giving a semblance of legitimacy to the offspring of its liaison with agribusiness. Item 7, therefore, has an important place in it. I have maintained, however, that item 5 and 6 are not well supported by scientific evidence. The conviction that they are well supported reflects that the narrative draws upon an impoverished conception of science, albeit one widely held in mainstream scientific institutions.[12] According to this, scientific research is conducted under 'decontextualizing strategies' (DSs) (Lacey 2012, 2016): theories and hypotheses are constrained so that they are able to represent things and phenomena as being generated from their underlying structures, their processes and interactions and those of their components, and the laws governing them; and empirical data that are sought for and recorded are largely quantitative, obtained by means of interventions with measuring instruments, and often of phenomena in experimental spaces. DSs dissociate the phenomena investigated from their human, ecological and social contexts, from any links they have with ethical and social value. If 'science' is limited to the use of DSs, then – provided that a sufficient array of SRAs (which deploy DSs) are passed – there is no 'scientific' evidence that serious risks (that cannot be contained by enforced 'scientifically' based regulations) are occasioned by using TGs (item 5). Moreover, 'science' cannot assess any alternatives to TGs other than those that rely on the input of research conducted under DSs; and of these alternatives (that include agrotoxics-intensive conventional agriculture) TGs may indeed be superior. That is what the 'scientific' backing for using TGs amounts to (see Lacey 2005, 230–235).

DSs are inadequate for investigating the consequences of using TGs, qua commercial objects and property, however, and so for investigating risks (potentially harmful consequences) for human beings, social arrangements and ecological systems that may be occasioned by socioeconomic mechanisms. Thus, the absence of 'scientific' evidence obtained under DSs for the existence of risks is never sufficient reason for denying that there are risks incurred by using TGs (Lacey 2005, ch. 9, forthcoming). Similarly, DSs are inadequate for investigating the possibilities of agroecosystems that are cultivated in accordance with the objectives of agroecology.[13] But, without investigating them, the questions raised about the space of agricultural options cannot be answered on the basis of empirical evidence, and item 6 could not become well established. The outcomes of research conducted more or less exclusively under DSs cannot be decisive in the context of actual agricultural practice – where risks are occasioned in virtue of all the kinds of things that seeds and plants are, and where there is plenty of empirical support for the benefits of practices like agroecology that are integrally connected with the contexts of their use.

To investigate risks and alternatives adequately, context-sensitive strategies (CSs) need to be adopted (complementing, not doing away with, DSs), where knowledge and understanding gained under them is held to the same standards of testing as those utilized under DSs. The strategies of research in agroecology provide exemplary instances (Lacey 2015a, 2016). Research conducted under CSs confirms that there are risks of using TGs, including that of undermining forms of agriculture that offer better promise of being responsive to food security issues throughout the world, and it underlies the claim that alternative approaches such as agroecology should not only not be discarded but given priority support (Lacey 2015a). The legitimating narrative is permeated through and through with the conception of science as using only DSs; and also with the widely shared presupposition that is linked more generally with technoscientific developments (Lacey 2012) that science (so conceived) not only is shaping the future, but is the key to a better future. However, this presupposition could not be supported without investigation that uses some CSs. Challenging item 7 of the legitimate narrative in this way shows that some of the items and presuppositions of the legitimating narrative can only be held dogmatically, and that the narrative lacks the categories needed to grasp fully and clearly what TGs are.

Concluding remarks

What are TGs? They are technoscientific objects that embody scientific knowledge gained under DSs. They are biological organisms, realizations of possibilities discovered in research conducted under DSs (in molecular biology, genetics and biotechnology) brought to realization by means of technical, experimental and instrumental interventions. As such, they are components of social and ecological systems that embody values of technological progress (Lacey 2005, 17–28, 2012). They are also normally components of 'n-tuple packages', whose other components are essential in the immediate agroecosystems in which they are grown, as well as parts of an agroecosystem (the market) with worldwide dimensions, in which they are commercial objects whose uses are constrained by claims of IPR. As such, they embody values of capital and the market.

TGs have effects on human beings, social arrangements and ecological systems in virtue of all of the kinds of things that they are. Using them consolidates breaking the traditional unity (seeds both sources and parts of crops; harvested grain both source of food and seed for new plantings), and destroys the network of social relations linked with maintaining it. What TGs are cannot be grasped on the basis of the scientific inquiry using only DSs and knowledge that underlies their coming into being and that explains the efficacy of their use. Furthermore, that there are fundamentally different alternatives to using them gets no recognition or empirically based rebuttal in the legitimating narrative, and so those immersed in it also cannot grasp the intelligibility and scientific foundation of the alternative possibilities (especially agroecology) from which the most significant criticism comes. The proponents of the widespread use of TGs cannot understand the turmoil that marks the life and times of TGs.

Notes

1 Techniques not based on DNA recombination are now being developed – for example, one that leads to suppressing the expression of genes that enable the reproduction of certain viruses involves inserting double stranded (ds) RNA into a plant's genome.

2 Altieri (1995) and Vandermeer (2011) are key sources on agroecology, evidence for its productive successes and further potential, and its special suitability for farming in impoverished regions. See Lacey (2005, ch. 10) for more details and references.

3 The analysis of the next two paragraphs derives largely from Kloppenburg (1987) and from Shiva (1991). See Lacey (2005, ch. 7).

4 The breakdown of the traditional unity is not complete, and efforts to strengthen it are growing throughout the world – especially fostered by movements who aspire to 'food sovereignty', for whom agroecology is the preferred approach to farming (Lacey 2015a).

5 These TGs were introduced in the mid 1990s, and by 2010 superweeds had begun to cause problems. See Benbrook (2012).

6 The critical narrative also includes allegations that the SRAs dealing with TGs that have actually informed the decisions of regulatory bodies have been marked by numerous scientific shortcomings, including that they do not take into account that TGs are parts of n-tuple packages (Lacey 2016, forthcoming).

7 The term 'transgenics' is intended in the critical narrative to connote that the kind of 'genetic modifications' involved in the production of these organisms could not occur in accordance with the mechanisms of natural or farmer-aided selection. Because the legitimating narrative highlights that they are, like those produced by farmer-assisted selection procedures and indeed any organism at all, modifications of already existing organisms, it refers to them as 'genetically modified organisms' (GMOs). As used in this article, 'TGs' and 'GMOs' are coextensive. I will use 'GMOs' only in this section.

8 The following claims are referred to in the text as 'item 1' to 'item 7'.

9 CGIAR, Consortium of International Agricultural Research Centers, http://en.wikipedia.org/wiki/CGIAR (accessed July 17, 2016). 'Golden rice' is the most celebrated instance of TGs being developed by CGIAR.

10 This claim is celebrated in high profile events – for example, in 2013, three scientists (connected with agribusiness) were awarded the World Food Prize for their significant contributions to the development of GMOs. See http://www.worldfoodprize.org/en/laureates/2010__2015_laureates/2013__van_montagu_chilton_fraley/ (accessed July 17, 2016). In contrast, in the same year the competing Food Sovereignty Prize, http://foodsovereigntyprize.org/the-honorees/ (accessed July 17, 2016), was awarded to three groups: Basque Country Farmer's Union, National Coordination of Peasant Organizations (Mali), and Tamil Nadu Women's Collective (India), that represent forms of farming present in the space of agricultural options that provide evidence for the productivity and sustainability of appropriately managed and developed traditional forms of farming, and that play key roles in addressing food security issues in impoverished communities.

11 This leaves aside questions about the alleged shortcomings of TGs; see also note 6.

12 This conviction is endorsed by many distinguished scientists (Lacey 2016, forthcoming).

13 See note 2.

References

Altieri, M. A. 1995. *Agroecology: The Science of Sustainable Development*. Boulder: Westview.

Benbrook, C. M. 2012. 'Impacts of genetically engineered crops on pesticide use in the U.S. – the first sixteen years.' *Environmental Sciences Europe* 24: 24–37.

Kloppenburg Jr., J. 1987. *First the Seed: The Political Economy of Plant Biology 1492–2000*. Cambridge: Cambridge University Press.

Lacey, H. 2005. *Values and Objectivity in Science; Current Controversy about Transgenic Crops*. Lanham, MD: Lexington Books.

———. 2012. 'Reflections on science and technoscience.' *Scientiae Studiae* 10: 103–128.

———. 2015a. 'Food and agricultural systems for the future: Science, emancipation and human flourishing.' *Journal of Critical Realism* 14 (3): 272–286.

———. 2015b. '"Holding" and "endorsing" claims in the course of scientific activities.' *Studies in History and Philosophy of Science Part A* 53: 89–95.

———. 2016. 'Science, respect for nature, and human well-being: Democratic values and the responsibilities of scientists today.' *Foundations of Science* 21 (1): 51–67.

———. forthcoming. 'Views of scientific methodology as sources of ignorance: Controversies over transgenic crops.' In *Agnotology: Ways of Producing, Preserving, and Dealing with Ignorance*, edited by Kourany, J. and Carrier, M.

Mitchell, S. D. 2009. *Unsimple Truths: Science, Complexity and Policy*. Chicago: University of Chicago Press.

Pollack, A. 2014. 'Altered to withstand herbicide, corn and soybeans gain approval,' *New York Times*, 17 September, http://www.nytimes.com/2014/09/18/business/altered-to-withstand-herbicide-corn-and-soybeans-gain-approval.html?ref=business (accessed July 17, 2016).

Shiva, V. 1991. *The Violence of the Green Revolution*. London: Zed.

Vandermeer, J. H. 2011. *The Ecology of Agroecosystems*. Boston: Jones and Bartlett Publishers.

11 Cardboard

Thinking the box

Cheryce von Xylander

"This is the record of a box man. [. . .] A box man, in his box, is recording the chronicle of a box man" (Abé 1974). These words introduce the protagonist of a novel set in modern Tokyo, the tale of a street person who finds himself contained by the containers that we use for the containment of goods. His fate implies an existential condition and suggests, specifically, that in making the cardboard box, we may have remade ourselves in its image.

The present essay engages this literary conceit from a technoscientific perspective. Is it possible that paper packaging brought a radically new spin to an age-old cultural practice, to wit the paper-based mediation of ideas and things? To the extent that science is vested in propositional representations of the world, it is also wedded – albeit not in techno-monogamy – to a mediation of ideas and things that relies on pen and paper. With a new paper product, however, namely the cardboard box as a container of goods for the industrial age, came a new way of relating ideas and things. The cardboard box may signify a form of materiality that is integral to the very idea of technoscience. If so, a philosophy of technoscience would disregard this object at its own peril.

Some facts and figures will help put matters in perspective. Together with paper, cardboard is the single largest source of trash in America: over 27 percent of the 251 million tons of municipal solid waste produced in 2012 (measured in bulk, the percentage would be even higher) (U.S. Environmental Protection Agency 2012, 5). This figure, impressively massive, does not address significant changes taking place within the pulp and fiber market. In Germany, where statistics for paper and cardboard are tracked separately, the direction things may be headed is easier to detect. According to the German Pulp and Paper Association (VDP), over eleven million tons of paper packaging materials (i.e., cardboard) were produced in Germany in 2014 – that is, 1.2 thousand tons over 2013 – while the amount of graphic paper produced over the same period decreased (Verband Deutscher Papierfabriken e.V. 2014). This shift in German consumption patterns reflects similar trends in other developed countries. With online commerce, deployment of paper packaging is on the rise worldwide (Chiang 2014). The expected rate of increase in this sector is likely to outstrip any savings concurrently prompted by the paperless office.

The box folds out of cardboard: it is, strictly speaking, an iteration of the history of paper (one materiality along a technological continuum). Yet semblances can deceive. The traditional 'republic of letters' stacks up quite differently than the emerging 'republic of cardboard' and, as to be shown, the citizens of these communities may be as unlike one another as the 'Flatlanders and Spacelanders' in Edwin Abbott's novel *Flatland* (1884).

In 2002, when *The Myth of the Paperless Office* appeared, the authors determined that paper, due to its exceptional 'affordances', was here to stay. Despite the rise of digital technologies, they anticipated no reduction in the amount of paper being used. Their argument hinges on its popularity, specifically on all the things "people can do with paper" (Sellen and Harper 2002, 17). But the proliferation of handheld, electronic devices tipped the balance; recent findings suggest that, contrary to this pre-Web 2.0 prediction, paper is indeed being phased out of office life (The Global Community of Information Professionals 2014). This need not diminish paper 'emissions' when operative logics stipulate that *less* means *more*: the paperless office is premised on an enabling mobility of things and that, in turn, relies on transmissions effected by means of the cardboard box. Items standardly complete a circuitous, if not irrational, route on the odyssey to a final destination. Untold box-junctions punctuate the meandering journey, repackaging and recombining takes place along the way. So, whereas the precise 'affordances' favored over time can vary – increasingly, they pertain less to sheet paper (used in managing office routines) and more to paper receptacles (used in managing the logistics of commerce) – for the time being, our profound reliance on paper's materiality persists. The authors adjudged rightly in getting it wrong: that the paperless office equates to a paperless lifeworld is a myth.

'Adult-to-kid transmogrifier'

Cardboard's pervasiveness seems remarkably unremarkable. Brown-gray, pedestrian and ever present, it's the canonical stuff-without-qualities – to the modern sensorium, a non-entity best bracketed. We deploy this man-made material almost wholly toward one end, namely to handle the logistics of exchange. It figures in all that is covetable and coveted (the stockpiling of commodities, long-distance haulage, mail-order delivery, retail presentation) but also, importantly, in that supra-mercantile context of dishevelment, the perpetual turmoil of re-location, be it personal or professional.

For clarification, 'cardboard box' designates a class of vessels having two manifestations: the slim, paperboard variant of the cereal box and the bulkier, padded variant of the hauling box. Both box types come in fixed and foldable formats. The tough, brown paper used to fabricate boxes of both types dates back to the 1870s, when the addition of sodium sulfide made it possible to produce a stronger, more resilient paper. Unlike lightweight sheet paper, whose functionality depends on the makeup of its surface – which for over two millennia (in China, less than half that long in Europe) has served as a carrier of symbols – heavyweight carton paper folds into three-dimensional shapes, capable of creating a protective, transitory space. For goods can only be shifted from here to there in receptacles, which form a third element in the transaction. This mobile placeholder is, literally, a holder-of-place in that it affords portable sanctuary and, thereby, suspends the trade-obstructing constraint of the fixed locality. Its de-contextualization is reminiscent of Bruno Latour's notion of the "immutable mobile" – a term he uses to explain how embodied, localized practice becomes black-boxed, figuratively speaking, and thus transferrable to other places – except that cardboard consigns context per se (not specific, regional practices) to a mobile neverland (Latour 1987, 223).

The manufacture of paper receptacles was not transformative from the outset. Made by hand, these receptacles were of marginal, commercial interest. The significant entrepreneurial breakthrough came with the ability to industrialize the making of paper packaging – this serviced and facilitated an emergent mass-market. Demand for ever more non-permanent receptacles grew rapidly, changing the dynamics of both consumption and production. Boxes in all shapes and sizes came to epitomize the market's transcience – each a microcosm of an increasingly commodified planet. Made-to-order deliverability necessitates the coordination of workflows across institutional divides and professional cultures, forcing the production sector to be in conversation with marketing departments, transportation systems, regulatory bodies, law-enforcement agencies, labor unions, vocational associations and so on. In short, boxing not only wraps goods; it intertwines the institutional loops in which those goods are manufactured and distributed.

Cardboard has proven an indispensable lubricant of global capitalism in the twentieth century. This consummate go-between escorts a constant conduit of things on orb-pulsing surges of supply and demand. The tally, in hindsight, is staggering. Consider the paper trail of trade transactions carried out since the box became a staple of modernity – mindboggling. Our species survival depends on consuming a much-ness sufficient to ensure further consumption: pandemic box failure would expose the geo-political pyramid scheme whose dialectic impels us to exacerbate the very crisis we strive to avert. From pizza-in-a-box to the infamous big-box store, cardboard structures the human-built environment. The big-box store may be a suburban abomination to some but the sobriquet speaks volumes about real-existing capitalism. Wikipedia once proclaimed that the term referenced the box-like shape of the stores themselves (Wikipedia 2012). Did the hive mind let slip that the big-box store had spearheaded a distinctive retail aesthetic – on-site merchandise presented in the trappings of its travel-worn delivery? Here the context of production engulfs that of presentation, the hauling box doubles as display case, and bargain-hunters on wheels make good on the *car* in cardboard.

Significant indicators suggest that the cardboard era of human history, or, if you will, the era of cardboard-humanity, may be as evanescent as previous ages of man. Paper-appliancing could prove but a transitional stage in the natural history of industrial evolution as other kinds of packaging win out. A 1998 study conducted by Germany's Federal Environmental Protection Agency (*Umweltbundesamt*) compared the environmental impacts of standard carrier bags and found that, contrary to popular belief, paper bags recycle less efficiently than plastic ones (Krieg 2004, 99). And, indeed, since 2003, the paper container seems gradually to be losing out over its plastic rival; a 2 percent lead of synthetic bags over paper ones by Europe's export champion may presage a global trend (Krieg 2004, 99). Palaeontologists, looking back from the future, could discern an 'Age of Cardboard', analogous to the Iron or Bronze Age in that it will be defined by the characteristic tool-making of the period but differing, significantly, in that, analogous to the Ice Age, it will correlate with a changing environment on earth. The relevant 'tools' having rotted away, cardboard may leave no physical trace of itself. Will this exonerate the sad biped that wrought planetary transformation and rendered *obscene* its self-styled Anthropocene (von Xylander 2015)?

Truly remarkable, all things considered, is that cardboard could ever have gone unremarked. After successfully hiding in plain sight for so long, it is now staking a claim in the pantheon of culture, rehabilitated as a card-carrying patron of the arts and sciences. Due, perhaps, to the variety of contents it daily conveys, the cardboard box is increasingly perceived to be a physical metaphor for creative possibilities – both cornucopia of and catalyst for imaginative thought – in short, a generalized idea stimulator. In 2005, this commonplace was inducted in the 'Toy Hall of Fame' of the National Museum of Play in Rochester, New York (National Toy Hall of Fame 2015). Viral uptake in 2012 of 'Caine's Arcade' – the video of a gambling arcade constructed from defunct cardboard boxes by a nine-year old boy in East Los Angeles – inspired the San Francisco Exploratorium to launch 'scientific creativity' competitions where the most innovative cardboard assemblages receive a prize (Isaacson 2012). These events have been franchised to participating schools worldwide (Imagination Foundation 2015). The comic series *Calvin & Hobbes* offers additional evidence of a sea change. Their pictorial pranks cast the cardboard box as the consummate empowerment tool: whether as a "cerebral enhancetron" (mind-enhancing instrument) or as "adult-to-kid transmogrifier" (a time-travel capsule and wish-fulfillment chest; see Wikipedia 2015a). Emphasis is invariably on the box's emancipatory thrust. But this same technology has historically functioned as a disciplining instrument, too. Industrial paper containment promotes a complex system of tutelage perpetuating the modern economic world order, a *dispositif* of consumption, to speak of Foucault. So, to euphemize, it functions philogenetically as a *kid-to-adult transmogrifier*.

The coercive power of industrial paper has entered the collective unconscious and spawned a design-led fairy tale modeled on *The Ugly Duckling*. Pictorially, this analogy was introduced in a 1940s advertisement for the cardboard box conceived by that modernist visionary of commercial promotion, Bauhaus graphic artist Herbert Bayer (1900–1985). His poster boasts a split image with an amorphous duck, made of discarded paper debris, on the upper left and a shapely swan, wrapped around a pristine, recycled cardboard box, on the lower right (Bayer 1939). A similar metamorphosis powers the ideology of design-think: *Out of the Box*, a lavishly illustrated, folio-sized paean to commercial packaging (by a Berlin-based design-book publisher), propounds the faith. Here the kingdom of signs matures into a beautiful swan conceived as the empire of embodiment: "Once upon a time, there was a retail landscape. It was a flatland of logos and print ads, a two-dimensional terrain that one day began to morph into a three-dimensional brandscape and this is where the new commerce began" (Klanten and Bolhöfer 2011, 4). Said brandscape's three-dimensional phenomenology implies not only a sign world of reference but also a haptic world of artifacts, contingent on the technoscience of carriage: the swan epitomizes the box and the unique economy of desire that the box has made possible.

Lewis Mumford (1895–1990), foremost American technology prophet of his time and self-proclaimed arbiter of New World modernism, saw the writing on the wall, if you will, in 1939, the early days of the box conversion. Of paper packaging, he wrote, it has "revolutionized modern society almost as profoundly, perhaps, as coal, steampower, and oil." The cardboard box – lightweight, robust, foldable, recyclable and cheap – has proven a true marvel of versatility. It exemplifies the very spirit of consumer society "to observe in detail the successive stages that marked the development of the paper container is to obtain insight into the rise of institutions so apparently remote as the advertising agency, the chain store, and the modern apartment house." Not only emblematic, it is positively constitutive of metropolitan life, a space-saving wonder that has made urban congestion tractable: "When we watch the painful creep of trucks through the side streets of New York, we should reflect that, if the folding box had not been invented, the congestion would be almost twice as great." His sights ever fixed on endangered technics (technological practice, in Mumford-land, tracks the rise and fall of civilizations), he did allow that such "paper foundations will not, perhaps, last forever" (Smith 1939, v–viii).

Packaging science

A technoscientific object can be both profound and profane. Besides being ordinary and quotidian, cardboard seems emphatically *low* tech. On strength of appearance, it might readily be subsumed in the recent spate of popular books (Henry Petroski, Bill Bryson, Roger-Pol Droit) on such quaint marvels of engineering as the pencil, paper clip, toothpick, sunglasses, suitcase and so on. These treatments delight in viewing the utensil under investigation in gloriously myopic isolation, impressed by how the objects in question have held their form. Such constancy seems profound in that it defies the flux of history. The regular slotted container (RSC) – that is, the standard corrugated box – "has changed little from that used in 1914" (Twede 2007, 243). As per Wikipedia, its specifications are as follows: "all flaps are the same length from score to edge" and, typically, "the major flaps meet in the middle and the minor flaps do not" (Wikipedia 2015b).

It also renders the cardboard box an improbable candidate with which to conceptualize technoscience. Whatever else one may take technoscience to mean, the notion implies cutting-edge developments in science and technology – a criterion that does not bode well for the technoscientific profile of cardboard. Of course, high-tech machinery and engineering expertise enter into its industrial production. But it shares that characteristic with mass-produced commodities in general. Like PlayStation, another object discussed in this book, cardboard manifests as technoscience-in-the-home. We encounter an unchanging object – but, behind the scenes, there is a continual reworking of the production process; sophisticated methods of manufacture are constantly optimized for efficiency. This increasing rationalization remains hidden from view.

To further complicate things, the meaning of technoscience itself is highly contested. Leading figures in science studies – Bruno Latour, Donna Haraway, Don Ihde, to name but a few – use it to describe a condition of profound entanglement between human beings and machines in which conventional distinctions nature/artifice, practice/theory, fact/fiction no longer apply. Others, like Michel Serres, provide a sophisticated conceptualization of the technoscientific connection – after we have mastered nature, he asks, "How can we master our own mastery?" – without, however, programmatically enlisting the term (Serres and Latour 1995, 172). The information society depends upon knowledge-systems. For Latour, these are co-constitutive of a vast military-industrial complex. To illuminate the extant power relations, he describes the world of research in actor categories that have "military connotations (trials of strength, controversy, struggle, winning and losing, strategy and tactics, balance of power, force, number, ally)" – expressions "rarely employed by philosophers to describe the peaceful world of pure science." The martial imagery underscores that "technoscience is part of a war machine and should be studied as such." Nor is this pervasive militarism restricted to "research into new weapons." It pertains, so Latour, to "the mobilisation of resources" in general, which comprises "research into new aircraft and transport, space, electronics, energy and, of course, communications"(Latour 1987, 172). Although cardboard is not mentioned in this connection, its quickening effect on commerce certainly facilitates the "mobilization of resources" that, for Latour, is the defining feature of technoscience.

Contrarians resolutely deny that the concept of technoscience does any useful work with respect to the conduct of scientists and engineers, theory and practice. Useless jargon, they say, an ugly buzzword, yet another stratagem for undermining the cultural authority of science. Red flag to some, indispensable terminology to others, little has changed since Ian Hacking's review of Latour's *Science in Action* remarked the contested terrain.[1] The term's deployment appears to be as controversial as the phenomenon to which it refers, highlighting, if nothing else, the political dimension of the problem. Indeed, that cold war reaction – the so-called policy of containment – epitomizes the tensions at issue. Nominally, it refers to a pledge by the U.S. government to provide military support as well as economic and/or technical backing to non-Communist countries and thus oppose the expansion of Soviet influence. Policy of *containment* is, perhaps inadvertently, a double entendre in that it could hardly have been put in place or had its deterrent force without the requisite arsenal of paper containers, some imprinted with the label C.A.R.E., some for moving supplies in case of conflict. To avoid getting embroiled in these partisan debates, focus will now shift to the production process at issue.

A major branch of packaging science concerns paper-based solutions to product delivery, a field mainly concerned with the design of the cardboard box. Engineers in this line of specialization learn to master one of the truly ingenious, lightweight-material innovations of the twentieth century: a paper vessel capable of holding many times its own weight in content. Airplane wings and nylon stockings belong

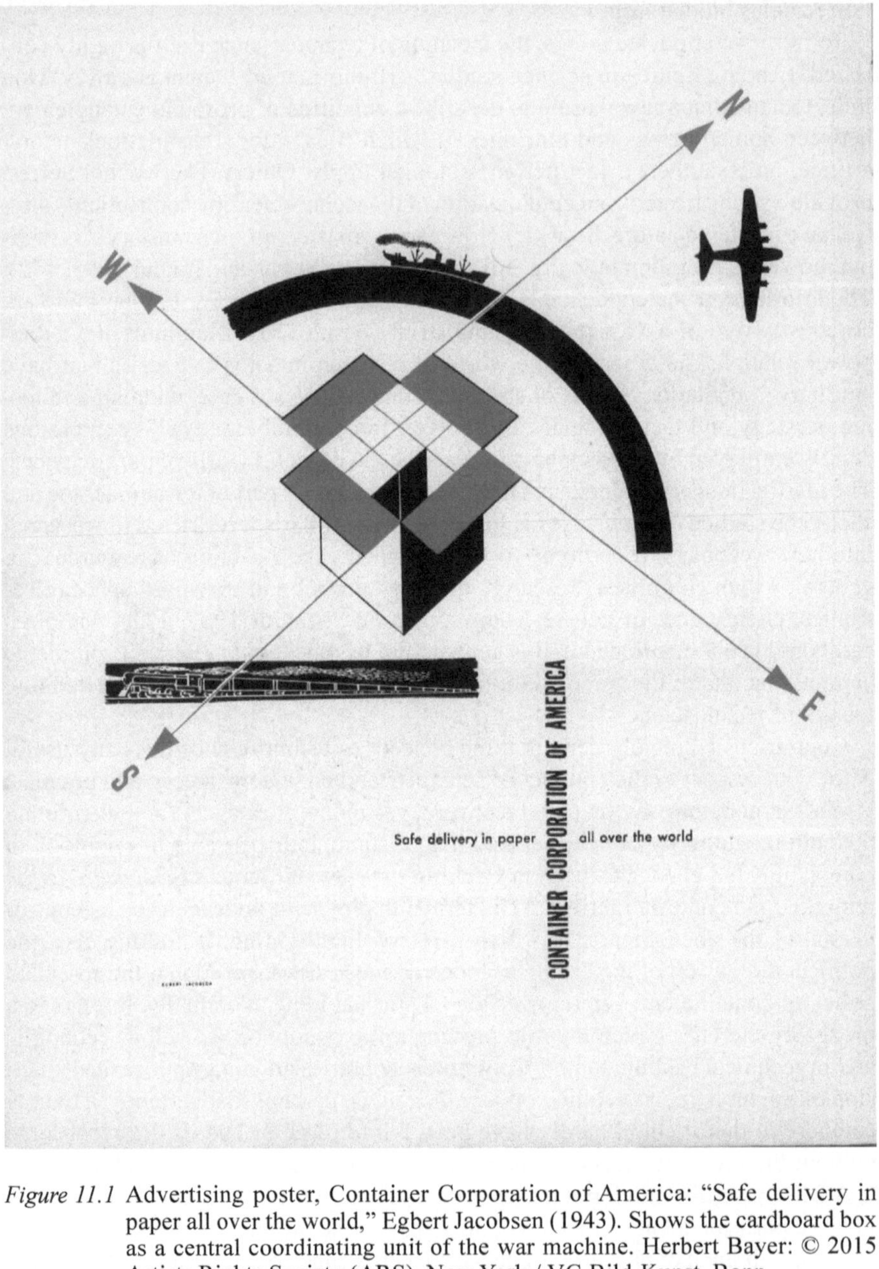

Figure 11.1 Advertising poster, Container Corporation of America: "Safe delivery in paper all over the world," Egbert Jacobsen (1943). Shows the cardboard box as a central coordinating unit of the war machine. Herbert Bayer: © 2015 Artists Rights Society (ARS), New York / VG Bild-Kunst, Bonn.

to the same category of inventions – objects made of materials that seem to defy their own physicality, materials able to outperform themselves, though savings in the weight of these lightweight materials tend to drive up cost of production

(McKinsey & Company 2012, 5). Whereas handling sheet cardboard hardly sounds like rocket science, the actual process of tailoring, with requisite precision, the supply of boxing solutions to the itemized demands of commerce represents a formidable logistical challenge. Suitable paper receptacles, except for generic readymades, are built to specification.

Box architecture demands a thorough understanding of the mathematics of stabilization over time, which includes changes of circumstance between dynamic and static load-bearing situations. Problem solving in this field of application draws on structural engineering expertise and product design competence. It demands a thorough grasp of the national sizing norms for paper products, international tariff-regulations, exigencies along planned delivery routes and operative transport systems – not to mention the psychology of desire and corporate brand management. Whereas digital modeling software can generate a first prototype, a box's life cycle is exposed to free variables that often only show up in on-site testing for real-world imponderables. The custom fitting of boxes to varied scenarios of protective packaging, from a properly sized product sleeve to the well-insulated bulk transport, demands skills of process engineering which belie the seeming simplicity of throw-away cardboard.

Strength is unevenly distributed: a box's vertical fluting is incomparably stronger than the same corrugated wall aligned in a horizontal direction. We've all been there trying to flatten a corrugated paper container that obstinately resists attack: "70 percent of its strength is found in the corners of each box, making squarely stacked cartons incredibly strong" (Vega 2015). Differences in functionality vary so greatly inside and outside the box that industry standards of box quality are set by two separate but complementary stress tests. The Edge Crush Test (ECT) determines how well a box will hold up during stacking; it is measured in pounds per square inch (psi). The Burst Strength Test (BST) or Mullen Test measures the weight a box can hold without failing. Both ratings – they have been in use since 1910 and were merged in 1919 – apply together as Rule 41 (Twede et al. 2014, 225–274). In the United States, these measures are stamped in the Box Manufacturer's Certificate located on the bottom flap of a corrugated cardboard box. The regulatory impact of this rule has kept U.S. box grades sturdier than European equivalents. Given the slim margins to be achieved in global turbo-capitalism, overpacking the freight (i.e., wasting resources) is as serious a miscalculation for realizing a profit as underpacking supplies and risking spoilage. Mumford had little truck with these basic max-min calculations. He approached the question of box quality as an aesthete. Boxes are built to highest standards, he mused, only to be filled with a glut of shoddy products. All the same, he found it heartening that, amidst the many vanities, pockets of real quality survive:

> and if at times we are more pleased by the appearance of a smartly designed paper package than by the quality of its contents, we may console ourselves with the thought that, in its most characteristic material and product, our metropolitan world maintains a standard of excellence.
>
> (Smith 1939, vi)

Recognizing that the cardboard box was reorganizing how humans relate, Mumford surmised that it marked the dawn of a new civilization: "it was, so to say, the swaddling clothes of our metropolitan civilization" (Smith 1939, vi). Language tells. "Swaddling clothes" was archaic at the time of writing. The only child in the modern American imaginary garbed in *swaddling clothes*, when Mumford composed this metaphor, was the baby Jesus. The biblical trope gives the box, and its putative contents, moral valence. *Swaddling clothes* wrap and protect what is most precious in Christendom, its Lord and Savior. By implication, what the cardboard box *swaddles* is of comparable sacredness. Goods bought and sold are the free market's holiest of holies, says this all-American take on the famous Marxist dictum that capitalism spawns *commodity fetishism*. The swaddling image also deflates that hackneyed trope of human progress, the *cradle* of civilization. Instead of the lullabies and coddling associated with 'cradle', Mumford evokes the physicality of a city's vital function. A decade before the disposable diaper made its appearance in American family life – an industrial 'paper' innovation of the 1950s (that falls, ontologically, somewhere between sheet paper and paper cartons) – Mumford reflected the paradox of the foldable box, which would enable some forms of cultural exchange while stifling others.

Another influential voice in American letters, contemporary with Mumford, was Elwyn Brooks White (1899–1985), better known as E. B. White, undisputed authority on the American usage of the English language. Both authors, who had witnessed the rise of cardboard in their lifetimes, deemed it a force to be reckoned with. Like Mumford, White observed that the cardboard box had immediate bearing on human relations. His best-selling book *Is Sex Necessary? Or Why You Feel the Way You Do* appeared in 1929 and devoted an entire chapter to the modern malaise of being "boxed-in" (Thurber and White 2004, 130–149). The book, a satirical marriage manual modeled on the sexological literature of the period, defines this condition as being "caught, or trapped, as a husband by a wife, or the delusion of being caught or trapped" (Thurber and White 2004, 174). The pathology is attributed to a common predicament afflicting modern marriage, namely increasing domestic claustrophobia due to the proliferation of private possessions. Home furnishings and domestic appliances, evidently agreeable to female humans, are seriously detrimental to the male, downright toxic, and in severe cases results in catatonia, "Total Loss, of Neuro-vegetative Reflexes" (Thurber and White 2004, 174).

Blindness to the pervasive influence of the cardboard box entails blindness to its ministrations, vaulted and mean. Mumford's *swaddling clothes* and White's *boxed-in* malaise make light of a serious point on which both authors are broadly in agreement – the cardboard box is a transformative social agent that transports not only commodities but also a moral economy, one conducive to mass consumer culture. The cardboard box epitomizes *compliance*. Made of lightweight yet robust material, it crunches down to a fraction of the volume to be contained in size and weight, standardly folding into a stackable variant that facilitates its own transport by untold orders of space-saving flatness. The cardboard box encourages *fastidiousness*. Its closed exterior, a boon to hygiene as advertising executives would extol, is safe from dirty hands and sneezes. The cardboard box is unapologetically *misanthropic*. It maintains a safety zone that travels with the contents it envelopes, like a sentinel, and allows itemized tracking. A box shell reinforced with tape (also made of *Kraft* paper) as well as validation slips, customs stamps, printed bar codes and inventory labels (measures taken against falsification, tampering and theft) communicate a bleak picture of warranted mistrust. In all, this highly purposive object reifies asserted givens of the socio-political order.

The commercial bias of box culture's moral economy has burgeoned, perhaps, most cynically in the karma of cardboard – that is, its planned recycling (*karma* being related to the proto-Indo-European root *kwer* – 'to make, form'). The modalities of a box's eventual pulping are factored into the economics of box making. Since World War II, box engineering has fathomed a continual reincarnation – cartons in perpetual cycles of renewal. The first nationwide recycling campaign in the United States was mobilized as part of the war effort in the 1940s. The Container Corporation of America, a cardboard box producing corporate conglomerate of the period, called for paper rationing and coordinated paper waste removal. Later, the high-profile campaign promoted forest conservation; wood fiber can be re-pulped up to eight times. This public education drive undoubtedly boosted ecological awareness; it also upped the company's profile. The use of salvaged cardboard in the manufacture of fresh cardboard cut costs (Anker 2007, 254–278). The Container Corporation's self-serving marketing strategy introduced the global recycling logo still in use today (Wikipedia 2015d). Perhaps it was at this point that the box moved conceptually closer to 'nature' and lost its social visibility: brown paper turned 'green' and, once naturalized, no longer registered as high tech.

'Paper foundations' of literacy

Cardboard, seen in the light of technoscience, can illuminate cultures of literacy. Advancement of literacy and accumulation of knowledge hinges on the dissemination of paper, the first mass medium. With paper came scribes and, after that, mechanical print, proliferation of books and spread of enlightenment, so the classical account – a pithy exposition (more is the pity that it distorts cultural realities). Sheet paper alone could hardly have yielded industrial society, let alone make it intelligible from within to the populations concerned. The connectedness that must be in place for that great chain of consumption, the modern, free-market mandala to function – linking, as it does, one to all and all to one – transcends the realm of letters. It would appear that another mode of paper mediation transpired as well.

Presumably, Mumford had a dialectical process in mind when he predicted the demise of modernity's 'paper foundations' – after all, *foundations* has both literal and figurative connotations. His comments appear in a preface to the biography of Robert Gair (1864–1927), the industrial packaging pioneer who invented both the corrugating process and the collapsible carton. A renowned comic author of the period, H. Allen Smith (1907–1976) composed the actual literary portrait of Gair. Both writers recognize that the implications of cardboard reach far beyond the economics of distance. They treat the cardboard box as a medium of cultural advancement akin to sheet paper but uniquely instructive for being functionally distinct. Smith notes that insertion of the cardboard box into the quotidian affairs of the marketplace unleashed a flurry of cultural activity similar to the historical developments imputed to Scottish philosophers in eighteenth century Edinburgh, Gair's city of birth – a veritable 'commercial enlightenment'. (Smith 1939, 71). His irreverent perspective blurs, productively, it must be said, a spurious distinction between writing paper and industrial paper that continues to dominate the discourse on literacy these many years later. Despite being charmingly dismantled by astute participant-observers of high modernism in America, celebrity authors whose views on topics other than cardboard proved tremendously influential; thus the exalting of sheet paper and its related textual practices persists.

Cardboard is a paper product with a twist. Virtually indistinguishable from sheet paper in material composition, it holds no claim to the cultural authority of the latter. It is not a fully recognized member of the paper dynasty. The legacies of these two agents of social conditioning could hardly appear more disjointed. Narrative embedding, dexterities furthered by their handling, propagation of 'content': along every variable of comparative practice, there is a functional rift between sheet paper and cardboard. This categorical disjunct is downright canonical hardwired in meta-narrative discourses that predate the invention of cardboard. Interestingly, this is the case even though the package is also the message, to paraphrase Marshall McLuhan, its surface presents a write-up of the contents (but that will be explored in a separate essay).

Sheet paper invites abstract reasoning and symbolic manipulation. Its literacy stands in a long and venerable tradition associated with formal codes of lettered exchange; these can be realized on a multitude of surfaces: clay tablets, papyrus, nano-cellulose. Digital literacy fits the pattern; duly tracing the flat expanse of screen real estate to a history that is staunchly conservative, its chroniclers tend to extend traditional notions of literacy to e-paper. According to this orthodoxy, scriptural routines have a resilience that migrates, in the civilizing process, from one platform of symbolic exchange to another. What these accounts do not address is another possible transition pertaining to paper, namely from the cardboard box to what, by analogy, might be called 'e-cardboard'. This is by no means a wildly speculative future. Although *e-cardboard* is at the time of writing these lines not a google-indexed term, corresponding phenomena already haunt our "virtual actuality." (Hubig 2006, 187). Cardboard furniture, cardboard housing, cardboard bicycle – the receptacle has mutated into the content it carries (see Green 2012; Fehrenbacher 2015; Karton Group 2014). With 3D printing, frictionless haulage seems possible: that objects, formerly trafficked through space as physical bulk, are instead digitally reconstituted via a print command.

A case in point is Lothar Müller, arts page editor of the *Süddeutsche Zeitung*, one of Germany's premier literary critics and cultural commentators. His magisterial study *Weiße Magie. Die Epoche des Papiers* ("White Magic. The Age of Paper") maintains that today's non-linear, multi-tasking e-skills were prefigured in the writings of authors who were far removed from the computer revolution, including Jean Paul, Thomas Carlyle, Hermann Melville, Henry James, Paul

Valéry and numerous others. Müller shows that the reading skills elicited by these texts anticipate how the computer desktop paradigm engages its users. His seamless *mise en scène* makes a powerful case for the traction of paper in cultural life. Against precipitate claims that sheet paper is in the process of being superseded by electronic paper, Müller pits a long history of semiotic practices. A dazzling panorama of changing social orders unfolds: rise and fall of empires, competition between knowledge regimes, fluctuation in trade cycles. Evidence enlisted to support one salient, historical fact – the persistence of paper in human affairs. Sheet paper appears in numerous guises as repository of knowledge (archive), ledger of transactions (account) and conduit of value (money) – anything and everything, it seems, except vessel of transport. Müller's conception of literacy assimilates electronic communications and canvases distinct realms of social activity. All the same, it stays wedded to a traditionalist understanding of knowledge and knowing. This bias, firmly entrenched in the history of thought, purports matters epistemological to be, inherently and in essence – cerebral, theoretical, abstract.

This point of view, medieval in origin, resolutely ignores another modality of paper that correlates with the spread of industry – cardboard. Of no less consequence for human affairs, it urges another inflection of literacy. Indeed, cardboard arguably instantiates a distinct order of literacy inseparably linked to that engineering way of being in the world that characterizes technoscientific virtuosity. Being able to pursue goals by manipulating physical objects is a learned competence that requires technological savvy. Competence in this sphere of activity is different in kind from the fluency in manipulating symbolic schemata that informs the received notion of scientific expertise. However, symmetry obtains: just as sheet paper facilitated classical literacy, the cardboard box can be said to foster a *literacy of things*. Without the mediating services of cardboard, technoscientific humanity could not have acquired the current aptitude it has in wielding sophisticated tools to accomplish complex routines. Whether or not the proliferation of these skills, which are associated with the circulation of commodities, are deemed commensurate – culturally, morally, and/or aesthetically – with the lexical dexterities of yore, it so happens that the cardboard box has furthered the exchange goods and wares with the same expediency as sheet paper the exchange of ideas.

As concerns these competing literacies – a *literacy of signs* (sheet paper) and a *literacy of things* (cardboard) – neither computes without reference to the other and never did, even before the invention of paper. Real-world literacy constitutes a single integrated phenomenon encompassing theory and practice; it engages the full spectrum of activity, which variously fosters and hinders human connectedness. Paper's praxis entails different ways of knowing: one trades on abstraction, representation, analogical reasoning and the virtual relatedness of objects in a realm of two-dimensional surrogacies; the other deals in boxed goods, the meaning of concretes, ready-mades, immanences, recalcitrance and all manner of handy commodities filtering our experience of the world. A comprehensive standpoint must fathom the dichotomous role of paper in shaping that perpetual work-in-progress – human self-formation.

Techno-sentience

To recall the opening question: can the cardboard box – an old-fashioned, crude-seeming trifle – be considered a technoscientific object of any real consequence? Cardboard's putative eco-friendliness – the belief that it is but a few pulpings away from the living tree – may, paradoxically, have eclipsed critical engagement with the systemic agency of this constant accompaniment to the commercial exchange. Moreover, a purposive, partial amnesia has expunged cardboard from view. Cultural preferment of the practices associated with sheet paper buttress an ideological rift between symbolic and manual labor. Entrenched hierarchies privilege certain forms of cultural expression over others and promote, for the select few, exclusive codes of sociality. An immersive, situated, embodied history of *Kraft* paper will, by contrast, de-familiarize the supposed *illiteracy* of cardboard. This perspective strives to spell out another dimension of cognitive wherewithal, beyond humanist learning, that might be termed *Kraft* literacy, or, alternatively, *spatial literacy* (understood not in the usual sense of a cartographic, mapping geographic awareness but rather as the ability to coordinate one's movements toward pragmatic ends) or *manual literacy* (to emphasize the critical function that the hand and digits play in this probing *savoir-faire*) or *ontic literacy* (meaning that the grammar of this mode of relating to the world and others is defined by design-led bits and bobs).

Ultimately, a technoscientific reading of cardboard urges a revised master narrative of paper's history. Like political science, social science and other cognate disciplines, 'technoscience' might be thought of as the *science of technology*. Only, the compound term implies its own reflexivity, an emergent techno-sentience, if you will, whereby the referent of each term conditions the other. Technology and science designate spheres in which creative ingenuity find expression; they connote different ways of knowing, being in the world, world-making. A long-standing battle has been waged between scholars whose outlooks differ over the pre-eminence and respective merits of these two domains of knowledge-production. Fusing these two, import-laden terms into one challenges us to rethink their dynamic relationship for present purposes and in its historical development. Today's emerging methodological protocols privilege expert practice with an engineering bias over knowledge-making as a theoretical endeavor and end in itself. My footnote to this big picture of knowledge production is that, if a technoscientific literacy of things is on the rise while the science literacy of signs is in decline, then, arguably, the cardboard box helped tip the balance.

A telling ambiguity adheres to the foldable cardboard box in that it conveniently shape-shifts between two- and three-dimensional objects, the topography of Flatland and the materiality of Spaceland, the republic of letters and the republic of cardboard – two complementary principles for parsing the universe. Recalling its genealogy, this should come as no surprise. The industrial production of cardboard deployed the machinery of mass printing. A historian at Michigan State University, School of Packaging, has shown that, "Three key technological inventions in the 1800s set the stage for the mass production of paper based packaging by the end of that century: the paper making machine, the process for pulping wood and

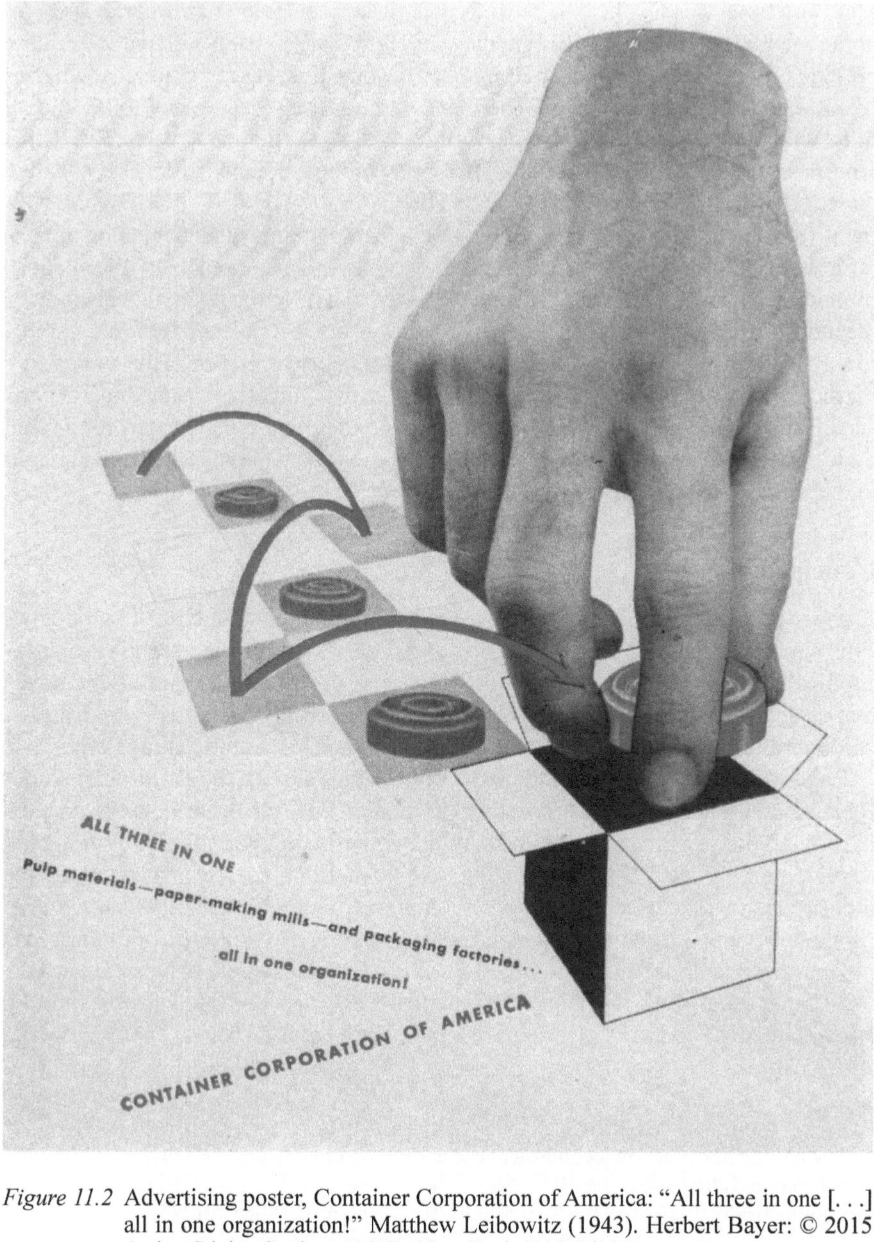

Figure 11.2 Advertising poster, Container Corporation of America: "All three in one [. . .] all in one organization!" Matthew Leibowitz (1943). Herbert Bayer: © 2015 Artists Rights Society (ARS), New York / VG Bild-Kunst, Bonn.

lithographic printing" (Twede 2007, 241). These are, of course, the very industrial innovations that set the stage for the expansion of print capital as well, through books, newspapers, magazines, posters and handbills.

Mediated opining morphs into industrial origami. Gair's key insight, as Smith and Mumford note, was that he delegated the industrial processing of the cardboard box to the machinery of mechanized literacy. In particular, he used a converted Aldine printing press to automate industrial folding; the machine was recalibrated to imprint creases instead of printing ink: "The blank came off the press ready to be folded, filled with merchandise, and locked" (Smith 1939, 66). After it had been retired, this repurposed press sat "on a pedestal in the lobby of the company's office," beneath a tablet bearing the proud inscription: "THE FATHER OF THE FOLDING BOX [*sic*]" (Smith 1939, 68–69). The section of the Gair plant in which the lineup of box-refurbished printing presses stood in a row was playfully referred to as "the allegory of industry" (Smith 1939, 58). This epithet – evocative of the lettered genealogy of the industrial process – suggests that the irony was not lost on the workforce, subject to shifts running night and day to meet demand for paper containers. They were using the technological apparatus developed to promote the quintessential medium of classical learning, the book, to manufacture, in overwhelming bulk, cartons for a retail trade that would, increasingly, deflect attention away from the sphere of pure thought to the messy nitty-gritty of physical engagement.

Conclusion

A key to succeeding in business, they say, is the ability to 'think outside the box'. This cliché perpetuates the self-effacement of cardboard while muddling a philosophical conundrum. The thinking done today is profoundly reliant on the boxes that structure the routines that make life happen. To think creatively is to think cardboard or, in short, to think the box in all its technoscientific entanglement.

Boxes beget boxes amidst cardboard's daily doings in the re-engineering of world. New realities incumbent on the consumer revolution have made it a matter of urgency to reframe the ruling conception of how the entire paper dynasty relates to human development. Since 1993, the prevailing medieval mind-set has lagged behind the industry standard. At the start of the e-commerce boom, a trade report found that: "Although paper is traditionally identified with reading and writing, communications has now been replaced by packaging as the single largest category of paper use at 41% of all paper used" (Pulp and Paper International 1994).[2] Business folk have turned the corner already. Theorizers of culture risk falling behind their coeval *commerçants*.

Notes

1 The polarized debate on technoscience is addressed in Ian Hacking's review of Latour's *Science in Action* (Hacking 1992, 510–512).
2 Figures listed are for 1993; see Monitor (2015) and Wikipedia (2015c).

References

Abbott, E. 1992 [1884]. *Flatland*. New York: Dover.

Abé, K. 1974. *The Box Man*, translated by E. D. Saunders. New York: North Point Press, Farrar, Straus and Giroux.

Anker, P. 2007. 'Graphic language: Herbert Bayer's environmental design.' *Environmental History* 12 (April 2007): 254–278.

Bayer, H. 1939. Ugly Duckling of the Office' Poster for the Container Corporation of America, http://americanart.si.edu/collections/search/artwork/?id=1544 (accessed July 17, 2016).

Chiang, J. 2014. 'Ship Shape: Paper and Packaging Prices on the Rise,' in IBIS World, http://media.ibisworld.com/2014/09/23/ship-shape-paper-packaging-prices-rise (accessed July 17, 2016).

Fehrenbacher, J. 2015. 'The Cardboard House,' in *Inhabitat – Design That Will Save the World*, http://inhabitat.com/the-cardboard-house (accessed July 17, 2016).

The Global Community of Information Professionals. 2014. 'Paper Wars 2014 – An Update from the Battlefield,' Iron Mountain (White Paper): AIIM, http://goo.gl/zcfx7Q (accessed July 17, 2016).

Green, S. 2012. 'The Cardboard Bike . . . and Some Indications for 3D Printing,' in Blog for a 3D World, http://blog.stratasys.com/2012/10/15/the-cardboard-bike-and-some-similarities-to-3d-printing (accessed July 17, 2016).

Hacking, I. 1992. 'Review of Latour's *Science in Action*.' *Philosophy of Science* 59 (3): 510–512.

Hubig, C. 2006. *Die Kunst des Möglichen*, vol. 1. Bielefeld: Transcript.

Imagination Foundation. 2015, http://imagination.is/our-projects/cardboard-challenge (accessed July 17, 2016).

Isaacson, A. 2012. 'The perfect moment goes perfectly viral,' *The New Yorker*, 24 April, http://www.newyorker.com/culture/culture-desk/the-perfect-moment-goes-perfectly-viral (accessed October 21, 2016).

Karton Group. 2014, http://kartongroup.com.au (accessed July 17, 2016).

Klanten. R. and Bolhöfer, K. 2011. *Out of the Box! Brand Experiences between Pop-Up and Flagship*. Berlin: Gestalten.

Krieg, T. 2004. *Verpackt und Zugeklebt, Das Praxis-Handbuch für professionelles Verpacken*. Dortmund: rokoko-Netzwerk für Kommunikation.

Latour, B. 1987. *Science in Action: How to Follow Scientists and Engineers through Society*. Cambridge, MA: Harvard University Press.

McKinsey & Company. 2012. Lightweight, heavy impact: How carbon fiber and other lightweight materials will develop across industries and specifically in automotive. McKinsey & Company Advanced Industries, http://www.mckinsey.com%2F~%2Fmedia%2Fmckinsey%2Fdotcom%2Fclient_service%2Fautomotive%2520and%2520assembly%2Fpdfs%2Flightweight_heavy_impact.ashx&usg=AFQjCNHB4rx8ixJcZX5o9tMFQ0zqqLPm_g&sig2=7403BlyT-NnC_Qbe7KTgSA&cad=rja (accessed October 21, 2016).

Monitor Business Machines Ltd. 2015. 'Facts about Paper and Waste', http://goo.gl/xLwkXJ (accessed July 17, 2016).

National Toy Hall of Fame. 2015, http://www.toyhalloffame.org/toys/cardboard-box (accessed July 17, 2016).

Pulp and Paper International 1994. 1995. *Fact and Price Book*. San Francisco: Miller Freeman, Inc.

Sellen, A. J. and Harper, R. H. R. 2002. *Myth of the Paperless Office*. Cambridge, MA: MIT Press.

Serres, M., with Latour, B. 1995. *Conversations on Science, Culture and Time*, translated by R. Lapidus. Ann Arbor: University of Michigan Press.

Smith, H. A. 1939. *Robert Gair: A Study*, with an introduction by Lewis Mumford. New York: The Dial Press.

Thurber, J. and White, E. B. 2004. *Is Sex Necessary? Or Why You Feel the Way You Do*, foreword by J. Updike. New York: Perennial Harper Collins.

Twede, D. 2007. 'The history of corrugated fiberboard shipping containers.' *Marketing History at the Center, Proceedings of the 13th Conference on Historical Analysis and Research in Marketing*, pp. 243–246. North Carolina: CHARM Association (Duke University, May 17-29).

Twede, D., Selke, S. E. M., Kamdem, D.-P. and Shires, D. 2014. 'Paper, Paperboard, and Corrugated Board Properties and Tests.' In *Cartons, Crates and Corrugated Board, Second Edition: Handbook of Paper and Wood Packaging Technology*, 225–274. Lancaster, PA: DEStech Publications.

U.S. Environmental Protection Agency. 2012. 'Municipal Solid Waste Generation, Recycling, and Disposal in the United States: Facts and Figures for 2012', https://www.epa.gov/smm/advancing-sustainable-materials-management-facts-and-figures (accessed July 17, 2016).

Vega, D. 2015. *Corrugated Cardboard: Plain Old Box.* http://goo.gl/KSmP99 (accessed June 3, 2015).

Verband Deutscher Papierfabriken e.V. 2014. 'Facts on Paper 2014', http://www.vdp-online.de/en/papierindustrie/statistik.html (accessed July 17, 2016).

von Xylander, C. forthcoming. 'Pictorialism (Prelude & Fugue).' In Aesthetics of Universal Knowledge, edited by Schaffer, S., Tresch, J. and Gagliardi, P. Basingstoke; Palgrave.

Wikipedia. 2012. 'Big-Box Store', https://en.wikipedia.org/wiki/Big-box_store (accessed July 17, 2016).

———. 2015a. 'Cardboard Boxes,' in Calvin and Hobbes, https://en.wikipedia.org/wiki/Calvin_and_Hobbes (accessed July 17, 2016).

———. 2015b. 'Currogated Box Design', https://en.wikipedia.org/wiki/Corrugated_box_design (accessed July 17, 2016).

———. 2015c. 'Paper Recycling', http://en.wikipedia.org/wiki/Paper_recycling (accessed July 17, 2016).

———. 2015d. 'Recycling Symbol', https://en.wikipedia.org/wiki/Recycling_symbol (accessed July 17, 2016).

Part III
Multiple temporalities

12 The multiple signatures of carbon

Sacha Loeve and Bernadette Bensaude Vincent

Figure 12.1 The 'coal tree'. By-products obtained from bituminous coal degasification. Courtesy of the Pharmazie-Historisches Museum, University of Basel.

Carbon is the fourth most abundant element by mass in the universe after hydrogen, helium and oxygen. The chemistry of carbon is both ancient and ubiquitous, with applications ranging from fine jewelry to heating, textiles, pharmaceuticals, energy and electronics. How can we present this familiar stuff, which has played a central role in the history of chemistry, as an exemplar technoscientific object? Wouldn't it be more adequately described as a typical scientific object?

To be sure, carbon is an object of 'pure research'. It emerged in the nineteenth century as a chemical *substance*, as an abstract albeit material substrate underlying a range of simple and compound bodies. However, this ontological perspective shows only one face of carbon. While scientific research was focused on what is conserved through change, the researchers' attention shifted to what might be changed: Carbon came back as a menagerie of allotropes – fullerenes, nanotubes, graphene and many more – which populate the 'nanoworld', making it a rich source of technological possibilities. The ontological status of carbon in a scientific perspective answers the question 'What is it?,' while in a technoscientific approach, the question answered is rather 'What might be performed with it?' or, more precisely, 'What might it afford?' We could thus portray carbon as a Janus and contrast its scientific and technoscientific identities.

However, this contrast provides a far too purified image of carbon. Carbon has assumed so many guises over the course of the centuries that it almost seems to play various *personae* moving with a momentum of their own, inspiring new adventures while exhibiting surprising physical and chemical properties. Diamond, charcoal, graphite, carbon compounds, mephitic air and CO_2 are among the characters that have played – and continue to play – significant roles in science and mythology, in industry and economy alike. Carbon is too ubiquitous, too polymorphic and too important in our history for being reduced to these Janus faces. Such a portrait would overlook the multiplicity of the *modes of existence* of carbon and their various inscriptions in daily life.

Multiple modes of existence

As the term 'allotropy' suggests (*allos*: other + *tropos*: manner, way or mode),[1] carbon displays a multiplicity of modes of existence. This chapter explores their coming into being in relation with human history. It tells many stories, ranging from legends about mephitic air and the key role of carbon in building the periodic table to the story of the carbon skeleton as a backbone of life and pillar of chemical industry. Due to its disposition for creating strong bonds with itself and other atoms, carbon affords robust and stable structures and a host of potential combinations, found both in nature and in human industry. At the nanometer scale, carbon writes a kind of fairy tale. As its bulk molecular structure shrinks to pure surface, it displays a wide range of novel properties and performs a variety of mechanical, electronic, optical and magnetic functions, presented as the promises of enchanted futures. In stark contrast to these futuristic tales, the story told by carbon deposits looks backward and is moving inexorably toward a sad ending. Charcoal and oil, the archives of life on earth, have been excavated and come back into existence as fire and power for human consumption needs. Two centuries after the beginning of the industrial revolution, the vestiges of life accumulated for billions of years in the soil are almost depleted. They have melted into air by a few generations of an ever-increasing population of humans under the form of carbon dioxide. As human societies realized the impacts of their carbon-based technologies, they elected carbon dioxide as a standard to regulate the flow of exchanges between ecology and economy.

Today, any activity correlated with greenhouse gas emissions – that is to say, most human activities – is translated in terms of 'carbon equivalent'. Carbon thus becomes a universal unit of currency. But while carbon trading is meant to get the economy 'back down to earth', it actually turns carbon into a speculative product in 'carbon trading'.

Yet the alliance of carbon with market economy is only one of its modes of existence. Instead, we want to consider all of them at once. Carbon is reducible neither to the fashionable abstraction of carbon equivalent, nor to its chemical identity. The simple and unique definition of carbon as an element in the periodic table, like the definition of water as H_2O questioned by Hasok Chang, is obviously too narrow to contain all its facets and roles in human history (Chang 2012). Just as Chang convincingly argues that the formula H_2O does not deserve a monopolistic dominance, we will see that carbon calls for multiple systems of knowledge. Not only is there a plurality of views of carbon in physics, chemistry, biology, materials science and environmental science; there are also many other relevant views of carbon in lay and traditional knowledge. Carbon proves the value of 'epistemic pluralism', and invites for more 'tolerance' among the various forms of knowledge. However, this chapter goes further. It claims that there is no reason to deduce the various facets of carbon from one single fundamental definition. Carbon as a substance, standing beneath all its modes of being, is only but one of these facets. Carbon invites *ontological pluralism* in addition to epistemic pluralism. Although each mode of existence has an 'imperialistic' propensity for configuring the other modes in its own template, all the modes of existence coexist. Kick one of them out the door; it comes back through the window. In paying attention to carbon's diverse modes of existence along the lines first developed in 1943 by Etienne Souriau and taken up by Bruno Latour (Souriau 2009; Latour 2013), this chapter makes the case for an *ontography*: it considers carbon as an agent of 'graphism' in the task of *writing* ontologies.

Carbon heteronyms

Carbon is polymorphic. Even in its elemental state it takes a variety of allotropic forms, some of them with well-ordered, crystalline structures (graphite, diamond, lonsdaleite),[2] others with amorphous structures (glassy carbon, carbon black, nanofoam), in addition to mixed structures displaying various degrees of order and disorder (charcoal, soot, coke).

Yet some of the allotropes of carbon have quite different physical properties: diamond is hard and translucent; graphite is friable, fragile and opaque. Diamond is abrasive; graphite lubricant. Diamond is an electrical insulator and thermal conductor; graphite an electrical conductor and a thermal insulator. Ever since the days of antiquity, diamond has been considered the hardest stone on earth. Its name, deriving from Greek αδάμας (*a*: 'not' + *daman*: 'to tame') highlights its remarkable property of resistance to all attempts to break it, as well as to the erosion of time: it is an 'everlasting stone', which literally 'cannot be tamed'. Diamond is adamant, refusing to change or to give in. Fittingly enough, it became the symbol of an unbreakable engagement between two persons: 'diamonds are forever'. What on earth could it have in common with the black and brittle substance that leaves a dark mark on a sheet of paper? In contrast to diamond, graphite (derived from the Greek *graphein*: 'to write') is so fragile and soft that it has to be inserted into a hollowed wooden stick to be used. The pencil patented by Nicolas Jacques Conté in 1795 was named 'lead pencil' because graphite was initially considered to be 'plumbagol' or 'black lead', a substitute for the lead formally used for writing since Roman times. Less surprising was the connection between 'black lead' (a residue from the distillation of animal or vegetable matter in a limited supply of air), and the use of the same material to draw the pictograms of cave paintings some 30,000 years ago. Yet this artists' material was also used to heat iron in metallurgy. What does coal – the fuel of the industrial revolution extracted by the ton from mines all over Europe and in the colonies – have in common with the precious diamond?

In their chemical mode of existence, diamond and graphite are two allotropes: they are made of the same carbon atoms and differ only in the bonds between atoms. However, from an ontographic perspective, they share more than chemical allotropy. From the graphite pencil to diamond-engraving tools and the 'code of life' based on a carbon backbone, from the periodic table to the 'carbon footprint', from the carbon black tattoos of Ötzi the Iceman (5200 years ago) to the 'tons of carbon equivalent', measuring the global warming and to radiocarbon dating (^{14}C), carbon is linked with the action of scripting (*graphein*). It exists in inscriptions, for conservation, standardization and circulation. Human history is written in carbon. If it is true that human civilization began in prehistory with the mastery of fire, then carbon, by its very name – derived from the Latin root *carbo*, 'burnt' – connects nature and culture. If it is true that history began with writing, then carbon in its form of graphite kicked off cultural history. Nowadays carbon traces allow humans to read their history against the background of a wide spectrum of timescales, ranging from a few years to geological eras.

This essay presents the various modes of existence of carbon as its *signatures*. It lends a voice to its multiple *personae* and allows them to write their own narratives. Just as the Portuguese poet Fernando Pessoa had various *heteronyms* sign his works of fiction, carbon has several signatures for writing ever-new scripts and telling various stories that interweave human history, natural history and cosmic processes. To what extent could we consider the name 'carbon' as a multiplicity of heteronyms, each having its own signature? In considering carbon as a scribe with multiple identities, this chapter will emphasize the ways in which each *persona* of carbon interacting with our material and symbolic practices create togetherness and sketch the lines of our common world.

Carbon as mephitis

In the *Aeneid*, the Roman poet Virgil describes the valley of the river Asanto in Campania as an entry to hell (Virgil 1909, 562–571). The foaming marsh, named *la Mefite* by local peasants after a fury called Mephitis (Di Lisio et al. 2010, 142),[3] has the largest natural non-volcanic CO_2 emission rate ever measured on earth (about two hundred tons per day). Virgil's scary description echoed and further propagated legends about the evil and pestilent nature of Mephitis. While the site is still considered potentially hazardous, especially in the absence of wind,[4] the shepherds of the neighborhood have continued to bring their sick sheep to *la mefite* during transhumance in order to cure them. Between Chthonian depths and Ouranian vapors, life and death, Mephitis 'stands in the middle' (Ibid., 143). She was a kind of *pharmakon*, evil and remedy at once, as well as a scapegoat. In the course of time, this ominous exhalation has been named 'mephitic air', 'pestilential vapor', 'lethal spirit', 'deadly spirit', 'fixed air', 'aerial acid', 'acid air', 'chalky acid', 'sylvester spirit' or simply, 'gas'. In fact CO_2 is the archetype of the notion of 'gas' (Soentgen 2010). This term, coined in the seventeenth century by Joan Baptista van Helmont, referred to both a deadly substance and a principle of life: on the one hand, it is a kind of *pneuma* animating a body; on the other, because of its impetuous and elusive character, it turns into a dangerous killer when released.

To be sure, the reform of the language of chemistry promoted by French chemists in 1787, was a serious attempt to purify carbon from the 'archaic symbols' conveyed by Mephitis's avatars. The name 'acide carbonique' erased the connotations of life and death, of good and evil, borne by the ancient denominations of CO_2. Let us see whether Mephitis's formidable power has been defeated or merely concealed by the *mêlée* of more modern carbon heteronyms.

Carbon as a standard

In nineteenth-century chemistry, carbon provided a taxonomic scheme that allowed chemists to establish order within the crowd of individual substances. When chemists undertook to set up a classification of chemical elements, carbon played a decisive role in defining what was to be classified. Dmitiri Mendeleev used the case of carbon as a kind of template to define the concept of chemical elements as distinct from that of simple substances. The allotropic forms of carbon (diamond, graphite and coal) are exemplars of the concrete stuff of simple substances.

> A simple body is something material, a metal or a metalloid, endowed with physical and chemical properties. The idea that corresponds with the expression simple body is that of the molecule. [. . .] By contrast, we need to reserve the name element to characterize the material particles that constitute the simple bodies and compounds and that determine the manner in which they behave in terms of their physical or chemical properties. The word element should summon up the idea of the atom.
>
> (Mendeleev 1953, 693)

While simple substances come into existence as concrete and physical entities at the end of a process of analysis and purification, elements are the material but invisible parts of simple and compound bodies. Carbon is a hypothetical abstract entity since it can never be isolated – in stark contrast to diamond (pure carbon) and anthracene (90 percent pure carbon). Nevertheless the element carbon is real and identifiable by a positive individual feature: its atomic weight. For Mendeleev the atomic weight was the signature of elements. The signature of carbon was 12.

How can atomic weight be quantified when it is impossible to measure the weight of atoms? The system of atomic weights had to be set up on the basis of relative and conventional values. Carbon again played a decisive role in this process by being chosen as the standard in the mid-twentieth century. The earliest systems of atomic weights based on a conventional standard unit H = 1 (hydrogen equals 1) or O = 16 (oxygen equals 16) no longer worked after the discovery of isotopes. For example, physicists attributed the value 16 to one single isotope of oxygen, while chemists attributed that same value 16 to the whole range of oxygen isotopes 16, 17 and 18. In 1959–1960, the International Union of Pure and Applied Physics (IUPAP) and the International Union of Pure and Applied Chemistry (IUPAC) agreed to put an end to this troublesome discrepancy by electing the isotope carbon-12 as the standard reference for determining the atomic mass of all other elements. In 1961, an IUPAC convention defined the Dalton or UMA (unified atomic mass unit) as one-twelfth of the mass of an atom of ^{12}C. Carbon thus acted as a diplomat, a mediator to settle a conflict between chemists and physicists.

It also provided the standard for defining the concept of mole. A mole, or N (Avogadro's number), refers to the amount of substance of a system, which contains as many elementary entities as there are atoms in 0.012 kg of carbon 12 (6.0220451023). N atoms of ^{12}C have a mass of 12 grams. As the basis for mole, carbon acts again as mediator. This time it mediates between the nanoworld of atoms and molecules and the macroscopic world of human operations on matter, making the chemist's life easier (Buès 2001).

Carbon as substance

The periodic system has been set up by Mendeleev in an effort to discover the unique and general law governing the irreducible diversity of chemical phenomena. This law stated that the properties of simple and compound substances are a periodic function of the value of the atomic weight of elements.

Mendeleev's emphasis on the centrality of elements was maintained and reinforced in the early twentieth century with the discovery of isotopes. Carbon came to be considered an underlying substratum (*sub-jactum*) that persists through chemical changes. This substantialist approach was conveyed by Friedrich Paneth, a German chemist who demonstrated that radio-elements could not be isolated from the most common isotope by chemical means, and thus could be considered as true elements. In trying to identify the mode of existence of what is preserved unchanged through chemical metamorphoses, Paneth adopted a metaphysical stance through his distinction between 'basic substance' (*Grundstoff*) and 'simple substance'(*einfacher Stoff*) (Paneth 1962, 144–160). *Stuff* thus became a central concept in chemistry (Ruthenberg and van Brakel 2008). In order to identify the 'something' that remains unchanged through chemical changes, Paneth revived the Aristotelian notion of *hupokeimenon* as a basic and general level of reality underlying the various particulars accessible to sense perception. Whereas the basic substance (G*rundstoff*) has no existence in space and time, simple substances are unique events located *hic et nunc.* Paneth was thus in a position to answer Aristotle's question: 'What is it (*to esti*)?' And he provided a dual answer, as Aristotle did: it is both a singular phenomenal entity, an individual that you can see and point your finger at (*ek-keimenon*) *and* a universal substrate (*hupokeimenon*) underlying and explaining particular and changing phenomena. Carbon as substance connects modern chemistry with ancient metaphysics.

Carbon as material

In chemical textbooks, carbon appears as a substance because of the ontological priority granted to atoms. However, in the material world around us, artisans and engineers do not find carbon atoms. Nature affords carbon molecules and compounds, which can be used as materials. Many of them have played key roles in human industries since prehistory. For instance, bitumen – a hydrocarbon containing 80 percent carbon and formed from plankton slowly accumulated in sedimentary basins – was already used by Ancient Egyptians for coating roads, boats, canals, dams and tanks. Asphalt – a mixture of bitumen and aggregates that coated the streets of London and Paris around 1820 – was used by seventeenth-century engravers for etching and by painters to bind their pigments. Coal and bitumen have also long been known for their antiseptic properties: the Phoenicians carbonized the wooden barrels of trading ships in order to preserve drinking water on their long journeys at sea, while the Egyptians treated teeth cavities with a mixture of bitumen and clay. Carbon fibers, which are widely used today in the manufacture of composites due to their remarkable properties – resistant, flexible and lightweight – enjoyed prior industrial success in lightbulbs, when Edison and Swan used filaments obtained by the 'carbonization' of cotton and bamboo.

Such industrial uses of carbon materials did not follow from the theoretical understanding of carbon's atomic structure and of the carbon-carbon bond. Rather, the potential uses of hydrocarbon molecules were gradually explored in a complex process of de-contextualization and re-contextualization of pharmacists' and manufacturers' knowledge and know-how about carbon materials (Travis et al. 1992, 91–111; Tomic 2010). Carbon materials bridge the gap between the phenomenal world of technological applications and the theoretical world of molecular structures and models.

Carbon as skeleton

The identification of the unique atomic structure of carbon (tetrahedral with four valence electrons) and its wide-ranging binding capabilities (C–C, C=C, C≡C, C–H, etc.) accounts for the remarkable *dispositions* of carbon (what it is capable of): stability, resistance and combinatory potential to form innumerable compounds.

In nature, biomolecules such as DNA, RNA, proteins, carbohydrates and lipid membranes are all structured around a strong carbon backbone.

In the realm of technology, millions of organic chemical compounds synthesized by the chemical industry are also made of a carbon backbone, albeit less complex than those made by nature. Whether natural or artificial, organic compounds are the products of the structural combinatorial game enabled by carbon binding dispositions. Thus synthetic products, usually presented as products of human design driven by social demands, are also examples of the actualization of carbon dispositions. Underlying the traditional divide between the natural productions of life and the highly artificial processes of industry runs an invisible thread, an elegant, robust but largely unnoticed microstructure: the carbon backbone.

Carbon as surface

Over the past twenty-five years, carbon has been the star of materials research, generating more and more exotic allotropes: nanotubes (Iijima 1991, 56–58), fullerenes (Kroto et al. 1985, 162–163), graphene (Geim and Novoselov 2007, 183–191), nanobuds (Nasibulin et al. 2007, 156–161), nano-onions (Rettenbacher et al. 2007, 1411–1417), nanotori (Cox and Hill 2007, 10855–10860), nanocones, nanohorns (Iijima et al. 2008, 605–629), nanobamboos (Koltai et al. 2009, 2671–2674), nanoribbons (Jiao et al. 2009, 877–880; Fasel et al. 2010, 470–473) and so on. This weird menagerie has garnered many awards: the 1996 Nobel Prize in Chemistry for the fullerenes, the 1998 Kavli Prize in Nanoscience for carbon nanotubes and the 2010 Nobel Prize in Physics for graphene, the individual sheet of graphite. The new avatars of carbon allotropes have prompted the coming into being of a kind of parallel world, which is invisible yet simultaneously amply visible in the media: the 'nanoworld'. Once again, after the great expectations generated by nineteenth-century synthetic chemistry, carbon appears as a cornucopia of futuristic products: screens, batteries, organic electronics, ultra-fast computers, ultra-thin sensors, traps for pollutants – and it even promises an elevator into space using a giant cable made of braided carbon nanotubes! However, these materials and their technoscientific mode of existence are far more bizarre and fascinating than their potential – and perhaps speculative – applications.

First, since these structures can be considered either as tiny crystals or as giant molecules, they challenge the boundary between organic and mineral. Second, their combinatorial potential no longer involves populations of molecules in chemical reactions; it rather refers to a game of construction involving the assembly of individual carbon molecules. Third, structure and properties matter less than the *functions* to perform (mechanical, electronic, photonic, chemical, magnetic). Finally, structures are viewed as configurations in an indefinite set of transformations.

Since these new avatars of carbon allotropes are single or multi-layered configurations of individual graphite sheets, graphene stands as the matrix of all of them, at the frontiers of matter. It is the thinnest and the strongest material known to exist (Salvage 2012, S30–S31). Its electrons behave like massless particles, travelling at a much higher speed than in silicon. Graphene is also flexible, light absorbent and is a chemical reactant. It is indefinitely malleable: it can be cut, bent, folded, rolled or zipped into a tube or a cone, just as in origami and kirigami art (García et al. 2009, 221–224). Graphene blurs the metaphysical divide between substance and phenomenon, between primary and secondary qualities.

In this technoscientific style of research, carbon is less a source of dispositions (i.e., substantial and permanent properties waiting to be actualized) than a source of *affordances* (i.e., relational and functional properties that can only be performed and instantiated by the combination of material dispositions and intentional contrivance; Harré 2003, 19–38). Indeed, graphene makes it possible to direct or confine electrons and photons, to change the polarity of a spin, to accelerate processes, to encapsulate molecules, to provide experimental systems for quantum field theory or 2D physics, to save information bits or to make sensors with the help of grafted molecule-probes. In brief, graphene affords an entire playground, displayed in the thousands of papers and patents relating to graphene that appeared in 2010.

Graphene is less the surface of something than a 'surface in itself', made of nothing else, a surface without bulk. Unlike carbon as substance, carbon as a pure surface does not stand as a permanent substratum securing an ontological identity at a deeper level of reality. Rather, it signs a new script of technoscientific research where superficiality matters.

Carbon as memory / carbon as fire

"Burning buried sunshine" (Dukes 2003, 31–44) – this sounds like a crazy project. It is no more a playful construction game with thin carbon sheets but rather a mass destruction of deeply buried underground carbonaceous materials. The fate of fossil hydrocarbons exemplifies another of the interactions between the material properties of carbon and human history and politics. The layers of biomass accumulated over millions of years as sediments of hydrocarbons are like a repository of the past. They make up the material memory of life on earth. By massively resorting to the energy of this 'buried sunshine', humans are consuming this memory at a rate of a few centuries per year. According to Alain Gras, the 'choice of fire' is the signature of our civilization, with its 'cowboy economy' and massive exploitation of fossil fuels (Gras 2007). The peculiar set of relations that have been set up and reconfigured between the material properties of carbon, international finance, expertise and democracy is well described in Timothy Mitchell's essay "Carbon Democracy." Coal played a critical role in forging democracy by encouraging social movements while oil has shaped and continues to shape the politics of Western and Middle East countries (Mitchell 2001).

Carbon's etymology, however, suggests that its association with fire is not limited to this contingent history. Carbon has been identified as a chemical element since through combustion in the experiments conducted by Lavoisier and Macquer in Paris and by Tennant in London (Lavoisier 1772, 591–616; Tennant 1797, 123–127). These experiments demonstrated that burning a certain amount of diamond and burning an equivalent mass of charcoal released the same volume of 'fixed air' (carbon dioxide).

Moreover, the association between carbon and fire does not necessarily mean irreversible consumption. Nineteenth-century chemists, investigating the circulation of carbon through the three realms of nature – mineral, plant and animal – described animals as combustion engines releasing carbon dioxide into the atmosphere, while plants 'fix' the carbon and release oxygen. Living beings do not 'burn' calories as an internal combustion engine burns gasoline. Instead, they convert calories into 'soft energy' (known today as adenosine triphosphate) and then reuse much of the waste. Ashes can be used as a fertilizer and can participate in cosmic cycles. Jean-Baptiste Dumas's 'chemical statics of organized beings' is reminiscent of carbon exhalations as Mephitis, dispensing Life and Death (Dumas 1842). Such 'archaic' symbols of carbon survive in today's climate science and policy.

Carbon as equivalent

In their efforts to mitigate the climate disruptions induced by the use of carbon-fire, humans have once again selected carbon as a standard. Carbon provides the 'rate of exchange' of all greenhouse gases (GhGs) according to their global warming potential (GWP),[5] just as gold or the U.S. dollar did for currencies. And carbon dioxide, the historical archetype of all gas, became 'carbon dioxide equivalence' (CO_2eq), the reference for devising 'green' markets and environmental finance devices.

As all equivalents, carbon affords *commensurability*. It allows quantitative comparison between various gases and between heterogeneous human actions. (How much carbon equivalent does your wedding party weigh? What is the carbon footprint of your alimentary diet?) Carbon thus bridges natural processes and human enterprises. It even allows to equate present and future GhGs emissions or reductions. For instance, planting trees or replacing a fuel generator with a wind turbine are supposed to compensate for global CO_2 emissions. A certain amount of GhGs emitted in one part of the world is thought to balance an equal reduction in another part of the world (Fragnière 2009).

Carbon markets are based on such equations (MacKenzie 2009, 440–455). The commensuration achieved by the carbon equivalent is turned into a mechanism of *compensation*, which has been compared to the system of *indulgences*, the 'market of pardon' once established by the Catholic Church (Adam 2006; Smith 2007). Buzzwords such as 'carbon neutrality', 'decarbonizing industry' or 'zero-carbon planet' make of carbon the 'bad guy'. This demonization of carbon revives the ancient identification of carbon dioxide with malefic power.

Carbon trading schemes wrongly put the blame on carbon itself. The problem is not that there is too much carbon (though there is, of course, too much CO_2 in the atmospheric air). In fact, the system of compensation nurtures the social construction of ignorance (Lohmann 2008, 359–365). It provides means for endlessly postponing effective measures aimed at reducing emissions. By dissociating actions from their consequences, it tends to dissolve the urgent issue of climate change into counterfactual and abstract speculation instead of considering how human activities have disrupted the carbon cycle over millions of years as described by paleo-climate scientists. By consuming centuries of 'buried sunshine' per year, we create a chasm between the historical time of human activities and the geological and biological timescales. Carbon trading increases the gulf between heterogeneous 'timescapes' (Lohmann et al. 2006). The alledged 'bad guy' that would have to be 'neutralized' or 'sequestered' should rather be rescued, because life depends on it. After all, carbon is life – not life in its 'essence', but life in its diversity, profusion and interdependence.

Toward an ontography of carbon

Carbon signs the rich narrative of a *persona* always on the move, always binding with itself as well as with others. Endlessly metabolized, exchanged, fixed and released, it forms as many heteronyms as there are ways for carbon to be an object. In surveying the multiple identities of carbon and treating them as heteronyms – real signatures of fictitious *personae* – this essay sought to highlight three major features of technoscientific objects.

First, the unbounded productivity of technoscientific objects may be due to their multiple identities. Carbon is much more than a scientific object. Without questioning the central importance of the definition of carbon as a chemical element, we argued that it is only one of the many ways for carbon to sign its name – as a permanent substrate underlying various phenomenological appearances. The hierarchy of levels of reality suggested in this metaphysical perspective, even when it is 'metachemically' refined in order to guarantee the notion of chemical individuality, results from the reductive entrenchment of many modes of existence into a single one. On the contrary, our approach was to spread and to maintain the constellation of carbon's modes of existence while resisting the 'imperialist' temptation to deduce their plurality from one of them.

Is it possible to characterize the scientific approach by the question 'What is carbon?' and distinguish it from a technoscientific approach interested in the question 'What does it do?' This is a false dilemma because both questions overlook the coexistence of multiple modes of being. Neither a substantialist nor an operational definition could take into account the varieties of carbon identities revealed through the circulation of carbon atoms over the centuries. The Janus portrait is only a caricature, which occults multiple personalities. Despite the broad diffusion of scientific names such as 'carbon' and 'CO_2', the purification attempted by means of the reform of chemical nomenclature never succeeded in eradicating the intimate relationship we have with carbon in our vital, technical and symbolic activities. Just as carbon forms many allotropes by binding with itself in specific and different ways, each of its modes of being is able to link up with the others via analogies, metonymies, metaphors and anaphors. For instance, carbon as a general equivalent in its economical mode plays a role analogous to the role of the mole in its chemical mode of existence, as a standard of commensuration.

Who is carbon, then? Here comes the second feature of this technoscientific object. To use the term Pessoa coined for referring to himself, the scientific name 'carbon' can be considered as the 'orthonym' of a multiplicity of heteronyms, which suggests a proper position, a degree of social 'correctness', but nothing like a deeper ontological level. But the orthonym carbon is neither object nor subject; it is a 'quasi-object' in the sense of Michel Serres (1982). Quasi-objects have no fixed essence; they are defined by the links and connections they create by circulating in local contexts. They build niches and make collectives. A striking feature of quasi-objects is that they continuously cross the boundaries between nature and culture and between the natural and the artificial. The narratives of carbon belong to natural history as well as to human history. Some of its modes of existence have been characterized as mediators, bridging the heterogeneous realms of the inorganic and the organic, of the natural and the synthetic, of ecology and economy. Through exchanges, combinations, uses and re-uses by humans and non-humans, carbon challenges the hierarchy of ontological levels by creating a sort of bio-pedo-geo-hydro-atmo-cosmo-sphere. Besides, carbon is a mediator establishing a common measure between heterogeneous regions of the world. Just as the balance in the hands of Lavoisier (Bensaude Vincent 1992, 217–237; Wise 1992, 207–256), carbon affords commensurability between the smallest and the largest scales, between the mineral, the vegetal and the animal, between the moral and the social, between the economic and the ecological. As a powerful mediating apparatus, carbon also encourages ambitious attempts at controlling the world. The balance of gains and losses, based on static equilibrium, generates a space of illusory rational control and power over the future, instantiated in carbon trading.

Finally, as carbon opens up the 'cosmopolitical' perspective of a common world shared by humans, other living beings and things, what kind of approach is the most appropriate to clarify its ontology? The narratives written by the heteronyms of carbon are at odds with the grand Promethean narrative forged by the champions of nanotechnology, synthetic biology and other ambitious technosciences. 'Shaping the world atom by atom' or re-engineering nature to improve it, are slogans that celebrate the power of human design over nature. Designers like to stress their enhanced capacity to control and shape materials thanks to their access to the molecular level. In their view, materials are no longer a prerequisite for design: fullerenes, carbon nanotubes, graphene are 'materials by design'. This phrase suggests that engineers are emancipated from all constraints of materiality. This essay, by contrast, emphasizes the affordances of materiality. As chemist Richard Smalley put it in his 1996 Nobel lecture, carbon has a 'genius wired within it':

> The discovery that garnered the Nobel Prize was the realization that the carbon makes the truncated icosahedral molecule, and larger geodesic cages, all by itself. Carbon has wired within it, as part of its birthright ever since the beginning of this universe, the genius for spontaneously assembling into fullerenes.
> (Smalley 1996, 90)

The physical and chemical properties of carbon afford opportunities for creating new lives and powerful concepts. Whether robust like the carbon backbone of DNA or delicate like the thin layer of graphene, all heteronyms of carbon are active and reactive in the hands of smart engineers. This is why an *ontography* seems more appropriate than an ontology for carbon. What is the difference?

We choose ontography as a genre for emphasizing the significance of 'graphism' and graphemes, of meaningful material traces printed by objects. Ontography offers a number of advantages. First, as Lynch argues, it is an attempt to dignify a descriptive approach based on empirical study (Lynch 2013, 444–462). Ontography deflates the quest for a unique substrate underlying a variety of materials. In stark contrast to ontology, it does not build a grand theory based on a hierarchy of the entities that make up the universe. Just as the term 'ontogenesis', ontography focuses on individual entities. It is compatible with pluralism. It seeks to give a voice to a plurality of beings rather than trying to silence them.

In addition, as an attempt to identify the modes of existence of particular entities, ontography does not assume a causal chain between levels of being. Nor does it single out any level of being. It opens up the possibility for humans to engage with objects as partners, thus co-creating their affordances. This process can be denoted as the 'instauration' of new identities (Souriau 1939). Finally, ontography allows space for a multiplicity of ontological modes. An entity such as carbon displays many different ways of being an object according to its relations, reactions and circumstances.

In this process, objects are not just materials. We have seen that carbon has become a matter of concern and a matter of affairs. May we suggest that the pattern of relations displayed by each heteronym of carbon has to become a matter of care as well? We would well be advised to take care of its multiple modes of existence in order to better inhabit the common world that we share.

Notes

1 Allotropy refers to the ability of an element to form different simple bodies made up of the same atoms bounded together in multiple specific ways.
2 Lonsdaleite, named in honour of crystallographer Kathleen Lonsdale, is a polymorph of diamond.
3 Furies or Erinyes are mythical entities personifying revenge and persecution.
4 In 1986, a giant CO_2 bubble explosion killed all animals – including hundreds of humans – in an area of a few square kilometers around the lake Nyos, in Cameroon's volcano region. The inhabitants attributed this stealth kill to Mazuku, the god of lakes (Soentgen 2010).
5 Saying that 'methane has a GWP of 23' means that 'methane creates 23 times more greenhouse effect than an equal volume of CO_2 would do over the same period of time'. Thus, to obtain the CO_2eq of a certain amount of methane, one multiplies the number of tons emitted by its GWP.

References

Adam, D. 2006. 'You feel better, but is your carbon offset just hot air?' *Guardian*, 7 October, https://www.theguardian.com/environment/2006/oct/07/frontpagenews.climatechange (accessed July 17, 2016).

Bensaude Vincent, B. 1992. 'The balance: Between chemistry and politics.' *The Eighteenth Century* 33 (3): 217–237.

Buès, C. 2001. *Histoire du concept de mole: à la croisée des disciplines physique et chimie*. Lille: Atelier national de reproduction des thèses.

Chang, H. 2012. *Is Water H_2O? Evidence, Realism and Pluralism*. Heidelberg, London and New York: Springer.

Cox, B. J. and Hill, J. M. 2007. 'New carbon molecules in the form of elbow-connected nanotori.' *The Journal of Physical Chemistry C* 111 (29): 10855–10860.

Di Lisio, A., Russo, F. and Sisto, M. 2010. 'Un itinéraire entre géotourisme et sacralité en Irpinie (Campanie, Italie).' *Physio-Géo* 4: 129–149.

Dukes, J. S. 2003. 'Burning buried sunshine: Human consumption of ancient solar energy.' *Climatic Change* 61 (1–2): 31–44.

Dumas, J. B. 1842. *Essai de statique chimique des êtres organisés*. Paris: Fortin, reprint 1972 in *Leçons de philosophie chimique*. Bruxelles: Culture & Civilisation.

Fasel, R., Cai, J., Ruffieux, P., Jaafar, R., Bieri, M., Braun, T., Blankenburg, S., Muoth, M., Seitsonen, A. P., Saleh, M., Feng, X. and Müllen, K. 2010. 'Atomically precise bottom-up fabrication of graphene nanoribbons.' *Nature* 466 (7305): 470–473.

Fragnière, A. 2009. *La compensation carbone: illusion ou solution?* Paris: Presses Universitaires de France.

García, J., Esparza, R. and Perez, R. 2009. 'Origami construction of 3D models for fullerenes, carbon nanotubes and associated structures.' *The Chemical Educator* 14: 221–224.

Geim, A. K. and Novoselov, K. S. 2007. 'The rise of graphene.' *Nature Materials* 6 (3): 183–191.

Gras, A. 2007. *Le choix du feu: Aux origines de la crise climatique*. Paris: Fayard.

Harré, R. 2003. 'The Materiality of Instruments in a Metaphysics for Experiments.' In *Philosophy of Scientific Experimentation*, edited by Radder, H., 19–38. Pittsburgh: University of Pittsburgh Press.

Iijima, S. 1991. 'Helical microtubules of graphitic carbon.' *Nature* 354 (6348): 56–58.

Iijima, S., Yudasaka, M. and Crespi, V. H. 2008. 'Single-Wall Carbon Nanohorns and Nanocones.' In *Carbon Nanotubes*, edited by Jorio, A., Dresselhaus, G. and Dresselhaus, M. S., 605–629. Topics in Applied Physics 111. Berlin: Springer.

Jiao, L., Zhang, L., Wang, X., Diankov, G. and Dai, H. 2009. 'Narrow graphene nanoribbons from carbon nanotubes.' *Nature* 458: 877–880.

Koltai, J., Rusznyák, A., Zólyomi, V., Kürti, J. and László, I. 2009. 'Junctions of left- and right-handed chiral carbon nanotubes – nanobamboo.' *Physica Status Solidi (b), Special Issue: Electronic Properties of Novel Materials* 246 (11–12): 2671–2674.

Kroto, H. W., Heath, J. R., O'Brien, S. C., Curl, R. F. and Smalley, R. 1985. 'C60: Buckminsterfullerene.' *Nature* 318 (6042): 162–163.

Latour, B. 2013. *An Inquiry into Modes of Existence*. Cambridge, MA: Harvard University Press.

Lavoisier, A. L. 1772. 'Sur la destruction du diamant par le feu.' In *Opuscules physiques et chimiques – Œuvres*, vol. 2, 1862, 591–616. Paris: Imprimerie impériale.

Lohmann, L. 2008. 'Carbon trading, climate justice and the production of ignorance: Ten examples.' *Development* 51: 359–365.

Lynch, M. 2013. 'Ontography: Investigating the production of things, deflating ontology.' *Social Studies of Science* 43 (3): 444–462.

MacKenzie, D. 2009. 'Making things the same: Gases, emission rights and the politics of carbon markets.' *Accounting, Organizations and Society* 34 (3–4): 440–455.

Mendeleev, D. 1953 [1871]. 'The Relation between the Properties of the Atomic Weight of the Elements.' reprinted in *Sourcebook in Chemistry 1400–1900*, edited by Leicester, H. M. and Klickstein, H. S. New York: Dover.

Mitchell, T. 2001. *Carbon Democracy: Political Power in the Age of Oil*. London and New York: Verso.

Nasibulin, A. G., Pikhitsa, V. P., Jiang, H., Brown, D. P., Krasheninnikov, A. V., Anisimov, A. S., Queipo, P., Moisala, A., Gonzalez, D., Lientschnig, G., Hassanien, A., Shandakov, S. D., Lolli, G., Resasco, D. E., Choi, M., Tománek, D. and Kauppinen, E. I. 2007. 'A novel hybrid carbon material.' *Nature Nanotechnology* 2 (3): 156–161.

Paneth, F. 1962. 'The epistemological status of the concept of element.' *British Journal for the Philosophy of Science* 13: 144–160.

Rettenbacher, A. S., Perpall, M. W., Echegoyen, L., Hudson, J. and Smith Jr., D. W. 2007. 'Radical addition of a conjugated polymer to multilayer fullerenes (carbon nano-onions).' *Chemistry of Materials* 19 (6): 1411–1417.

Ruthenberg, K. and van Brakel, J. (eds.). 2008. *Stuff: The Nature of Chemical Substances*. Würzburg: Verlag Königshausen & Neumann.

Salvage, N. 2012. 'Super carbon.' *Nature* 483 (7389): S30–S31.

Serres, M. 1982. *The Parasite*, translated by Lawrence R. Schehr. Baltimore and London: Johns Hopkins University Press.

Smith, K. 2007. *The Carbon Neutral Myth, Offsets Indulgences for You Climate Sins*. Amsterdam: Carbon Trade Watch.

Soentgen, J. 2010. 'On the history and prehistory of CO_2.' *Foundations of Chemistry* 12 (2): 137–148.

Souriau, E. 1939. *L'instauration philosophique*. Paris: Alcan.

———. 2009. *Les différents modes d'existence*, with an introduction by Bruno Latour and Isabelle Stengers. 2nd ed. Paris: Presses Universtaires de France.

Tennant, S. 1797. 'On the nature of diamond.' *Philosophical Transactions of the Royal Society* 87: 123–127.

Tomic, S. 2010. *Aux origines de la chimie organique – Méthodes et pratiques de pharmaciens et des chimistes (1785–1835)*. Rennes: Presses Universitaires de Rennes.

Travis, A. S., Hornix, W. J., Bud, R. F and Homburg, E. 1992. 'The emergence of research laboratories in the dyestuffs industry, 1870–1900.' In *British Journal for the History of Science*, 25 (1): 91–111.

Virgil. 1909. *Aeneid*, translated by John Dryden. New York: P. F. Collier and Son.

Wise, N. 1992. 'Mediations: Enlightenment Balancing Acts, or the Technologies of Rationalism.' In *World Changes, Thomas Kuhn and the Nature of Science*, edited by Horwich, P., 207–256. Cambridge: MIT Press.

Internet Resources

Lohmann, L. et al. 2006. Carbon Trading: A Critical Conversation on Climate Change Privatization and Power, http://www.thecornerhouse.org.uk/sites/thecornerhouse.org.uk/files/carbonDDlow.pdf (accessed September 2014).

Smalley, R. 1996. 'Discovering the Fullerenes,' Nobel Lecture, 7 December, http://www.nobelprize.org/nobel_prizes/chemistry/laureates/1996/smalley-lecture.html (accessed September 2014).

13 Monitoring and remediating a garbage patch

Jennifer Gabrys

Located across the world's oceans are several sizeable concentrations of plastic debris that have variously earned the title of 'garbage patches'. The Great Pacific Garbage Patch in particular has become an object of popular and technoscientific interest. It is an environmental anecdote to confirm our worst fears about overconsumption, and it is an imagined indicator of what may even outlive us, given the lengths of time that plastics require to degrade. The garbage patch is in many ways an amorphous object, drifting through media spaces and environmental narratives as an ominous form that focuses attention to the ways in which oceans have become planetary-sized landfills.

'Discovery' of the 'garbage patches' is often attributed to Charles Moore, a captain turned scientist who deployed and publicized the term to describe his observations of a high concentration of suspended plastics in the clockwise currents of the North Pacific Gyre, and so brought the phenomenon of plastics in the Pacific to greater public attention (Moore et al. 2001; Moore and Phillips 2011). However, oceanographer Curtis Ebbesmeyer originally coined 'garbage patch' as a term to describe the tendency for flotsam to collect in sub-orbiting gyres (Ebbesmeyer and Scigliano 2009, 167). Although scientific observations of the circulation patterns of gyres and the accumulation of debris had taken place previously (Carpenter and Smith 1972), Ebbesmeyer and Moore both suggest that it was the naming of the ocean debris as 'patches' that eventually galvanized attention for this issue (Ebbesmeyer and Scigliano 2009; Moore and Phillips 2011). Anecdotally, the garbage patches have become one of the most potent figures of public concern about environments. The imagining of vast stretches of oceans choked by plastics is at once a media device for expressing the worst of the destructive impacts of humans on the planet, and an attractor for scientific study into ocean plastics, since it is a topic about which citizens most frequently make inquiries to environmental agencies.

Geomythologizing a garbage patch

Popular imaginings of the Pacific Garbage Patch have included comparisons of its size to the state of Texas, or even as an island that might be named an eighth continent formed of anthropogenic debris. Upon hearing of the concentration of plastic wastes in the Pacific, many people search on Google Earth for visual evidence of this environmental calamity. Surely a human-induced geological formation of this magnitude must be visible even from a satellite or aerial view? However, because the plastic wastes are largely present as microplastics in the form of photo-degraded and weathered particles, the debris exists more as a suspended soup of microscopic particles. While Google Earth may be a platform for visualizing and locating ocean data,[1] this visualization technique is much different than seeing the patch as an actual photographic object. The inability to locate the garbage patches on Google Earth, a tool for scanning the seas through a conjugation of remote sensing, aerial photography and online interfaces even gives rise to popular controversy about how to locate the patch and whether the plastic conglomerations are actually present in the oceans, and if so, how to address the issue. The relative invisibility and inaccessibility of the patches render them as looming imaginative figures of environmental decline, and yet amorphous and un-locatable and so seemingly resistant to environmental action.

The difficulty of visually locating the patch as an identifiable object reveals how the garbage patch is on one level a 'myth' about how plastics accumulate in the oceanic gyres. While plastic exists in considerable quantities in these areas where currents converge into still expanses of oceans, the form that the plastic takes is often in varying stages of decomposition, suspended within water columns, and even filtered through various organisms that ingest these particles. Several scientific entities such as the U.S. National Oceanic and Atmospheric Administration (NOAA) have gone to lengths to dispel the 'myth' of the garbage patch by clarifying that the patch is not literally a surface coating of plastics but more of a zone with higher concentrations of suspended plastics and especially microplastics.[2] Yet the term 'patch' persists in use, not least because it is seen to bring increased public attention to an environmental issue somewhat removed from everyday experience. Scientific agencies such as NOAA may explain that the patches do not assemble as islands of plastics, but may also continue to use the term as shorthand to describe plastics concentrations; while the media may use images of accumulated concentrations of plastics in urban harbors, for instance, to stand in for the more distant and difficult to visualize garbage patches; and artists may focus on sites such as the Midway Atoll to capture the effects of macroplastics that wash up on islands proximate to oceanic gyres, and which are often taken to be representative of the general constitution of the garbage patches (Jordan 2009). The patch is a concept that accumulates uses, images and imaginaries, where the more complex and amorphous garbage patches resist easy identification.

But what sort of 'myth' might the garbage patch embody? And what sort of technoscientific object is this? Rather than seek to 'dispel the myth' of the patch, in this chapter I am interested to take up the ways in which the indeterminate and changeable qualities of the plastic garbage patch focus practices for determining how and what this concentration of plastics in the ocean consists of, how to monitor these plastics, and how excess plastic marine debris might be remediated. I consider two primary aspects of the garbage patch that pertain to its materiality and ongoing circulation. In the first instance, I look at how microplastics or small-scale plastic particles are the primary material strata of the garbage patch, and how these materials influence the processes and form of this oceanic debris-space. I then look at techniques for mapping the circulation of ocean debris, and focus on the Global Drifter Program as an ocean observation project that has deployed buoys equipped with sensors that communicate with satellites to study the drift of plastics and other debris in the oceans.

Based on these accounts of monitoring plastics in the oceans and locating debris concentration zones, I consider how the garbage patch could be termed a geomythology, or a tale of uncertain earth events and forces. Geomythology is a term coined to describe the ways in which distinct earth formations are often explained by myths that capture how they came to be (Vitaliano 1973). The way in which environments and earth features form is not just a matter of geologic process, but is also a social, cultural and narrative process that conveys imagined or actual accounts of how earth formations come to be recognizable objects. From volcanoes to floods, these formations and events have been generative of stories that describe their emergence and ongoing significance. Google Earth has even been used as a tool to identify these formations as features, and to provide legitimacy for these stories as attached to actual earth objects. I adopt and adapt this term to consider how the uncertain and indeterminate aspects of the garbage patches give rise to environmental narratives and technoscientific practices for bringing these newer geological objects into view.

While geomythology might typically exist as a narrative form explaining earth events and formations of indeterminate origins, this inquiry is less focused on explaining the *origins* of the garbage patch. Instead, the geomythology developed here considers how much of the uncertainty around the gyres involves exploring what kind of earth or ocean object the garbage patch is and, even more, what *potential* events and effects may unfold through this shifting formation. Plastics have inevitably been present in the oceans for many decades, but at some undefined moment the concentration of plastics in oceans became high enough to constitute an object of study. Not only is the genesis of this object indeterminate, but so too is the way in which multiple other objects of study emerge within this space.

This chapter attempts to craft a geomythology about the garbage patch in order to consider how the potentiality of this amorphous and changeable object informs technoscientific practices for monitoring and remediating this environmental phenomenon. Such an approach works with a generative understanding of the garbage patch as a *technoscientific object in process*, which further gives rise to new objects and societies of objects to come (Whitehead 1929). The genesis of the garbage patch is not singular, but constituted through multiple objects that emerge within these ocean ecologies, and which are of indeterminate and ongoing duration, since plastics appear to persist and transform in environments for indefinite periods of time.

Locating a technoscientific object

In more current scientific literature, the Pacific Garbage Patch is often referred to as the Eastern Garbage Patch. The patch area is located between Hawaii and

California within the North Pacific Subtropical High, a shifting zone of high pressure and relatively calm water. The Eastern or Great Pacific Garbage Patch is not the only location where marine debris collects in the Pacific, moreover. A Western Garbage Patch has since been identified near Japan, and the Subtropical Convergence Zone at the transition zone between the Subpolar Gyre and the Subtropical Gyre is also noted to collect larger amounts of marine debris (Pichel et al. 2007; Howell et al. 2012).

These debris collection zones are also connected to five identified oceanic gyres, including two in the Pacific, two in the Atlantic, and one in the Indian Ocean (Lebreton et al. 2012). As Howell and collaborators write, the Pacific subtropical gyre is the 'largest circulation feature on our planet, and the earth's largest continuous biome.' Gyres tend to spiral or converge inward, and the North Pacific Subtropical Gyre – the larger system of which the Eastern Garbage Patch is but a small part, or a 'gyre within a gyre' (Howell et al. 2012) – has been estimated to be roughly between seven to nine million square miles.[3]

Of the many forms of marine debris floating through oceans and seas, plastics are the primary form of waste moving through and collecting in oceans. Sixty to eighty percent of all marine debris in oceans is composed of plastics (Gregory and Ryan 1997; EPA 2011). Plastic fragments sifting through ocean waters most often travel from land-based sources, typically migrating from urbanized areas, wastewater, landfills and plastics manufacturing sites into oceans. A smaller proportion of plastics derive from marine-based sources, including offshore shipping and fishing activities (Derraik 2002). Yet with all of these forms of primarily plastics-based marine debris, plastics circulate from manufacturing, use and disposal to wayward and often unidentifiable objects congealing in the shifting spaces of oceanic gyres.

The accumulation of plastics in oceans and seas is increasingly remaking oceanic materialities and environmental processes (Gabrys et al. 2013). However, plastic in gyres including the Great Pacific Garbage Patch assembles less as an identifiable mass of plastic, but more as a suspended soup of finer plastic fragments and microplastics. The 2001 *Marine Pollution Bulletin* article in which Moore and his collaborators describe their findings of plastic to plankton comparisons in the North Pacific Gyre indicates that up to 98 percent of the plastic material gathered through trawls of the Pacific gyre were composed of finer plastic particles. Of these finer particles, 'thin films and polypropylene/monofilament line' were present as identifiable plastic, while 'unidentified plastic' in the form of plastic fragments were the three main types of plastic sampled.[4]

If Google Earth or a satellite view of the garbage patch proves to be an impossible undertaking, it is because the plastics suspended in oceans are not a thick choking layer of identifiable objects, but more of a confetti-type array of plastic bits. Practices of sampling plastics in areas of high concentration of marine debris involve working with fine-mesh trawls. These trawls are able to collect microplastics across a range of visible and relatively invisible sizes. Establishing a universal standard for microplastics smaller than 5 mm has been seen as an important step in regularizing the study of microplastics in seas since plastics break up into such a wide range of forms and sizes (Thompson et al. 2004; Arthur et al. 2009). Some plastic fragments and objects are large enough to pose ingestion hazards to marine organisms – these are often termed macroplastics (Gregory 2009). Other plastic fragments are too small as to be invisible and undetectable, or to be readily ingested by many marine organisms without immediately obvious effect. Size is an important indicator in assessing ocean plastics, since on the one hand there is the risk of entanglement and ingestion hazards, and on the other hand there are more unknown issues as to how smaller particles of microplastics may transform ocean ecosystems. The technoscientific object of the garbage patch in this way generates additional objects that wander into view as the complex processes of plastics in the oceans begin to unfold. Microplastics were settled at a certain size in order to study the effects of this distinctive and pervasive category of plastics, which does not apparently cause immediate harm, but does influence oceans and ocean life in often yet-to-be-known ways.

The impacts of these microplastics are in many ways still somewhat uncertain, and may have a web of effects that may range from adsorption, chemical transfer of persistent organic pollutants (POPs) among other substances, endocrine disruption, alterations in plankton feeding habits, decreasing biodiversity and shifts in climate change (Barnes et al. 2009; Takada 2009; Thompson et al. 2009; UNEP 2009). Numerous 'data gaps' exist in relation to microplastics (EPA 2011, 1506). Plastics do adsorb and concentrate chemicals such as POPs from seawater, and transport these substances to other locations (Takada 2009). But how do chemical substances migrate into and across organisms, and what effects do these substances have on organisms over time? How do they cause endocrine disruption within marine organisms and humans (Di Lorenzo et al. 2008; Boerger et al. 2010)? What effect do microplastics have on plankton and insect populations, and how might these also affect food webs, biodiversity and climate change, by altering the composition and source-sink dynamics of oceans (Di Lorenzo et al. 2008; Goldstein et al. 2012)?

Attempting to establish the matters of fact related to garbage patches in the oceans is an experimental process that could be understood to be less about how to demystify the garbage patches and more about how the ongoing attempts to make sense of the garbage patch and the effects of plastic are bound up with these complex constellations of objects that pose pressing matters of concern. Here, attempting to establish matters of fact in relation to plastics in oceans is not about dispelling fictions, but about experimental modes of narrating, testing and sensing that bring an object of concern toward a space of workable interpretations and engagement.[5] Experimenting within ecology is often understood as the process of testing interventions in order to form new hypotheses (Underwood 1997). Here and in relation to technoscientific objects, this experimentation of matters of fact and concern is understood as an intervention of a different sort, which questions how technoscientific objects are made evident and how they may proliferate other objects as part of the processes of identifying, locating and monitoring their actuality.

Data gaps are an important part of the way in which matters of fact in relation to plastics are experimented through matters of concern. These gaps serve to mobilize technoscientific experiments on how plastics transform and rematerialize in oceanic spaces. Gyres with higher concentrations of marine debris, and in particular the North Pacific convergence zone, are primary sites where questions related to the effects of plastics unfold. Yet from this perspective, the garbage patch is one of several objects that emerge in this site of technoscientific study and concern. Microplastics are primary objects within the garbage patch, and they must also be studied in order to gauge the characteristics of marine debris gyres. Locating the garbage patch as an object is not a simple delineation in space, but a topological unfolding of ongoing potential effects. The garbage patch is thus not one technoscientific object, but a *"society" of objects* (following Whitehead) that in their interaction (or intra-actions, as Barad suggests) give rise to new and ongoing relations, formations and actual occasions (Whitehead 1929; Barad 2003). Locating the Pacific Garbage Patch is not a matter of demarcating a stable continent of plastic on a satellite map. Instead, the garbage patch requires locating technoscientific objects within objects – objects that are intra-acting and giving rise to new potential effects and environmental conditions. The geomythology of the garbage patch must grapple with the very ways in which the object-ness – and the *potential* object-ness – of the garbage patch is never stable or given, but in process and giving rise to new technoscientific engagements.

The genesis and ontology of technoscientific objects shift from one of establishing origins and stabilizing uses to one of anticipating new effects and relations through conditions of becoming. Rather than a technoscientific ontology of control or stabilization, plastics in oceans indicate how the material agency and intra-action of the multiple objects within this garbage patch society are continually generating new conditions, objects and societies of objects. The potential technoscientific ontology that emerges here is multi-agential and more-than-human. Adsorption of chemicals may alter the habits of some marine organisms; degradation of plastics may shift the composition of source-sink dynamics. The garbage patch is a site where objects proliferate. New potential effects arise in relation to these new objects and intra-actions. As plastics fall apart, they generate new occasions of becoming and new processes of materialization (Gabrys 2011; Gabrys et al. 2013). Plastics as they persist in environments are characterized by *plasticity* less as a condition of frictionless adaptation, but as more as a condition of material persistence and transformation within and without organismal boundaries. The experimental ontologies that emerge here are necessarily material, distributed and multi-agential. Not one object eventually fixed for study or exploitation, but multiple objects in shifting and changeable conditions that assemble within a geomythological topology: this is the garbage patch.

Monitoring a technoscientific object

Locating the garbage patch is on one level bound up with determining what types of plastic objects collect within it, and what effects they have. Yet on another level, locating the garbage patch involves monitoring its shifting distribution and extent in the ocean. As has been discussed so far, the garbage patch is not a fixed or singular object, but a society of objects in process. How does the garbage patch come into view as detectable while also in process? The composition of the garbage patch consists of plastics intra-acting across organisms and environments. But it also moves and collects in distinct and changing ways due to ocean currents, which are influenced by weather and climate change, as well as the turning of the Earth (in the form of the Coriolis effect) and the wind-influenced direction of waves (in the form of Ekman transport). As an oceanic gyre, the garbage patch moves as a sort of weather system, shifting during El Niño events, and changing with storms and other disturbances (Howell et al. 2012).

Techniques for studying marine debris coincide with techniques for studying ocean circulation. In some cases, flotsam is directly observed and modeled as a way of gauging likely movements of debris across ocean currents. A well-known study by Ebbesmeyer focused on the movement of bath toys (ranging from ducks, frogs, beavers and turtles) that spilled from a container ship, which was washed overboard during a storm in 1992 (Ebbesmeyer and Scigliano 2009). Based on beachcomber efforts, as well as identifying by serial number the bath toys, and mapping and inputting coordinates into the Ocean Surface Current Simulator (OSCURS) computer program, Ebbesmeyer developed a circulation model that gave the locations of gyres (which corresponded with related gyre studies), and the likely time that objects spend in gyres.

His "flotsametrics" technique drew on decades of studies that have attempted to discern patterns in ocean circulation by mapping the pathways of flotsam. Here, the drifting message in a bottle, or MIB, is a classic reference point for studying and experimenting with objects as they travel in oceans, with a MIB recently having been found in Scotland that dated to 1914 (Madrigal 2012). But numerous other experiments have been developed in this space besides, including a 1976–1980 experiment by NOAA that replaced "tens of thousands of plastic cards in response to significant oil spills along the East Coast from Florida to Massachusetts" (Ebbesmeyer n.d.). Working on behalf of NOAA, Ebbesmeyer collected these plastic spill cards over time, some of which have drifted for over twenty-five years. Given the length of these drifts, Ebbesmeyer estimates the plastic spill cards may have circled between seven to nine times around the North Atlantic Subtropical Gyre – the Atlantic version of the Pacific Garbage Patch (Ebbesmeyer n.d.; Law et al. 2010).

'Traceable Drifter Unit', or TDU, is the term that Ebbesmeyer uses to describe flotsam that is released en masse (with releases sometimes exceeding 100,000 drifters) in the ocean and which yields data relevant to ocean surface currents. These TDUs have ranged from Guinness beer bottles to MIBs with biblical or governmental messages, as well as material from known container spills. As Ebbesmeyer writes of flotsam drift studies, 'By their endurance for as long as a century, flotsam provides a tool for tracing long planetary drifts. Drifters riding the global conveyor belts, for example, require twenty years to circle the earth'.[6] The ways in which flotsam travels, drifts and collects in oceans may be studied over long periods of time, and the different exit points for flotsam to head toward coasts, or extended times in which it takes to reach coasts, may indicate just how long marine debris remains within oceans and in particular how many times debris circulates around ocean gyres. The convergence zones are not just collections of primarily plastic stuff, but are also metamorphosing oceanic repositories that include items from the early boom years of plastics, sporadic spills from container ships and passing fashions in consumer goods. The packaging, films, fragments and assorted objects that cycle around gyres may remain there for many decades to come, eventually forming new oceanic environments and influencing organisms and food webs.

Many monitoring practices are used to observe ocean circulation and the likely movement of marine debris, including airborne sensors, coastal webcams, drifter buoys and tracers, remote sensing via satellites and even iPhone apps (IPRC 2008). One project working in across techniques of drifter tracing and sensor communications, the Global Drifter Program, has deployed tracking buoys that communicate with satellites to establish patterns in ocean currents. Along the way, the drifters have also become devices for establishing the likely movements of marine debris, since where the drifters collect is likely to indicate the same locations in which other flotsam collects (Maximenko et al. 2012).

The Global Drifter Program consists of a platform of more than 1250 drifting buoys that have been deployed over several decades, spanning from initial development in 1979 to current annual mass deployments to monitor the oceans (Lumpkin et al. 2012). The buoys monitor the upper water column, and provide information on ocean surface and atmospheric conditions, as well as fluxes between air and sea. Detecting and sensing sea surface temperature, barometric pressure, wind velocity, ocean color, salinity, and sub-surface temperatures, the buoys monitor ocean conditions primarily to determine weather and climate patterns. As they circulate, the buoys consistently send 140-character messages on location and ocean conditions – what physical oceanographer Erik van Sabille has referred to as 'Twitter from the ocean' (UNSW 2013). Part of the Global Earth Observation System of Systems (GEOSS) of monitoring technologies, the Global Drifter buoys link up with earth models to provide forecasting data. Run through the Atlantic Oceanographic and Meteorological Laboratory (AOML) in Miami, Florida, the drifters are deployed in study sites, and then circulated through the oceans.

In addition to functioning as weather, climate and circulation observation devices, the drifters have provided detailed and longer-term data on the likely movement of debris in oceans. A high proportion of drifters have gravitated toward the five gyres, and in this sense have provided further data for establishing where these gyres are located and how long drifters or debris may converge in these areas (Dohan and Maximenko 2010; Lebreton et al. 2012; Maximenko et al. 2012). Through studies that use Global Drifter data, the emergence of a sixth Arctic gyre has been identified, as well as the ways in which patches are 'leaky' and circulate debris across regions over centuries (Van Sebille et al. 2012). The drifters are in many ways proxies for demonstrating how debris travels over time in oceans, how debris converges in gyres, and the length of time it may take debris to exit convergence zones (if at all) and wash up in coastal regions. The drifters were not necessarily originally developed as monitoring devices to study the accumulation of debris directly, since they focused on circulation patterns. But the drifters became an imported technique for studying how debris circulates and settles in ocean spaces in relation to the study of ocean circulation. The drifters also eventually become debris, as they have a limited (five-year) battery life, and cease to function due to mechanical error, environmental stress and more (Lumpkin et al. 2012).

The Global Drifter Program potentially not only corroborates or qualifies prior and differing studies on ocean circulation, but also provides a more real-time observation platform for understanding how gyres may shift – and debris concentrations along with them. In many ways, the ongoing deployments, shifting oceanic trajectories and real-time communication of the drifters are practices that emerge in relation to and through a fidelity to the shifting technoscientific objects under study. Debris concentrations – whether differently termed and identified as the garbage patch or gyre or convection zone – exist as technoscientific objects within objects, and these objects change the other objects with which they are intra-acting. New object conditions are continually generated here, from changes in chemical and biological conditions, to alterations in habitats, shifting locations of the garbage patch due to ocean and atmospheric circulation, and even changes in climate and wider environments (EPA 2011). The sensing and satellite-linked drifters enable a mode of technoscientific practice that is able to more continually monitor these shifting object conditions and processes.

Remediating and transforming a technoscientific object

Such a shifting society of objects, which are in process and so oriented toward further potentialities (Whitehead 1929, 214–215), gives rise to distinct technoscientific practices for engaging with these emerging and generative technoscientific ontologies. The garbage patch is an entity – more or less accurately defined – where plastics are in process, rerouted and circulating across oceanic systems. As discussed here, the plastics that drift through oceans and debris patches are indeterminate objects of study that are often approached obliquely and through their unknown potentialities. Given its material constitution, the garbage patch is also not external to that which inhabits it, but occupies the many different organisms that live amid it, where many organisms filter plastics through their bodies. As an ocean-in-the-making (Harvey and Haraway 1995; Gabrys 2013), the geomythological force of these object-events unfolds as a space of potential objects to come, and of indeterminate environmental events to make sense of.

These oceans- and objects-in-the-making are also sites for technoscientific practices- and models-in-the-making. Drifters and sensors, together with studies of particle movement and ocean currents, are *both* abstract approaches to understanding the garbage patch, as well as concrete things in the world that mobilize matters of concern (Stengers 2008; Helmreich 2011). The garbage patch on one level could be seen as a geomythological model, an abstraction that is also a "lure for feeling" that experiments toward matters of fact (Whitehead 1929, 88). Such abstraction, as a lure, is not separate from concrete events, but instead is an attractor for identifying that which matters, and how to make sense of that experience (Stengers

2008, 96). In this geomythology of the garbage patch, the *genesis* and *ontology* of this technoscientific object is bound up with a genesis that is always in the middle of things, as processual ontology (Bensaude Vincent et al. 2011). Experimental practices and compound objects converge in these oceanic gyres. Being alert to the garbage patch and debris concentrations in the oceans requires developing an attention to the generative and potential materialities that may continue to unfold through these objects.

In many ways, this geomythology finally shares a sideways correspondence with that earlier plastic *mythology* rendered by Roland Barthes. In his concise postwar account of plastics, he describes plastics as "the stuff of alchemy," through which the "transmutation of matter" takes place (Barthes 1957, 96). His description charts the 'transit' of plastics from raw material to any number of objects. This study of the garbage has in a related but different way dealt with the transit of plastics from discarded object to environmental and oceanic agent. As plastics break down in oceans, they assemble in oceanic gyres, filter through marine organisms, alter environmental conditions, and turn up as technoscientific objects of concern. The same plastic changeability – or plasticity – that Barthes expounded on as giving rise to an infinite array of consumer goods redirects toward a material changeability that influences and transforms environments on a planetary scale. The 'transmutation' that takes place here is equally subject to speculation: what potential events and objects will emerge through these plasticized oceans, marine organisms, and technoscientific studies?

The garbage patch emerges across multiple registers as a technoscientific object. On at least one level, it is present as the product of technoscientific advances in materials, where plastics give rise to new environmental and technoscientific problems as a result of the solutions they initially presented. On another level, monitoring the plastic waste requires new technologies of observation, from remote sensing to distributed sensor buoys, to bring plastic as marine debris to attention. Such technoscientific observation techniques focused on marine debris in the gyres inevitably also mobilize responses for remediating and managing the issue of plastics in the seas. In this sense, the garbage patch in its intractable plasticity gives rise to technoscientific practices not just to monitor but also to repair, control or otherwise manage this object of study and concern.

How does the relationship between monitoring and intervening in the garbage patch influence this technoscientific object and the practices employed to study and respond to it? Intervening within and developing strategies for addressing the garbage patches may on one level appear to require a beach-cleaning effort or anti-litter campaign. Yet on another level, as a matter of concern, the garbage patch indicates how designing a philosophy for engaging with the indeterminate and ongoing intra-actions and societies of objects may be one way to engage with the garbage patch as a space in which to experiment the matter-of-factness and concernedness of plastic objects as they transform in oceans. Within this technoscientific ontology, new understandings of responsibility may also proliferate, in terms of how the relations between objects are articulated abstractly and concretely, how societies of objects mobilize distinct types of technoscientific and environmental practices, and how the material occasions of oceans are not even an external object of study, but an event in which we are now participating and through which we will continue to be affected.

A key question arises from this study of the garbage patch as a generative technoscientific object, which is what other forms of technoscientific engagement and *politics* might be necessary not just to articulate a project of environmental awareness (which is what the project to identify the patch as an explicit and *visual* aberration perhaps demonstrates), but as an object that provokes new forms of environmental participation and attention to the *eventual* effects of our material lives. What experimental forms of politics and environmental practices might we develop that are able to attend to these processual matters of concern? A repurposed geomythic discourse of the garbage patches might then attend to the indeterminate edges of technoscientific objects, and to the modes of engagement yet to be experimented and generated at these sites. Perhaps this geomythology gives rise to the need for a cosmopolitical approach to technoscientific objects such as this (Stengers 2010), which do not emerge as much through their performative or instrumental capacities, as they do through the debris of useful applications (plastic) that have other capacities and material effects beyond that which was anticipated. Here, new technoscientific objects come together as the remains from original technoscientific pursuits, and which necessarily give rise to new practices for studying these residual and yet emergent objects – as societies of objects, with more-than-human effects.

Notes

1 National Center for Ecological Analysis and Synthesis (NCEAS), "A Global Map of Human Impacts to Marine Ecosystems," http://www.nceas.ucsb.edu/globalmarine.

2 National Oceanic and Atmospheric Administration (NOAA) "De-mystifying the 'Great Pacific Garbage Patch,'" at http://marinedebris.noaa.gov/info/patch.html.

3 Meanders and eddies are areas where most visible 'patches' of garbage emerge, whereas gyres tend to have overall higher concentrations of microplastics that are quite often invisible or difficult to detect (NOAA n.d.).

4 "Several limitations restrict our ability to extrapolate our findings of high plastic-to-plankton ratios in the North Pacific central gyre to other areas of the ocean. The North Pacific Ocean is an area of low biological standing stock; plankton populations are many times higher in near-shore areas of the eastern Pacific, where upwelling fuels productivity [. . .] Moreover, the eddy effects of the gyre probably serve to retain plastics, whereas plastics may wash up on shore in greater numbers in other areas. Conversely, areas closer to the shore are more likely to receive inputs from land-based runoff and ship loading and unloading activities, whereas a large fraction of the materials observed in this study appear to be remnants of offshore fishing-related activity and shipping traffic" (Moore et al. 2001, 120).

5 According to Isabelle Stengers, "the production of the matter of fact that could operate as a reliable witness for the 'adequacy' of an interpretation is always an experimental achievement. As long as this achievement remains a matter of controversy, the putative matter of fact will remain a matter of collective, demanding, concern" (Stengers 2008, 94). See also Latour (2004).

6 In "Using Flotsam to Study Ocean Currents," Ebbesmeyer further writes, "We must develop networks, which remain vigilant to collect this flotsam." This is something he has supported through the efforts of Beachcombers Alert, http://www.beachcombers.org.

References

Arthur, C., Baker, J. and Bamford, H. (eds.). 2009. *Proceedings of the International Research Workshop on the Occurrence, Effects and Fate of Microplastic Marine Debris*, September 9–11, 2008, University of Washington–Tacoma, Tacoma, WA (NOAA Technical Memorandum NOS-OR&R-30).

Barad, K. 2003. 'Posthumanist performativity: Toward an understanding of how matter comes to matter.' *Signs: Journal of Women in Culture and Society* 28 (3): 801–831.

Barnes, D. K. A., Galgani, F., Thompson, R. C. and Barlaz, M. 2009. 'Accumulation and fragmentation of plastic debris in global environments.' *Philosophical Transactions of the Royal Society B: Biological Sciences* 364 (1526): 1985–1998.

Barthes, R. 1957. 'Plastic.' In *Mythologies*, translated by Annette Lavers, 97–99. New York: Farrar, Straus and Giroux, 1972.

Bensaude Vincent, B., Loeve, S., Nordmann, A. and Schwarz, A. 2011. 'Matters of interest: The objects of research in science and technosciences.' *Journal for General Philosophical Science* 42 (2): 365–383.

Boerger, C. M., Lattin, G. L., Moore, S. L. and Moore, C. J. 2010. 'Plastic ingestion by planktivorous fishes in the North Pacific Central Gyre.' *Marine Pollution Bulletin* 60 (122): 275–2278.

Carpenter, E. J. and Smith Jr., K. L. 1972. 'Plastics on the Sargasso Sea surface.' *Science* 175: 1240–1241.

Derraik, J. G. B. 2002. 'The pollution of the marine environment by plastic debris: A review.' *Marine Pollution Bulletin* 44: 842–852.

Di Lorenzo, E., Schneider, N., Cobb, K. M., Franks, P. J. S., Chhak, K., Miller, A. J., McWilliams, J. C., Bograd, S. J., Arango, H., Curchitser, E., Powell, T. M. and Rivière, P. 2008. 'North Pacific Gyre Oscillation links ocean climate and ecosystem change.' *Geophysical Research Letters* 35: 1–6.

Dohan, K. and Maximenko, N. 2010. 'Monitoring ocean currents with satellite sensors.' *Oceanography* 23 (4): 94–103.

Ebbesmeyer, C. n.d. 'Using Flotsam to Study Ocean Currents,' NASA Ocean Motion and Surface Currents, http://oceanmotion.org/html/gatheringdata/flotsam.htm.

Ebbesmeyer, C. and Scigliano, E. 2009. *Flotsametrics and the Floating World: How One Man's Obsession with Runaway Sneakers and Rubber Ducks Revolutionized Ocean Science*. Washington, DC: Smithsonian Books, Collins and HarperCollins.

EPA (U.S. Environmental Protection Agency). 2011. *Marine Debris in the North Pacific: A Summary of Existing Information and Identification of Data Gaps*, EPA-909-R-11-006. San Francisco: EPA.

Gabrys, J. 2011. *Digital Rubbish: A Natural History of Electronics*. Ann Arbor, MI: University of Michigan Press.

Gabrys, J. 2013. 'Plastic and the work of the biodegradable.' In *Accumulation: The Material Politics of Plastic*, edited by Gabrys, Jennifer, Hawkins, Gay, and Michael, Mike, 208–227. London: Routledge.

Gabrys, J., Hawkins, G. and Michael, M. (eds.). 2013. *Accumulation: The Material Politics of Plastic*. London: Routledge.

Goldstein, M. C., Rosenberg, M. and Cheng, L. 2012. 'Increased oceanic microplastic debris enhances oviposition in an endemic pelagic insect.' *Biology Letters* 8 (5): 817–820.

Gregory, M. R. 2009. 'Environmental implications of plastic debris in marine settings -entanglement, ingestion, smothering, hangers-on, hitch-hiking and alien invasions.' *Philosophical Transactions of the Royal Society B: Biological Sciences* 364 (1526): 2013–2025.

Gregory, M. R. and Ryan, P. G. 1997. 'Pelagic plastics and other seaborne persistent synthetic debris: a review of Southern Hemisphere perspectives.' In *Marine Debris Sources, Impacts and Solutions*, edited by Coe, J. M. and Rogers, D. B., 49–66. New York: Springer Verlag.

Harvey, D. and Haraway, D. 1995. 'Nature, politics, and possibilities: A debate and discussion with David Harvey and Donna Haraway.' *Environment and Planning D: Society and Space* 13 (5): 507–527.

Helmreich, S. 2011. 'Nature/Culture/Seawater.' *American Anthropologist* 113: 132–144.

Howell, E. A., Bograd, S. J., Morishige, C., Seki, M. P. and Polovina, J. J. 2012. 'On North Pacific circulation and associated marine debris concentration.' *Marine Pollution Bulletin* 65: 16–22.

IPRC (International Pacific Research Center). 2008. 'Tracking ocean debris.' *IPRC Climate* 8: 14–16, http://iprc.soest.hawaii.edu/newsletters/iprc_climate_vol8_no2.pdf.

Jordan, C. 2009. Midway: Message from the Gyre, http://www.chrisjordan.com/gallery/midway/.

Latour, B. 2004. 'Why has critique run out of steam? From matters of fact to matters of concern.' *Critical Inquiry* 30: 225–248.

Law, K. L., Morét-Ferguson, S., Maximenko, N. A., Proskurowski, G., Peacock, E. E., Hafner, J. and Reddy, C. M. 2010. 'Plastic accumulation in the North Atlantic Subtropical Gyre.' *Science* 329 (5996): 1185–1188.

Lebreton, L. C. M., Greer, S. D. and Borrero, J. C. 2012. 'Numerical modeling of floating debris in the world's oceans.' *Marine Pollution Bulletin* 64: 653–661.

Lumpkin, R., Maximenko, N. and Pazos, M. 2012. 'Evaluating where and why drifters die.' *Journal of Atmospheric and Oceanic Technology* 29 (2): 300–308.

Madrigal, A. C. 2012. 'Found: World's oldest message in a bottle, part of 1914 citizen-science experiment,' *The Atlantic*, 5 September, http://www.theatlantic.com/technology/archive/2012/09/found-worlds-oldest-message-in-a-bottle-part-of-1914-citizen-science-experiment/261981/.

Maximenko, N., Hafner, J. and Niiler, P. 2012. 'Pathways of marine debris derived from trajectories of Lagrangian drifters.' *Marine Pollution Bulletin* 65: 51–62.

Moore, C. J., Moore, S. L., Leecaster, M. K. and Weisberg, S. B. 2001. 'A comparison of plastic and plankton in the North Pacific Central Gyre.' *Marine Pollution Bulletin* 42 (12): 1297–1300.

Moore, C. J. and Phillips, C. 2011. *Plastic Ocean: How a Sea Captain's Chance Discovery Launched a Determined Quest to Save the Oceans*. New York: Avery.

National Oceanic and Atmospheric Administration (NOAA). n.d. 'De-mystifying the "Great Pacific Garbage Patch,"' http://marinedebris.noaa.gov/info/patch.html.

Pichel, W., Churnside, J., Veenstra, T., Foley, D., Friedman, K., Brainard, R., Nicoll, J., Zheng, Q. and Clemente-Colon, P. 2007. 'Marine debris collects within the North Pacific Subtropical Convergence Zone.' *Marine Pollution Bulletin* 54: 1207–1211.

Stengers, I. 2008. 'A constructivist reading of *Process and Reality*.' *Theory, Culture & Society* 25 (4): 91–110.

———. 2010. 'Including Nonhumans in Political Theory: Opening Pandora's Box?' In *Political Matter: Technoscience, Democracy and Public Life*, edited by Braun, B. and Whatmore, S., 3–34. Minneapolis, MN: University of Minnesota Press.

Takada, H. 2009. 'International Pellet Watch: Global Distribution of Persistent Organic Pollutant (POPs) in Marine Plastics and Their Potential Threat to Marine Organisms.' In *Proceedings of the International Research Workshop on the Occurrence, Effects and Fate of Microplastic Marine Debris, September 9–11 2008*, edited by Arthur, C., Baker, J. and Bamford, H., 371–428. Tacoma, WA: University of Washington–Tacoma (NOAA Technical Memorandum NOS-OR&R-30).

Thompson, R. C., Moore, C. J., vom Saal, F. S. and Swan, S. H. 2009. 'Plastics, the environment and human health: Current consensus and future trends.' *Philosophical Transactions of the Royal Society B: Biological Sciences* 364 (1526): 2153–2166.

Thompson, R. C., Olsen, Y., Mitchell, R. P., Davis, A., Rowland, S. J., John, A. W. G., McGonigle, D. and Russell, A. E. 2004. 'Lost at sea: Where does all the plastic go?' *Science* 304: 838.

Underwood, A. J. 1997. *Experiments in Ecology: Their Logical Design and Interpretation Using Analysis of Variance*. Cambridge: Cambridge University Press.

UNEP (United Nations Environment Programme). 2009. 'Marine Litter: A Global Challenge,' Nairobi: UNEP, http://www.unep.org/pdf/unep_marine_litter-a_global_challenge.pdf.

UNSW (University of New South Wales, Australia). 2013. 'Our Plastics Will Pollute Oceans for Hundreds of Years,' http://newsroom.unsw.edu.au/news/science/our-plastics-will-pollute-oceans-hundreds-years.

Van Sebille, E., England, M. H. and Froyland, G. 2012. 'Origin, dynamics and evolution of ocean garbage patches from observed surface drifters.' *Environmental Research Letters* 7 (4): 1–6. doi:10.1088/1748–9326/7/4/044040.

Vitaliano, D. B. 1973. *Legends of the Earth*. Bloomington: Indiana University Press.

Whitehead, N. 1929. *Process and Reality*. New York: The Free Press, reprint 1985.

14 Polar ice cores

Climate change messengers

Aant Elzinga

In a 2013 book, *The White Planet*, three prominent internationally acclaimed scientists deal with the shrinking world of ice and researchers' efforts to make sense of what is happening and why. They present a dramatic account of the struggles with ice whereby knowledge has developed during the past fifty years or more, a process wherein they are still very much a part. In the very last paragraph of their book they say:

> Let us repeat: climate warming is one of the great challenges facing our civilization today, and the polar ice is a witness to and an essential actor in it. These are good reasons for ice researchers to be concerned, well beyond the recent International Polar Year, with the state of the Polar Regions, those sentinels of our environment.
>
> (Jouzel et al. 2013)

In the plot of their story regarding the icy world a main character is the message-bearer, the ice core.

In polar regions and on high mountains, snow precipitates to form annual layers that accumulate over many years. As time goes on, the bulk of the snow at the top presses down and compacts the snow at lower levels, eventually turning it into ice. In Antarctica and Greenland, in particular, there thus exist extensive ice sheets that extend down to bedrock. In some cases as in the Antarctic, these ice sheets are several and even up to four kilometers thick. In recent times scientists have used special drills to extract long cylindrical columns of ice from Antarctic and Greenland ice sheets. These polar ice cores are used to study traces of the past history of variations in the earth's climate. These studies are nowadays linked to an ambition to better understand past climates with an eye to comparing them to a notable current warming that is attributed to anthropogenic causes of emissions of greenhouse gases. Therefore the field of analyzing and interpreting variations in the record of past ice sheet temperatures and attendant variations in the composition of air locked in the ice as bubbles has become highly relevant in discussions that address the issue of climate policy.

Knowledge of past climate change has a bearing on models constructed to predict global warming over the next one hundred years and finding appropriate measures to slow down the enhanced greenhouse effect by reducing emissions of carbon dioxide and other greenhouse gases into the atmosphere, as well as finding ways to adapt to warmer climates in the future. Analyses of the ice cores from Antarctica show that the curves for carbon dioxide and methane match the temperature curve – when the climate is cold, there is less carbon dioxide and methane in the atmosphere. The measurements indicate a strong relationship between the atmospheric content of these gases in relation to the Earth's path around the Sun as well as the inclination and direction of the Earth's axis. However nowadays due to emissions from extensive use of fossil fuels the level of carbon dioxide is considerably higher than in those ancient climates. From having been objects of curiosity that might reveal how weather patterns and local climates varied in the past ice cores, in connection with the debate about global climate warming, ice cores have become technoscientific objects that figure centrally in policy making.

Techniques of recovering, preparing and managing ice cores

An ice core is a long cylindrical plug that is pulled out of an ice sheet. This operation occurs with a drill, usually a hollow metallic tube with a set of rotating knives at its bottom to grind into the ice, thereby producing a core that is pulled up in sections. These sections may vary from two to six meters in length. In the case of a very deep drilling operation, the successively deepening hole left by the ice core is filled with fluid to prevent pressure deep down in the circular wall of the hole from narrowing. This is to avoid the drill – suspended on a long cable (up to four thousand meters) moved by a winch – getting stuck on its journeys up and down into the ice sheet, an event that would jeopardize the whole project as has sometimes happened. As the drill proceeds deeper and deeper to pull up sections of ice core, the operators on the surface of the ice sheet have to wait longer and longer for the drill to reach its destination and then reappear at the surface where it is tipped at an angle for the core section to slide out into a trough for subsequent treatment. The actual cutting of a core section may only take five or six minutes, but the average waiting time sending the drill down and then bringing up a core section from a depth of say two kilometers may take an hour. So a lot of time is spent hanging around the drill and waiting, stamping one's feet to keep warm.

Normally an ice sheet moves over the bedrock toward coastal areas where the ice extends into the sea, floating as an ice shelf that breaks off or calves into icebergs. If one were to drill deep into this moving ice, the drill hole would bend in the direction of the flow that twists the annual layers of ice and moreover makes it impossible to bring up a core since the drill would also tend to drift away from the central point where it is positioned at the surface of the ice sheet. Deep ice cores are therefore recovered from sites called domes. These are places on the ice cap where there is no horizontal movement of the ice. At these sites, the annual layers of snow/ice are ideally deposited along a straight vertical line from top to bottom. The only thing that happens is that the annual layers of ice get thinner and thinner, the further down one goes.

There are different kinds of drills that are used for recovering an ice core. A drill that has frequently been used both in Greenland and Antarctica is the ISTUK, a name that combines the Danish word for ice ('IS') with the Greenlandic word for spear, awl or drill ('TUK'). It was constructed in 1978 and built on experience and ideas gained in connection with the first shallow lightweight electromechanical drill used in Greenland in 1974–1975. The new drill was the outcome of a partnership between two leading laboratories in ice core science. It combined an electromechanical mechanism constructed in Copenhagen to drive a drill head, cutters and shoes designed at the physics department of the University of Bern in Switzerland. Instead of using an earlier technique where an electric current flows through a rather thick and therefore very heavy cable suspending the drill to power its cutters, the new design used a battery to power the cutters; the battery gets recharged as the drill passes down and up through the drill hole. It proved to be a lighter and more efficient technology than older drills and has served well in both Greenland and Antarctica.[1] The drill and barrel assembly is suspended from the top of a tower designed to keep the cable on a straight vertical trajectory downward into the shaft. The chips of ice that are produced as the drill cuts its way around the core are propelled up through a chip transport channel in the outer housing of the barrel that comprises the core tube to fill chip chambers that are emptied each time the drill surfaces back at the top.

Ice core drilling does not involve high technology of the kind used in producing some other kinds of technoscientific objects. Rather, it involves more mundane things like modes of transport to get to the drilling site, living quarters for scientists and technicians, building a shelter to house the drill, digging into the snow a deep covered rectangular cavity (the 'science trench') for preliminary laboratory work adjacent to this and a further cavity (the 'core storage trench') to store successive sections of the core prior to their transportation to the scientists' home countries. It is a case of 'little science' in the field that at an aggregate level becomes part of big science, viz. an interplay of a distributive assembly of humans, instruments, snowmobiles, airplanes and ice cores, large-scale logistics, multi-governmental state support and political agendas.[2]

Owing to the steep costs, only a few countries have their own major ice coring projects: Australia, Japan and the United States. To compete with these three, the EU together with national funding agencies in ten European countries have come together to form a technoscientific consortium. Most recently China has come onto the stage as a new actor that has just set up its own ice coring facility ('Dome A') at a very high and dry site far away from Antarctica's coasts, endeavoring in a few years to bring up sections of ice that will perhaps reach back one million years into the past. How large and what kind of slice of an ice core a particular laboratory gets usually depends on two factors: one is the amount of funding a country puts into the project in the first place; the second concerns in which disciplinary fields relevant laboratories excel. Consequently, there is a fair amount of both macro- and micropolitical negotiation that comes into play when it comes to determining where various slices of a deep ice core find their ultimate home.

How ice became a marker of time

In the 1930s glaciologists began to hand-dig snow pits and measure the annual snow accumulation cycles revealed in the walls of such pits to a depth of perhaps fifteen meters. It was not until the early 1950s that a few bore holes were created in which cores of over one hundred meters were taken up. One of these first sizeable ice cores was born on glacial ice off the coast of Dronning Maud Land, Antarctica, during the years 1950–1951. It was under the auspices of the Norwegian-British-Swedish Antarctic Expedition. The drill used was a mining drill converted for the purpose. It was found that the character of the core along its depth changes form, snow at the top, to compressed but still porous snow (called firn), and then at about sixty to seventy meters there is a transition zone beyond which the annual layers of precipitation has turned into ice. At this level the air diffused from the atmosphere into the pore spaces is closed off and is encapsulated in air bubbles in this ice. The age of the ice can still be determined by counting the number of annual strata clearly visible in alternating strands associated each with a distinct difference between the summer and winter season.

The International Geophysical Year (1957/58) rekindled an interest in polar ice.[3] It is at this stage ice cores became 'time markers'. The first deep polar cores were drilled in the 1960s. Later investigations both in Greenland and Antarctica revealed that the depth of the transition phase varies across drilling sites. Near the Antarctic coast, for example, precipitation is much greater than in the distant high and dry interior of the continent. In the latter case, the annual layers are much thinner and the point at which air gets encapsulated in the ice as bubbles is at a much lower level – for example, 100 meters below the ice cap surface. At the South Pole firn turns to ice about 122 meters down and represents the climate of 100 years ago, while the bottom 100 meters of ice from the oldest Antarctic site represents the oldest 100,000 years of that record. This also means the air in the bubbles is much younger than the ice that surrounds it in the core, therewith implying elements of uncertainty. Such uncertainties have to be taken into consideration when engaging with the core by various techniques to analyze and interpret information about ancient air that is taken to be representative of a time period, not a single annual age.

After ice cores had been turned into time markers, in the early 1980s scientists used them to determine the effect of human activities on the earth's climate. It is in this second stage therefore that ice cores became technoscientific objects. In particular it was the Danish paleoclimatologist Willi Dansgaard who first recognized that the earth's climatic history was stored in the Greenland ice cap. His research together with that of Claude Lorius of France and Hans Oeschger of Switzerland revolutionized scientific knowledge of how temperature and composition of the atmosphere have changed over the last 150,000 years, demonstrating a clear link between carbon dioxide and methane concentrations and global temperatures. Thus, it may be said that the birth of the ice core as a technoscientific object did not really occur until the early 1980s, and with it the complexity of the subject as well as the uncertainties attending it began to be articulated. Before the early

1980s, the ice core was essentially still an object of scientific study that was not yet encapsulated into the overt policy domain. This shift in the kind of questions asked about ice cores and the kinds of answers expected can be tracked by calling upon some famous ice cores.

Drilling further and further back into the depth of time

The first deep core was taken at Camp Century, Greenland, a military research facility created by the U.S. Army in 1958. At the time there was a strong military interest in the physical properties of ice in connection with the development of a chain of fifty-eight radar stations – the Distant Early Warning (DEW) line – stretching from northern Alaska across Canada, over Greenland and finishing in Iceland. Glaciology became a prioritized discipline, and there was no clear boundary between basic research and military projects. Located near the big U.S. Air Force base Thule, Camp Century was built under the snow largely hidden from view. Drilling began in 1961 to recover for the first time ever a long ice core, an operation completed three years later. The core reached down continuously to about 1,390 meters, corresponding to some 115,000 years into the past, thus reaching snow deposited before the last Ice Age. Core sections were stored in the United States, supervised by Chester Langway, who arranged for Willi Dansgaard to receive samples for stable oxygen isotope analysis carried out in Copenhagen. The method was based on his earlier groundbreaking discovery of the seasonal variations in climate over short intervals by mass spectroscopic technique to measure variations in stable oxygen ratio data ($^{18}O/^{16}O$). The application of the method gave spectacular results, producing the first climate curve from an ice core.

It was a pioneering result published in both *Nature* and *Science* in 1969. At the time, the primary interest was not directed to understanding the causes of climate change but rather one of testing and developing the method of dating the core and recording temperature variations in order to apply the methodology on new cores in the future (Dansgaard 2005).

In the mid-seventies Claude Lorius and colleagues at the glaciological laboratory in Grenoble gained the support of U.S. logistics to start some drilling at 'Dome C' in the Antarctic. After several setbacks, the team reached a depth slightly greater than nine hundred meters, and seven tons of ice were brought back to several French labs where analysis eventually showed that more than forty thousand years of archives were now available from a first drilling at 'Dome C'. This encouraged the idea of going deeper, a mission that was finally realized under the auspices of EPICA twenty years later (discussed later). In the meantime, as a method of extracting air bubbles trapped in the ice was perfected, collaboration between the French teams and Hans Oeschger and Bernhard Stauffer at the physics laboratory in Bern contributed, as Lorius et al. put it,

> to a major discovery and confirmed the prediction made by S. Arrhenius at the end of the nineteenth century; the contribution of carbon dioxide was indeed

at the LMG about 30% less than that of the of the preindustrial period before human activity began to change it.

(Jouzel et al. 2013, 89–90)

This finding was reported around 1980. By this time, new and deeper drilling operations were already afoot in Greenland. In what follows, some major operations are highlighted, first for Greenland and then Antarctica.

Greenland

Over the years cores have been extracted at several sites in Greenland, four of which will be mentioned here: GISP, GRIP, GISP2 and North GRIP. The Greenland Ice Sheet Program (GISP, 1979–1982) was a site of convenience situated at 'Dye 3', a former DEW line (six stories high) radar station in southern Greenland where collaboration between American and Danish scientists was extended to include physicists from Bern University, Switzerland (Langway 2008). The latter supplied expertise in the analysis of air bubbles in ancient ice. A core (10.2 centimeters in diameter) of just over two thousand meters was retrieved. For the analysis, it was split into two parts along the axis, an 'American' and a 'European' part. The upper part of the core gave a relatively good record of annual layers into the past ten thousand years. For the time beyond that, though flow-induced layer thinning and diffusion of isotopes required modeling for further calibration, the next ninety thousand–year period the found pattern of climate variations corresponded fairly well to the relevant part of the Camp Century core record, including abrupt shifts during late glacial time. Introducing a new ice dating method based on radioactive isotope studies and using gas analysis the Swiss analysts mapped the variation in concentration levels of atmospheric composition, for example carbon dioxide and methane, in air bubbles encapsulated along a timeline over the same period. This opened up for the study of correlations between changes in temperature and those in relative amounts of greenhouse gases. Also a novelty was the installation of an on-site under-snow research laboratory (the science trench) to examine, record, measure, log and photograph many of the physical, mechanical and optical properties and to prepare tens of thousands of oxygen isotope and other samples.

Subsequently tensions between American and European funding agencies led to two separate but coordinated and in part competing operations, a European one called GRIP (Greenland Icecore Project, 1989–1992) and a U.S. one called GISP2 (1989–1993). GRIP reached down to just over 3 kilometers, corresponding to 200,000 years or more. The two sites at the ice sheet's highest point called Summit in the middle of Greenland were only 30 kilometers apart, which proved to be an advantage since it allowed close comparison for about 110,000 years into the past and as such provided strong support of climatic origin for even the minor features of the ice annals. In particular the isotopic temperature records verified twenty-three interstadial (or Dansgaard/Oeschger) events first recognized in the GRIP record. These findings now suggested scenarios of catastrophic changes triggered by rapid climate warming with retreating glaciers and rise of global sea levels were a possibility even in the future if nothing was done to curb the enhanced greenhouse effect in our own day. A heightened sense of urgency took hold of discussions at the policy table.

Soon a new European campaign, coordinated by Denmark, led to a more truly multinational effort 300 kilometers further north, now also involving teams from the United States and Japan. It was called NGRIP (North Greenland Icecore Project, 1996–2003), reaching a depth of 3,085 meters before subglacial water flooded into the hole. Analysis of the core has provided a more detailed picture of rapid climate changes (Dansgaard-Oeschger events) and stimulated the introduction of a common timescale for comparing relevant results from the Vostok and other Antarctic ice cores.

Antarctica

The Vostok research station is located near the center of the Antarctic ice sheet. Already in the 1980s and early 1990s, important publications emanated from ice coring based on data covering 120,000 years, thus including one whole glacial cycle in the past. Collaboration between Russian, U.S., and French scientists led, finally in 1996, to an ice core of 3,623 meters spanning over 400,000 years. Important correlations were found between CO_2 and temperature variations with CO_2 levels ranging from 180 ppm during periods of lower temperatures to 280 ppm during warmer periods. This gave a natural baseline in the pre-historical record that can be compared to current CO_2 levels (more than 385 ppm) that are influenced by human activity. Publication of results in *Nature* 1997 and 1999 represented climate curves stretching back over four glacial/interglacial cycles; the last three such periods count for roughly 110,000 years each into the past (Petit et al. 1997, 359–360, 1999, 429–436). It now became clear that knowledge of long series of past climate variations obtained from the icy archives embedded in the world's ice sheets allows the teasing out of natural and anthropogenic drivers of global climate change.

Once a broader contingent of European scientists had gained experience and capability in Greenland to take the lead in ice core drilling, this momentum led to a major program called EPICA (European Project for Ice Coring in Antarctica) run by a consortium of ten European nations. Drilling was conducted at two different Antarctic sites, 'Dome C' (1996–2004) where the French-Italian Concordia station is located, and on Dronning Maud Land (DML, 2000–2006) at the German Kohnen station. The 'Dome C' core, reaching down to 3,270 meters, set a new record with climate information for the past 740,000 years. For the four most recent glacial cycles, the data agree well with the record from Vostok. The earlier period, between 740,000 and 430,000 years ago, was characterized by less pronounced warmth in interglacial periods in Antarctica, but a higher proportion of each cycle was spent in the warm mode. The transition from glacial to interglacial conditions about 430,000 years ago (called Termination V) resembles the transition into the present interglacial period in terms of the magnitude of change in temperatures and greenhouse gases, but there are significant differences in the patterns of change (Augustin et al. 2004).

The DML operation reached a depth of 2,770 meters where the ice is 300,000 years old. Because of its relative nearness to the Atlantic coast, annual snowfall in the region of this station is much higher than at 'Dome C'. Thus the core gives a higher resolution of variations in temperature and greenhouse gas concentrations and it contains climatic signals from the Atlantic Ocean that allows for comparison to rapid climate events in the northern hemisphere witnessed in the Greenland ice cores.

Japan and the United States joined in the competition at two other sites, respectively at 'Dome Fuji' in East Antarctica (1995–2007) and the West Antarctic Ice Sheet (WAIS, 2005–present). The Japanese reached a terminal depth of just over 3,000 meters where relatively warm temperatures were found; key publications on the Fuji core report on climatic cycles in Antarctica over the past 360,000 years with possible couplings to past events in the northern hemisphere. The WAIS effort situated at an ice flow divide includes forty separate but synergistic projects (including cryobiology) to analyze the ice and interpret the records down to a depth of 3,405 meters. This corresponds to roughly 100,000 years ago and is important since it lies nearer a coastal area and has traces of subglacial marine activity. The West Antarctic Ice Sheet has been the subject of some debate, owing to predictions that in a warming world it might rapidly lose ice by melting and iceberg formation off the ice sheet into the ocean and that this disintegration would cause a substantial rise in sea level in a foreseeable future (O'Reilly et al. 2012). Policy concerns presently make the WAIS an important object of study targeting abrupt climate change events.

Engaging with the ice core

Engagement with the ice core involves a chain of activities, first at the drill site, then transportation to home country laboratories for storage and scientific analysis, generating data utilized in climate modeling and international publications that in turn project images and scenarios which also enter climate policy discourses and dramaturgical representations in mass media for public consumption. In its interplay with humans, ice gains multi-dimensional agency as it is successively reconfigured from a physical to a technoscientific object – the core – whose contents, translated into science, help generate socio-political or cultural imaginaries.

Once a section of the core is brought up out of the drill hole it is taken to the 'science trench' where it is logged. The top of the newest core piece is fitted to the bottom of the previous core piece to see if they fit exactly or whether some loss of ice has occurred that in such a case is recorded and taken into account. The length of the new core piece is then measured. It is the total sum of the series of core pieces that gives the total length of the core.

The core pieces are cut into manageable sections that are then cut parallel to the core into two halves, one of which is sent to a cold room depository destined to be a reference object in an ice core 'library' at some major research center. The other half is put under a line scanner in the 'trench' to reveal visual layers or stratigraphy and then cut into four parallel vertical slabs for further analysis in the home

Figure 14.1 Working in the science trench. Courtesy of the University of Copenhagen.

laboratories using a variety of different techniques. These are to determine concentrations of greenhouse gases, secondly impurities at different levels in the ice, thirdly amounts of sea salt that vary with the seasons, fourthly the content of stable isotopes. The latter is done with a mass spectrometer to reveal past surface temperatures at the drill site. Continuous Flow Analysis (CFA) of successive melted samples of ice occurs to produce proxy-based reconstructions on important climate variables (including natural and anthropogenic forcing).[4] Instruments used include gas chromatography to separate gases and measure relative concentrations of each. Another method is to use laser absorption spectroscopy. Instruments are used to provide raw data which is further analyzed; trends are interpreted and visualized.

The penultimate outcome of the various modes of engagement with the ice core is the scientific paper. By now there are hundreds of ice cores that have yielded data sets that may be accessed at various data centers. Publications stemming from ice cores have proliferated – hundreds of papers from the EPICA project alone – and they form core ingredients in the assessments on climate change undertaken by the 'Intergovernmental Panel on Climate Change' (IPCC). Behind the scientific paper, there is always considerable work of transplanting, describing, melting, reconstituting and re-inscribing various elements of the ice core as parameters in graphs and visualization of curves correlating temperature changes with variations in levels of greenhouse gases over time. The chain of practical and analytical activities has all the characteristics of experimental science that gets incorporated into the technoscience, whereby the ice core itself is turned into an instrument, a 'paleothermometer'. This virtual instrument is not only used to measure natural phenomena but also to trace human behavior and gauge its effects in the climate record of the more recent past. Science and technoscience, then, are complementary. Technoscience does not replace science but is a manifestation of its extended character as a social-natural science of climatology in which the earlier purification of nature from culture is replaced by a mutual interpenetration.[5]

Generally speaking, ice coring technology has successively developed and the science of interpreting their message gained in sophistication as the core graduated from being a scientific object as 'time marker' to a technoscientific object that tells us about co-variation between past changes in atmospheric temperature and levels of greenhouse gases at many local sites in Greenland and Antarctica. Depth in the ice sheet – the spatially vertical – converts into a temporal dimension, timelines along which in turn at various levels characteristics of past levels and compositions of greenhouse gases in the atmosphere are probed. Analytically a shift occurs back and forth between the horizontal and the vertical. Presence of sea salt or dust particles recorded in ice cores tell of storms associated with very active atmospheric circulation during the last glacial maximum twenty thousand years ago and serve as a natural baseline for assessing changes during the recent past of an interlinked human-and-earth-system. Rapid climatic variations (Dansgaard-Oeschger events) associated with transitions from a colder to warmer climate regimes in the past suggest warming builds up to a tipping point whereupon catastrophic shifts tend to follow, with a rapid collapse of glaciers and steadily rising sea levels.

By 'revealing' details in such scenarios, ice cores as technoscientific objects serve as cryospheric sentinels that alert us to danger. Thus, there is also an emotional dimension that comes into play, raising levels of anxiety about the future on our planet; later research suggests conditions prevailing on the ocean surface are also involved and need to be factored in (Jouzel et al. 2013, ch. 9). The conclusion elicited from ice cores and other paleo-data is that concerted policy and action is needed at global, national, regional and individual household levels to modify human behavior in our own day.

The findings of ice core science thus have an import that goes far beyond the laboratory and touches on the symbiotic nature-culture-economic-politics relationship that has come to have a strong bearing on global climate change. The ice core's earlier role as time marker continues in elementary ice core science, but in its added role as a technoscientific object, the accent shifts to its redefinition as a marker of human damage to the planet. These two definitions live side by side in discourses that also affect each other. When addressing the whole spectrum of practices from ice core drilling, data construction, interpretation, analysis and visualization in scientific papers, to the activities the IPCC regularly undertakes to strengthen an evidential base for political decision making, the historian and philosopher of science consequently requires a bifocal lens. The following section touches briefly on this question of translating a spatial relation into a temporal one.

An astute and recalcitrant instrument

Attending ice core technology, there are both practical and epistemological issues that should be noted. The former has to do with fieldwork; the latter with uncertainties that arise when a spatial relation is translated into a temporal one.

Life in the (ice) field involves both glaciological field-work and ice core drilling. Both are time consuming, sometimes adventurous, and less visible than the more dramatic statements made by the gas analysts or computer modelers. But the trend analysts need for their work both the glaciologists' feasibility studies and the materials that the humbler field workers and drillers make available. It is the stuff from which ultimately are constructed proxy accounts of parallel changes in atmospheric temperature and greenhouse gases, as represented in elegant graphs published by leading scientific periodicals. It is also the stage in which ice goes over from having been a facet of nature in its own right to becoming an object for us, to be interrogated by humans – that is, as an ice 'core'. In as far as it is forcefully pulled up in cylindrical lengths of a given diameter, the 'recovery' of the ice core is essentially a human effort, depending for its success on costly logistics. It is also constrained in practice in the field by attention to factors both of human safety and protection of the polar environment itself. In this context ice is far from being a passive or malleable entity; it is capricious and puts up plenty of resistance. Many things can 'go wrong' in the field (e.g., drills get stuck, a cable gets lost, a new drill hole has to be started nearby and the process starts over again); trial and error is a prevalent factor in many different dimensions of an expedition.

It is here we find the immediately more physical aspect in the researcher's engagement with the ice core: its capriciousness as a physical object recovered from an ice sheet by humans to be cared for and coached. It is the stage before greenhouse trace gas analysis and other sophisticated modes of engagement with arrays of advanced instruments and computers come into play.

When it comes to epistemological questions, there are several. One of these arises in the construction of proxy climate data based on subsequent engagement with ice cores both in 'the trench' and in home laboratories. When visual observation of seasonal variation and therefore identification of annual strata is no longer possible because the 'layers' or 'bands' in the core get too thin, a model has been introduced to construct a chronology for the deeper part of the core. For mid- and high-latitude precipitation and mean annual temperature in polar regions, there has been a simple formula according to which $\delta^{18}O$ regularly decreases by 1 per mil every time the temperature drops by 1.5°C when going across the ice field (e.g., from coastal to central Greenland). This is a relation that featured in the paleothermometer used to translate information on changing ^{18}O-isotope ratios at different depths into variations of temperature over time. When the GRIP and GISP2 cores were analyzed, the formula was found to collapse for short periods of apparent rapid climate change recorded in the 'natural archive' of the Greenland ice sheet.[6]

Apparent rapid fluctuations (Dansgaard-Oerschger events) in ancient climates at crucial times in transitions from ice age to warmer interglacial periods challenged the traditional assumptions of fairly simple linear correlations between deuterium and oxygen-18 isotope ratios and atmospheric temperatures near polar plateaus at times of past precipitation in cases of abrupt climate change. To make sense of thermal anomalies, isotopic analysis of other gases (in this case 15N/14N and 40Ar/36Ar ratios) were brought into play in much more sophisticated calculations that take into account differences between the age of the trapped air bubbles and the age of the surrounding ice matrix containing them. Therewith a more sophisticated virtual paleothermometer was constructed.

Jean Jouzel, who headed the overall organization of EPICA, summarized:

> In using this (the aforementioned traditional) relation as a paleothermometer, researchers have assumed that the present-day spatial relation does not change with time; that is, spatial and temporal slopes are assumed to be similar. Simple models show that this assumption holds only if such factors as the evaporative origin and the seasonality of precipitation remain unchanged between different climates, which is not at all guaranteed. These limitations have long been recognized and examined through simple and complex isotopic models.
>
> (Jouzel 1999, 910)

Surprisingly the more sophisticated analysis also seems to lend itself to what may be rhetorical overtones in a debate regarding the relative neglect of climatological studies that have a different geopolitical frame of reference. In concluding his review of the revised analysis of events punctuating the glacial period 14,650 years ago, and the rapid onset of warming, Jouzel says, 'this finding constitutes a breakthrough which will be extremely useful for deciphering mechanisms of abrupt climate changes and already suggests a North Atlantic rather than a tropical trigger for the climate event' (Jouzel 1999, 911).

Models involve epistemic uncertainties, both at the outset when interpreting proxy data using the virtual paleothermometer and later when reworked data sets are 'plugged into' simulation models (Edwards 2010, 344f.). Further data uncertainty derives from the difference in age between air bubbles encapsulated in the ice at a given point of the sample and the encapsulating ice itself. The air is older than the ice and the difference varies from site to site in accordance with the conditions obtaining at the time of precipitation (more snow near oceans, less and more fine grained snow on high and dry polar plateaus in the interior of Antarctica). Heavy snow accumulation in coastal regions helps in dating the ice because the years can be counted, while thin annual deposit levels require indirect methods based on models that again employ reasonable assumptions. The record contained in air bubbles is also more readily interpreted for regions of high accumulation than for regions where the compacting of the ice that contains them blurs the annual stratification lines. Further, there is the problem that in some cores the very bottom ice near the bedrock is subject to flow and other effects that have confused ice analysts in their efforts to construct chronologies.

Degrees of uncertainty may be reduced to some extent afterward by revisiting the data and reviewing assumptions made in ice core modeling and fine-tuning in simulation experiments (Edwards 2010, 342).

A hot political issue

Ice cores from Greenland and Antarctica have achieved iconic status in scientific advice to policy makers, not least through the assessment reports of the IPCC. In scenarios projected into the future they are mobilized to issue warnings and enrolled as witnesses to tell us why we humans should be concerned about the enhanced greenhouse effect.[7] IPCC assessment reports have systematically incorporated the latest findings as in the peer reviewed literature, including visualizations of co-variation curves representing changes in temperature in tandem with changes in concentrations of greenhouse gases over many millennia. Working Group I (covering the scientific basis) of the IPCC Fourth Assessment Report had 619 contributing authors. In a ranking of these authors in terms of how often they are cited in the international literature, the French ice core scientist Jean Jouzel tops the list (Prall 2008–2016). This attests to the centrality of the ice core in the scientific discourse on global warming. Apart from the knowledge base in natural science, the IPCC also reviews the literature on the social and economic consequences of global climate warming and what policies should be adopted to mitigate the human induced factor and at the same time adapt to a warmer world in the future.

As new cores are retrieved and analysis becomes more sophisticated, 'surprises' surface and get picked up in the media to briefly become hot topics with political implications. The Dansgaard-Oerschger events signifying rapid climate fluctuations and concomitant extremes in regional and local weather patterns or storms in the past is an example, another is the possibility of a rapid disintegration of the West Antarctic Ice Sheet. A further surprise came in January 2013 when the Danish group at the Copenhagen Centre for Climate Change published new research results from a multinational ice core drilling project that followed on the heels of NGRIP in Greenland. Called the North Greenland Eemian Ice Drilling (NEEM) project, it focused specifically on the earth's climate during the last interglacial warm period, the Eem 130,000–115,000 years ago. Upon analyzing the new core (2,540 meters long, drilled in the years 2010 to 2012) it was found that when the earlier Greenland ice sheet melted during the Eem, it actually contributed far less to global sea level rise than predicted. A press release from the Copenhagen Centre in 2013 summarized:

> The good news is that the Greenlandic ice sheet is not as sensitive to global warming as earlier assumed. The bad news is that if Greenland's ice sheet did only contribute with more than a few metres Antarctica must have been responsible for a significant part of the sea-level rise.
>
> (Hansen 2013; cf. NEEM Community Members 2013)[8]

In other words, the surprise is that Greenland ends up telling us a lot about Antarctica; ice sheets down there must have a much greater potential to raise sea levels than previously surmised.

The iconic role of the ice core has gained visibility in the media both through leading scientists as popularizers, and debunking campaigns organized by so-called climate skeptics who reject scientific evidence. Since drama sells well in mass media, 'skeptics' are often presented on par with knowledgeable climate researchers as if there is a real scientific controversy (when there is not). Still, when rejectionists promote their speculations, they also draw attention to the ice core whose symbolic value they seek to deflate when depicting the reigning scientific consensus as a conspiracy of climatologists and environmental researchers hungry for continual funding and comfortable careers. A lot of creativity goes into efforts to cast doubt on science and scientists in the mind of the public (Oreskes and Conway 2010). Such campaigns on the other hand also work the other way, spurring more scientists in various countries to come forward and engage in explaining how they arrive at their findings in an effort to increase the public understanding of ice core science and paleoclimatology. Two striking examples are open letters defending the integrity of climate science and informing the public as an active manifestation of a social responsibility of scientists (e.g., Gleick et al. 2010).

Concluding remarks

The authors of the book cited at the outset of this chapter argue that we now live in a new era called the Anthropocene. They date its origin back to the beginning of the nineteenth century when there was an appreciable increase of pollution due to the activities of mankind. The first phase of the industrial revolution brought steam power and smokestacks, coal-fed blast furnaces and gas lighting, and in a second phase expansion of manufacturing, new dyestuffs and chemical industries. By the end of the nineteenth century, smokestack industries and smokestack towns had proliferated. Humankind began to leave an indelible imprint on the environment and therewith on the climate of our planet and the future. The Anthropocene, the three authors write, "signifies the passage in the evolution of the environment of our planet from a balance that depended essentially on natural causes to a state in which the influence of human beings becomes the governing factor" (Jouzel et al. 2013).

Behind this conclusion stands a generation of research in which the ice core lives as a prime object, polar ice transmuted first into a chronograph and then additionally as a climate messenger. In the language of the present volume, the ice core became a technoscientific object. In this capacity its fortunes are closely interwoven with predictive climatology, policy making and media happenings, its icy annals both praised and contested. Numbers of ppm of carbon dioxide in fossilized atmosphere in the Vostok core and other deep ice cores when compared to present-day levels of greenhouse gases serve as important anchoring devices in these discourses.

By drilling further and further into ice sheets, ice core scientists probed deeper into the depths of time. They constructed two chronologies, one for the annual layers in the ice and another for fossilized atmosphere encapsulated in the ice at the various levels. Habituation prompted phrases like 'the archive shows' or 'reading the record' of the past. The ability to read glacial archives, 'frozen annals', as one scientist called them, involved the knack first of all of translating spatial parameters into markers of time. But ice cores soon revealed themselves to be more than mere chronographs in a global natural history. When researchers subsequently cracked the code of the air bubbles and atmospheric molecules in the ice matrix, they were also able to interpret more localized and qualitatively different climate change triggers. Therewith the stakes were raised. What until then had been speculation amongst a few now became evidence of anthropogenic climate drivers. Rather than just recounting the atmosphere's natural history, ice cores now began telling us about our common history with the climate, and the message travelled from small circles of climate specialists to the broader scientific community, plus policy makers and publics.

There has been no pretence of certainty since surprises emerged time and again. Antarctic and Greenland ice is capricious, putting up resistance when ice cores are extracted and treated for transport. The cores in turn too are recalcitrant in their role as technoscientific objects. They play tricks on the readers and interpreters of their 'annals' regarding the histories of climatic turning points under the surface of the high and dry dessert-like plateaus, right down to the bottom where the ice sheet sits on bedrock. Depending on their sites of origin, individual cores may also differ when they speak to us through various instruments after entering laboratories for closer examination. Although there is convergence on a general storyline, the archives also differ in stories of change from one polar locality to another. Currently the complex interrelationship of some distinctive past climate events at specific junctures along the cryospheric timelines of Greenland compared to those of Antarctica are still being teased out.[9]

Nevertheless the main storyline is on the table both of scientists and politicians, with the ice core now also acting as a marker of human damage to the globe. It is no exaggeration to say that ice cores have become 'common concerns' at the international policy table. The paleothermometer is not only an instrument now in the hands of humans to measure natural phenomena. It also indicates the imprint of humans on nature and what may happen in the future. It provides a means to gauge the extent of human effects; in a sense the ice cores are now also 'measuring us'.

Notes

1 For a detailed description of this drill, see Johnsen et al. (1994).
2 For a longer term historical perspective on changes of scale in polar research from little science to big science, see Elzinga (2009). A discussion of criteria characterizing big science appears in Elzinga (2012).
3 For a history of the four international polar years, see Barr and Lüdecke (2010).
4 Since the year 2000 a count of articles only dealing with ice cores published by authors at or affiliated with the Bern Centre gives a total of well over one hundred during the twelve year period.
5 Already in the 1970ies, philosopher Gernot Böhme raised the question, "Alternatives in Science – Alternatives to Science?" (1979). The editors of the present book offer a conceptualization that assigns a philosophy of technoscience as a complement to the philosophy of science (Bensaude Vincent et al. 2011).
6 As yet this has not affected interpretations of the Antarctic ice archive that appears to be less affected by seasonality, but it does call for caution in conceptual reconstructions of key correlations.

7 For Lorius et al.'s now classical graphic curves, see also, for example, Street-Perrot and Robert (1994, 47–68); for a discussion of the elements of chemical histories preserved in ice, see, for example, Graedel and Crutzen (1993, 223–229).
8 Behind the name 'NEEM Community', there is a network of some 130 authors.
9 At first it seemed there might be a kind of bipolarity where climate events in the North seesawed with those in the South or vice versa. Later when with better resolution various details in many episodes could be 'seen' more clearly in the ice core annals, each Dansgaard-Oeschger event in Greenland displayed a counterpart in Antarctica but of a different form – the warming phase and that of cooling was relatively symmetrical in the Antarctic, whereas in the North the warming was rapid and the cooling slower and less intense. The extremes in Greenland could reach as high as a 16°C fluctuation, but in the Antarctic they kept in the range of 2–3°C. The South began to warm up when the North was very cold, and the rapid warming occurred in Greenland when the temperature reached its maximum in Antarctica. It was not a matter of a 'seesaw' relationship but one where warming in Greenland displayed many more jerky effects because of influence from the neighboring Atlantic and could be delayed by more than one thousand years after the onset of warming over the Antarctic. But on all this, it seems, the final word has not yet been said (Jouzel et al. 2013, 148).

References

Augustin, Laurent et al. (EPICA Community Members). 2004. 'Eight glacial cycles from an Antarctic core.' *Nature* 429 (10 June): 623–628.

Barr, S. and Lüdecke, C. (eds.). 2010. *The History of the International Polar Years (IPYs)*. Berlin and Heidelberg: Springer Verlag.

Bensaude Vincent, B., Loeve, S., Nordmann, A. and Schwarz, A. 2011. 'Matters of interest: The objects of research in science and technoscience.' *Journal for General Philosophy of Science* 42 (2): 365–383.

Böhme, Gernot. 1979. 'Alternatives in Science – Alternatives to Science?' In *Counter-Movements in the Sciences: The Sociology of the Alternatives to Big Science*, edited by Nowotny, H. and Rose, H., 105–125. Dordrecht: D. Reidel.

Dansgaard, Willi. 2005. *Frozen Annals: Greenland Ice Cap Research*. Copenhagen: Niels Bohr Insitute.

Edwards, P. 2010. *A Vast Machine: Computer Models, Climate Data, and the Politics of Global Warming*. Cambridge: The MIT Press.

Elzinga, A. 2009. 'Through the lens of the polar years: Changing characteristics of polar research in historical perspective.' *Polar Record* 45 (3): 313–336.

———. 2012.'Features of the current science policy regime: Viewed in historical perspective.' *Science and Public Policy* 39 (4): 416–428.

Gleick, P. H. et al. (255 Members of the American Academy of Sciences). 2010. 'Climate change and the integrity of science.' *Science* 328 (10 May): 689–690.

Graedel, T. E. and Crutzen, P. J. 1993. *Atmospheric Change: An Earth System Perspective*. New York: W. H. Freeman & Co.

Hansen, L. H. 2013. 'Important Results from NEEM Community', http://www.isogklima.nbi.ku.dk/nyhedsfolder/uk_with_dk_companion/january-2013-article-in-nature-neem-community (accessed July 2, 2016).

Johnsen, S. J., Gundstrup, N. S., Hansen, S. B., Schwander, J. and Rufli, H. 1994. 'The new improved version of the ISTUK ice core drill.' *Memoirs of National Institute of Polar Research*, Special Issue No. 49: 9–23.

Jouzel, J. 1999. 'Calibrating the isotopic paleothermometer.' *Science* 286 (29 October): 910.

Jouzel, J., Lorius, C. and Raynaud, D. 2013. *The White Planet: The Evolution and Future of Our Frozen World*. Princeton, NJ: Princeton University Press.

Langway Jr., C. C. 2008. *The History of Early Polar Ice Cores*. Hannover, NH: U.S. Army Engineer Research and Development Center (ERDC/CRREL) (TR 08-1).

NEEM Community Members. 2013. 'Eemian interglacial reconstructed from a Greenland folded ice core.' *Nature* 493 (24 January): 489–494.

O'Reilly, Jessica, Oreskes, Naomi and Oppenheimer, Michael. 2012. 'The rapid disintegration of projections: The West Antarctic ice sheet and the Intergovernmental Panel on Climate Change.' *Social Studies of Science* 42 (5): 709–731.

Oreskes, N. and Conway, E. M. 2010. *Merchants of Doubt*. London: Bloomsbury Press.

Petit, J. R., Basile, I., Leruyuet, A., Raynaud, D., Lorius, C., Jouzel, J., Stievenard, M., Lipenkov, V. Y., Barkov, N. I., Kudryashov, B. B., Davis, M., Saltzman, E. and Kotlyakov, V. 1997. 'Four climate cycles in Vostok ice core.' *Nature* 387: 359–360.

Petit, J. R., Jouzel, J., Raynaud, D., Barkov, N. I., Barnola, J.-M., Basile, I., Benders, M., Chappellaz, J., Davis, M., Delayque, G., Delmotte, M., Kotlyakov, V. M., Legrand, M., Lipenkov, V. Y., Lorius, C., Pépin, L., Ritz, C., Saltzman, E. and Stievenard, M. 1999. 'Climate and atmospheric history of the past 420,000 years from the Vostok ice core, Antarctica.' *Nature* 399: 429–436.

Prall, J. 2008–2016. 'Most-Cited Authors on Climate Science', http://www.eecg.utoronto.ca/~prall/climate/AR4wg1_authors_table.html (accessed July 2, 2016).

Street-Perrot, F. A. and Robert, N. 1994. 'Past Climates and Future Greenhouse Warming.' In *The Changing Environment*, edited by Roberts, N., 47–68. Oxford: Blackwell.

15 Nuclear waste

An untreatable technoscientific product

Sophie Poirot-Delpech[1]

This chapter is based on exploratory fieldwork conducted in the underground laboratory of the *Agence nationale pour la gestion des déchets radioactifs* (ANDRA, National Agency for the Management of Radioactive Waste). The purpose of this laboratory, located in the town of Bure in Lorraine, France, is to prepare the way for an underground nuclear waste disposal, called the *Centre industriel de stockage géologique* (CIGEO, Industrial Centre for Geological Storage), where caverns excavated at a depth of 500 meters with a surface area of 30 km^2 are to be used for burying long-lived nuclear waste. The decision to build such a facility was established by the so-called Bataille Law of 1991, and construction work began in 2000. Over time, the project managed by ANDRA prompted a number of controversies, and so a nationwide public debate was conducted over three months, from September 15 to December 15, 2013, in order to assess its relevance, its goals and its characteristics (Commission particulière du débat public Cigeo 2013).

The aim of our research was to understand how nuclear waste as a radically new and disturbing object, along with the options currently available for its geological storage, makes sense to our contemporary societies and to the local communities that are involved in and concerned with the practical aspects of managing nuclear waste. Beginning with a brief presentation of the technical aspects of the project, this paper emphasizes the anthropological dimension of the problem and raises the question of whether we humans can share our world with nuclear waste – or, to put it differently, whether the existence of nuclear waste calls into question the very habitability of our planet, at least as we experience it today. This radical question will be discussed here in conjunction with a set of issues that have been raised not (only) by social scientists but by various practitioners encountered in the field.

The future, or rather the possible futures, of nuclear waste as a hazardous technoscientific product for generations and civilizations to come is a major matter of concern for those individuals and institutions tasked with the management of nuclear waste. How can we warn future generations about the long-term risks of the substances that will be passed on to them, given that some of these substances will remain hazardous for hundreds of thousands of years? How can we 'convey' to future civilizations the message that one such site should be kept out of human reach for eternity? In addition to this issue of *the memory of the future*, there is another problem that concerns us, namely, that of *the future of memory*. Given that the work of memory is always done in the present, it may be extremely hard for us to imagine how future generations and civilizations will understand and interpret the signs and leftovers we bequeath to them. It may be equally hard to imagine the inexorably slow dissolution of radioactive substances into a state – an inconceivably remote state – of harmlessness.

A further aspect of the problem raised by nuclear waste is less obvious but nevertheless underlies the discourses and practices surrounding it, namely, the ontological status of nuclear waste with regard to the great divide between the natural and the artificial, between nature and society. The identification of 'natural reactors' by CEA (Atomic Energy Commission) researchers in the 1970s leads us to interrogate the notion of a 'natural analogue' along with its imaginary and symbolic functions.

An object awaiting a solution

'Long-lived nuclear waste', known as HAVL in French,[2] comprises the (according to our current state of knowledge) untreatable residues from nuclear energy generated for industrial, military and civil purposes. For decades, nuclear scientists and engineers did not concern themselves much with nuclear waste, even if a few whistle-blowers saw early on how nuclear waste could lead the development of nuclear energy to a dead end.[3] Waste was treated in many different and inappropriate ways: it was buried in former uranium mines, or even mixed with other waste in landfills. Thousands of tons of nuclear waste were dumped into the sea before public opinion and some sections of the scientific community began to feel concerned about the potential effects on the natural environment.[4] In the heyday of the conquest of space and the thirty-year boom period known as '*les trente glorieuses*' in France, the solution proposed by some people was to launch nuclear waste into space.[5] This 'futuristic myth' is so robust that it resurfaces from time to time even today. Until relatively recently, there has been no systematic monitoring of the location and composition of radioactive waste in the United States, France, Britain, the former Soviet Union or indeed anywhere else in the world. Large marine and terrestrial environments across the globe have already been contaminated and declared 'no-go' areas because of such practices, while most of these packages, whose location has been 'forgotten', are unrecoverable. Such chaotic management of nuclear waste will leave the indelible 'mark' of twentieth-century inconsistencies on future times.

The reprocessing of used nuclear fuels has been designed to produce plutonium for nuclear weapons. Reprocessing is presented today as a solution to the waste problem as well as a guarantee of the independence of uranium non-producing countries. Today, its industrial implementation in a number of 'nuclearized' countries is often presented as a way of transforming nuclear energy into 'renewable' energy, as it enables used fuels to be 'recycled'. In France, used fuels 'cool down' for a few years near nuclear plants before being transported to The Hague in huge packages of 110 tons, of which 10 tons is nuclear material. These packages are known as 'castles' ('*châteaux*' in French), indicating the dimensions and scales so characteristic of the nuclear world. The 'castles' are then placed in containers for further cooling in pools. Finally, the uranium rods are sheared and dissolved by means of chemical treatment, thus separating uranium and plutonium from non-recyclable fission products. This new waste will eventually be vitrified and stored. In the end, each ton of waste results in 10 kg of plutonium, 950 kg of depleted uranium and 40 kg of waste, which retains high levels of radioactivity. After reprocessing, the nuclear ores thus produced are transported again across France to the southern cities of Marcoule and Pierrelate for the manufacture of MOX (mixed oxides, i.e., a mix of depleted uranium and plutonium) that will be used in a number of nuclear plants. Once it has served its purpose, this fuel turns into waste that is even more toxic than the waste produced by 'conventional' fuels. There is currently no sustainable solution for such residues. Moreover, the reprocessed residues of uranium ('yellow cakes') must be enriched. There are some enrichment plants in Germany, the Netherlands and France, but 'yellow cakes' are, for the most part, shipped or transported by train to Russia for enrichment, and then returned – in principle – to their country of origin. In reality however, it seems that a very large volume of these residues is stored in Russia.[6] One major problem is that the fuels extracted in Kenya or Australia remain much more competitive than MOX or enriched uranium, which are highly processed products. In addition (and this is a key factor), these two countries are less concerned about safety rules and social norms (Topçu 2013).

Reprocessing, the pride of French technology, does not enjoy unanimous approval. The United States, among other countries, opposes it. It is neither easily manageable nor particularly 'profitable' at this time. Reprocessing requires the deployment of secret or 'exceptional' convoys of hazardous materials on air, sea or road routes, thus increasing the risk of accidents or even disasters. Even though it seems economically advantageous since it would allow countries to 'stockpile' uranium, reprocessing chiefly serves an ideological purpose: to shift the image of nuclear power from being that of a linear, irreversible and god-like power to appearing as a form of energy-based on cycles and renewability. It conveys the image of a technoscience seeking to return to nature. The attractive power of the concept of renewability can be seen as a form of denial of the irreversibility and the depletion of resources.

In sum, although reprocessing seems to solve the problem of waste by creating a 'cycle', in reality it re-ignites the problem because it generates yet more toxic waste. The only true 'solution' would be to shift from nuclear fission to nuclear fusion. Only fusion would turn nuclear power into a renewable form of energy.[7] However, the feasibility of this option is still hypothetical, despite the huge investments already made to examine it.[8] In any case, the problem of nuclear waste that has been produced and is still being produced today would remain. What to do with the thousands of tons of highly toxic waste sealed in packages and stored in the vicinity of nuclear plants, pending a more sustainable and, possibly, safer solution?

Objects of hatred

Nuclear waste is legally defined in France as 'final waste'. Toxicity is only one of its properties; the definition also specifies that final waste cannot be recovered or reintegrated into human activities or natural cycles (unless considered from a very, very long-term perspective). Nuclear waste is untreatable. All final waste is a material in which the present is embedded and which projects onto the future a variety of temporalities. This is not specific to contemporary times. Waste is distinctive of human existence on earth, or, more particularly, of industry in general. For instance, rivers in the French region of Morvan, a place with emblematic significance in Gallic history, still contain traces of mercury resulting from industrial activities undertaken by the Celts in the first century AD according to recent archaeological findings. The 'production' of final waste, whether viewed as a hazard or merely as a risk factor, has significantly increased with industrial development, especially with the advance of the chemical industry, and has reached a critical level today. This 'historical' argument is often voiced to minimize and, to some extent, even to trivialize the issue of nuclear waste as well as the type of energy involved in its production. This argument, however, is weak. First, and generally speaking, our era differs from the previous one precisely because of our widespread awareness of our impact upon the world. Second, while being a 'technological product' of human activity, the distinctive feature of nuclear waste, and especially of highly toxic long-lived waste, is that it represents a danger to all forms of life precisely because of the lifetime of such metals.

In addition to its technical and material dimensions, the issue of nuclear waste has also a strong symbolic dimension. The term 'waste' – *déchets* in French is etymologically close to *déchu* (fallen from honor) – evokes the notion of dirtiness. The concern about waste is an invariant aspect of all cultures and all societies, one taken up in the anthropological study of cleanliness and dirtiness. Mary Douglas has developed the hypothesis, based on comparative anthropological studies, that we consider as dirty what is not in its proper place (Douglas 1966). In every culture, the dirty is that which does not fit into established categories. According to Douglas, this gives rise to hatred or disgust. That which is considered dirty (e.g., excrements, suburbs, no-man's lands) is both dangerous and terror-inspiring, but it is also a factor in a society's creativity – hence the shared etymology of dirty (French: *sale*) and the sacred (French: *sacré*; Latin: *sacer*). If the dirty refers to disorder, disorder is the condition of creativity, of creation, of circulation and of renewal. Yet as we will see, nuclear waste defies classification, not only in the material order but in the symbolic order as well.

Objects of containment

It appears that the era in which nuclear waste was ignored has passed. The management of waste requires ever more stringent norms and processes, and huge investments are being made to support research into its future. Of all the options that have been proposed, the burial of waste in the depths of the earth is held to be one of the most 'reasonable' answers to a problem that even now has yet to be finally resolved. From the 1980s onward, the concept of deep geological burial of nuclear waste has become the most consensual 'answer' in scientific circles.[9] It is not, however, really 'new'. In the early days of the nuclear age, waste was buried in former uranium mines or deposited hastily in 'depositories' (as ANDRA experts call them today), otherwise known as dumps. The advocates of deep geological disposal, however, make a clear distinction between this type of storage and simple, reckless burial. They claim that deep geological storage is a technically and politically responsible act that is aimed at protecting future generations from the consequences of the present. This project materialized in the creation of in situ laboratories in a number of nuclearized countries such as the United States, Sweden, Finland and Germany. Here we focus on the French case.[10]

At our field research site, Bure (Lorraine, eastern France), geologists have been commissioned to define the geological conditions necessary to implement this 'solution'. For this purpose, a deep underground laboratory has been built. It is a pilot site to test the behavior of 'packages' composed of vitrified waste and designed to protect people from radioactivity while minimizing the possibility that future generations will come into contact with the radioactive substances. The containment consists in burying artificial containers in a natural environment (argillite, a sedimentary rock composed of clay particles in the case of Bure). In this scenario, the natural environment is expected to contain the radioactivity after the artificial container is degraded. However, because of the concentration, in one place, of rare sources of energy and of increasingly rare and precious metals used to construct the containers (for example, copper in Sweden), the major risk posed by this strategy is not just the failure of predictability; it is also, and perhaps above all, uncertainty about the future of human action through intentional or unintentional intrusion. Terrorism, theft and random accidents are all scenarios that have been systematically modeled by professionals in charge of prospective studies.

Most experts originally thought that, for reasons of cost and security, the best option would be to forget about the geological sites where nuclear waste is 'stored' and consequently to plan irreversible storage (which is, as we will see, an oxymoron). Thus, several in situ laboratories have been established throughout the world since the end of the 1980s to experiment with geological storage. In France, the Bataille Law made provisions to establish three such laboratories in different geological sites in order to help select the best storage space. However, only one of them – the underground laboratory in Bure – has been effectively implemented. It is located in the northeast of a scarcely populated area geographers – the so-called empty diagonal of France[11] – close to Verdun, a territory indelibly marked by twentieth century history. The choice of this site is not only motivated by the properties of the rocks but also by economic concerns associated with an economically depressed, de-industrialized region and the expectation of political support from the local municipal government.

In fact, the process of dialogue with local people and members of environmental movements proved to be even more difficult than the selection of an appropriate geological site. The experts tried to convince the local authorities and residents to accept that a specific area of the world, 'their' world, would henceforth be dedicated to storing nuclear waste. Interestingly, it turned out that the creation of a zone cut off from the rest of the world designated to house waste prompted more reservations and protests than the building of nuclear power plants itself, even though its associated risks are more tangible, far-reaching, and better known.

Finally the construction of the Bure laboratory turned out to be acceptable only on condition that the nuclear waste stored in it would be retrievable. The 1991 Bataille Law – named after Rapporteur Christian Bataille – introduced a margin of reversibility in irreversibility. It stipulates that 'storage' should be made reversible within 100 years. In other words, it is expected that, if a new process of treatment were discovered, the method of storage will allow the recovery of packages. According to French sociologist Yannick Barthe (Barthe 2006), the Bataille Law is a landmark in the history and politics of science for two reasons: (1) the uncertainty of our relationship to the future is acknowledged and institutionalized in a legal act; (2) it is the first political intervention in a question hitherto considered relevant to technical experts only.

Objects of storage

The term 'storage', which has been used since the very beginning of nuclear waste burial projects, requires some discussion. It reveals a paradoxical or at least unprecedented dimension of material and temporal confinement. After being warehoused near the reactors, the waste has to be 'stored'. The language of warehousing and storage pertains to commercial and industrial settings, and usually refers to a resource (a product or an object) waiting to be used; moreover, these operations evoke the notion of traceability, of transparency – in other words, of 'memory'. Storage, here, comes after warehousing; but over a certain period of time, the stock is no longer intended to be available. On the contrary, nuclear waste must be 'forgotten' in the depths of the earth so that the earth takes over from human activity to protect human beings, living beings and the environment.[12] Two different intimations of time are thus intertwined in this project, two different ways of relating to the future and to memory: (1) withdrawing forever the underground site (i.e., a part of the world) from the influence of human history through a once-and-for-all containment conveys the idea that future generations might not be able to solve a problem created in the present; (2) allowing the containment to be reversible is more attuned to the representation of a promising future that is commonly embedded in technoscientific discourses. Yet all the geologists interviewed at the Bure laboratory were concerned with the impacts of reversibility on the efficiency of containment technologies. The period of 100 years – hardly significant in comparison with the timescales associated with radioactive activity – appears as a compromise between these seemingly incompatible visions of the future. Furthermore, the imperative of reversibility requires markers that will keep the memory of a storage site alive for future times, while the production of such time markers requires that the memory of the future be anticipated and controlled.

Each package of final waste, each with its own timescale, projects us into the future by generating a becoming that is both natural and human. The events that will influence its fate, whether anticipated or not, will be the product of both human activity and terrestrial conditions – the evolution of society and meteorological or cosmic events. The 'half-lives' of HAVL project a future that may have meaning for geologists, chemists or physicists; yet it remains intangible for a layperson or a politician.[13] However, all of them have to cope with (or at least to envisage) these long timescales.

Should we remember? Should we forget? Actually, the very terms through which oblivion is programmed (e.g., 'storage') convey a duty of memory. In a documentary film about nuclear waste storage in Finland, with the evocative title *Into Eternity*,[14] one of the interviewees summarized this paradox: "it is necessary to remember to forget." This dilemma between memory and oblivion appears to be inherent in the question of nuclear waste. How to mark the sites where nuclear waste is buried is a crucial concern for ANDRA and other national institutions. What kind of material traces should we leave behind for the future (spoken of in terms of hundreds of thousands of years – no one dares speak of generations anymore!) so as to preserve the memory of a site and its dangers (Galison 2010, 304–305)?[15] The question is less how long these time markers will endure physically than how long they will remain meaningful.

Most ANDRA officials rely on a notion of memory produced in the present 'for' the future. During our interviews with them in Bure, many of them mentioned the 'greatness' of nuclear energy and considered themselves as 'pioneers'. They claim to be the 'first' people to take full responsibility for the effects of Man's imprint upon the physical world by retrieving it from the future for the sake of future generations. Remarkably, the main actors involved in this technoscientific enterprise express ethical considerations that are not so dissimilar to those of Hans Jonas.

Objects of memory

Although our interviewees acknowledged the uncertainty associated with the desire of sparing future generations, they seemed to be unaware of a well known phenomenon in disciplines such as archaeology and anthropology: that memory is always a reconstruction of the past not just *by* but also *for* the present.

This suggests that forgetting is not the opposite of remembrance. Maurice Halbwachs once said (and Paul Ricoeur later elaborated) that there is no memory without forgetting and that to know *again* is to know *anew* (Halbwachs 1992). How will the beings that will populate the world in the future 'know anew' the traces we have left? For this, an interdisciplinary team has launched a search for universal terror-inspiring 'shapes' or 'symbols' that could prevent future accidental intrusion onto the burial site (Sandia National Laboratories 1991, F-49). Interestingly enough, the shapes supposed to repulse any future visitor have much in common with contemporary art and architecture (Fig. 15.1). The message of these 'markers for the future' is thus mainly addressed to the present.

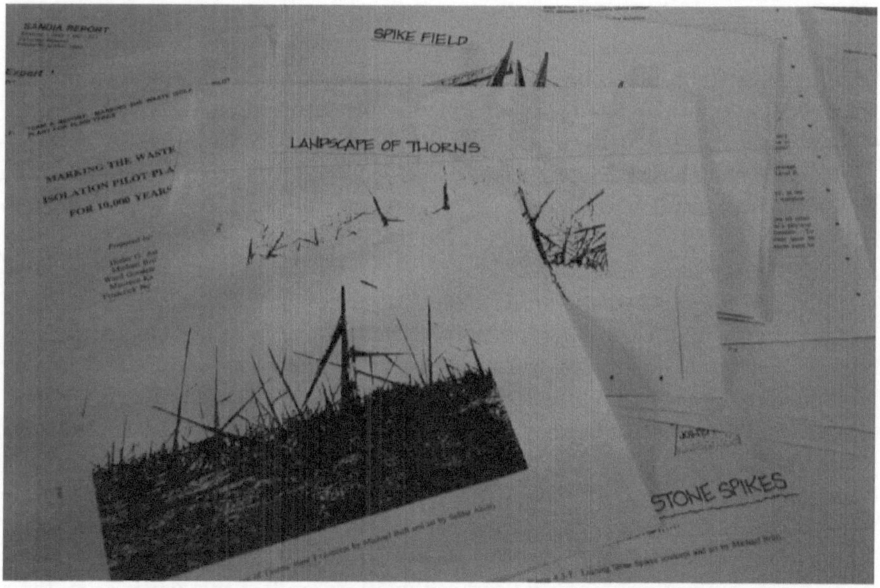

Figure 15.1 Projects of creating warning markers to deter inadvertent human intrusion into underground waste burial sites. 1991 Sandia National Laboratories report. Concept by Michael Brill, art by Safdar Abidi.

Waste and residues, these 'objects of memory', are the bedrock of archaeological research. In a book containing reflections on his discipline, archaeologist Laurent Olivier argues that vestiges and traces of the past are changing witnesses to a past to which they can give us direct access (Olivier 2008). Rather, such traces are constantly 'recast', as Maurice Halbwachs noted, in a process as natural as it is cultural and through which we represent our past to ourselves. Olivier thus regards each vestige as a 'palimpsest' which, thanks to its ability to 'survive', offers a substantial foundation for archaeological work. Archaeology, Olivier argues, deals more with memory than with history. Yet because of the intractability and intrinsic danger of nuclear waste, their containment prevents this technoscientific object from becoming an object of memory.

Techno-socio-natural objects

Closely linked to the problem of the memory of the future is the ontological status of nuclear waste. Indeed, nuclear waste is considered to be 'final' because it can no longer be transformed, re-treated or re-employed. Accordingly, nuclear waste can hardly be viewed as a technoscientific object because it has no use value for human societies. On the contrary, humans strive to protect themselves from it. Nuclear waste is not a natural object either, because it cannot be re-inscribed in the economy of nature and it is an extreme threat to all known forms of life today.

Neither technological nor natural, this waste is a monster. It has become untreatable, and this very 'untreatability' condemns such an object to be contained indefinitely, pending the reduction of its toxicity. The only way for nuclear waste to escape this destiny would be transmutation, a line of research prioritized by the Bataille Law in France but one that remains extremely hypothetical, according to the actors we spoke with during the course of our research.

Hannah Arendt has characterized the contemporary era by referring to the changing ontological status of technological objects (Arendt 1958). While such objects belong to Arendt's category of *work* (as products of *Homo Faber* that contribute to the durability of the world), 'waste as a resource' would fall within the domain of the *Animal Laborans* because of the way it reinforces the incessant renewal of the 'vital process' (i.e., the cycle of production/consumption). With nuclear waste, however, we are witnessing the emergence of a radically new kind of object – a 'technoscientific product' that does not fit into either of Arendt's categories. As the final element in the process of treating radioactive substances, nuclear waste cannot be reintegrated – 'recycled' – in the unceasing work of the *Animal Laborans*. On the other hand, even though it is durable and clearly the product of human work, it threatens the durability of our common world in the short, medium and long term. To make a sustainable world, we have to discard such products of human industry from our common world (natural and artificial), to protect this world from their toxicity and to eject them from the future of our memory. A novel figure emerges from this dilemma, that of the 'monstrous' object. It is a monster because it cannot occupy any category through which we usually make sense of the world. The monstrous character of nuclear waste accounts in part for the repulsion that the project of burial inspires in the immediate surroundings. This character also explains the repeated emphasis that scientists and 'communication specialists' place on the 'naturalness' of nuclear energy, and the special attention they pay to natural analogues at the different stages of their development.

Natural analogues?

It is no surprise, therefore, that the 1970s discovery of a natural analogue to a nuclear reactor in a Gabonese uranium mine aroused great interest among the proponents of nuclear energy and researchers studying waste burial. The idea that phenomena akin to 'nuclear fission' in uranium and thorium could exist in nature has been around since the 1950s (Wetherill 1953). This hypothesis was 'confirmed' by a surprising discovery in the 1970s in Pierrelate, a nuclear site in the south of France where uranium for the atomic bomb was being enriched. A routine check of the enriched material revealed that a sample of uranium hexafluoride contained 0.717 percent of U_{235} instead of the expected 0.702 percent. Tests on other samples at Pierrelate yielded the same results, and researchers traced the pattern all the way to the Gabonese mine of Oklo where this 'abnormal' mineral had been extracted. Fifteen other similar natural 'nuclear reactors' were 'found' using this kind of inquiry. They are approximately two billion years old, originating in a time when a number of critical conditions came together over the course of hundreds of thousands of years to initiate a process of fission.

In stark contrast to the familiar technoscientific strategies of mimicking nature, the notion of natural analogues to nuclear reactors proceeds from the retrospective recognition of a prototype in a natural phenomenon, which is itself described in technological terms (e.g., 'reactor,' 'start-up', 'regulation'). Natural analogues are of interest because they vindicate an artifact (i.e., a man-made product), rather than because they may increase scientific knowledge about nature. By naturalizing nuclear technology, the Oklo site could also 'legitimize' nuclear waste containment and burial, since Oklo's soils may contain radioactivity that has been present for billions of years. As Jules Horowitz, a physicist who wrote a foreword to the book authored by the 'discoverer of Oklo', Roger Naudet, states, "We had to acknowledge that natural reactors had 'worked' naturally for two billions years before Man, in turn, was able to achieve a controlled chain reaction" (Naudet 1991, XI).[16] Nature becomes the analogue of the artificial rather than the artificial imitating nature. The world is thus reconfigured as a kind of Oklo site. In fact, Oklo could have served as a laboratory for observing in vivo – and to some extent for validating over the course of an extremely long period of time – the 'solution' of burying nuclear waste. Unfortunately, the opportunity to observe a living nuclear lab has been 'buried' since the mine pit in eastern Gabon became flooded in 2011.

However, Oklo appears to make it possible to re-inscribe nuclear waste, this so contemporary object, in a natural memory and in geological timescales. Natural analogues tell us something about the current relationship of technoscience with 'nature'. Severely criticized by sociologists and philosophers,[17] the notion of nature, however, remains a heuristic for scientists engaged in nuclear waste burial projects. Unlike Bruno Latour, nuclear scientists and engineers still believe in nature! As technoscience fails to plan the future beyond imaginable timescales, nature takes over since she has the experience of such timescales. Geologists, in particular, refer to storage sites as places where the human imprint has not (or has hardly) impacted upon natural processes.

Conclusion

The nuclear waste 'problem' can be considered a 'reverse salient'. This term, used to describe a phenomenon in a military strategy that is likely to reverse a situation completely, has been used by Thomas Hughes to characterize the moment when an innovation can fail (Hughes 1983; Gras and Poirot-Delpech 1993). Waste is the Achilles heel of nuclear power. Geological storage is not only an answer for a non-solvable problem; it can be seen as a way of 'burying the problem'. In this perspective, nuclear waste is not just a problem or a risk factor awaiting a technological solution: it is the very touchstone of nuclear power.[18]

Notes

1 This chapter, translated from the French by Philippe Bardy, is the outcome of the fieldwork I conducted jointly with Laurence Raineau. Many of the data and views presented here are the products of our joint effort, but the responsibility for the writing of this paper is mine alone.

2 HAVL stands for *Déchets de haute activité à vie longue* ('high-level radioactive waste that is long-lived').
3 Consider, for example, the environmental movement Friends of the Earth (Les Amis de la Terre 1978).
4 This may be one more case of 'agnotology' (Proctor and Schiebinger 2008). The negligence with nuclear waste, whether an obstruction, more or less deliberate, or an 'oversight' linked to present imperatives, has been a key factor in the development of nuclear energy. French philosopher of science Michel Serres presents this negligence as a symptom of the similarity between human development and the propensity of animals to mark their territory to colonize it. The possibility of a 'contract' between Man and Nature, suggested by Serres in order to establish a more harmonious relationship between men and their milieu, implies distancing ourselves from such practices (Serres 1995, 2008).
5 This proposal came at a time when manned space flights programs were first envisioned – a fact which, by the way, might be worth of interest for the advocates of waste as a fundamental anthropological component; see Monsaingeon (2014).
6 See the chapter on Muslimovo (Russia) in Noualhat (2009).
7 The issue of greenhouse gas emissions has represented a 'windfall' for nuclear power by touting it as a form of 'clean' and renewable energy.
8 ITER (International Thermonuclear Experimental Reactor) is a joint mega project involving around thirty countries with aim to prove the feasibility of a nuclear reactor based on fusion. Its cost is enormous but, as all large-scale nuclear power facility implementation, it is very attractive for the territory that welcomes it due to jobs and the economic activity it creates. The extent of its promises lives up to the critics of which it is the object. Without getting into the detail of these controversies, it suffices to say that the ITER project renews the myth, today demolished, on which nuclear industry had been established, that of an eternally renewable power source.
9 Most actors agree that there is no 'solution' to the nuclear waste problem, but only 'answers'.
10 Two observations, however, might be useful here. First, it must be noted that Nevada's Yucca Mountain Nuclear Waste Repository was finally closed by the Obama administration as a result of the successful but 'unholy alliance' between environmental activist groups and the gaming and tourism industry that felt threatened by the proximity of the site (the story is told in Noualhat 2009). Although beyond the scope of this chapter, based on the experience with some actors of nuclear waste storage in France, we must underline, again, the importance of a cross-country comparative study on waste storage. We have indeed emphasized the cultural dimensions of nuclear waste, which is considered differently depending on many contextual differences (of culture, of development, of democratization level of society, etc.). Like conventional waste, and, more generally, as in every discussion on the clean and the dirty, or the healthy and the unhealthy, nuclear waste – the fuels which become the 'log' after consumption – may not be 'waste' in countries where a recycling process does exist. Second, the waste issue is not as pressing in countries like the United States or Russia, where large areas of land are available, as it is in smaller countries such as Switzerland or Finland.
11 It has been argued that nuclear waste puts us in a special relationship with time – they are these technoscientific products which are placed beyond human time (Adam and Groves 2007). However, they also pose problems with regard to *space* in a much broader context (i.e., of the increasing disqualification of areas of land due to the long-lasting effects of radioactivity). Not only sites in the vicinity of Fukushima and Chernobyl nuclear facilities have been doomed by nuclear energy, but also zones such as the Hanford site in the United States, also known as the 'nuclear craddle'. Nuclear power requires space, bare lands or lands out of the realm of human activity.
12 This was emphasized by Raineau (2012).

13 Borrowed from pharmacology, the notion of half-life refers to the amount of time required for a substance to lose half of its activity. In nuclear physics, it measures the average period at the end of which the nucleus of an atom has a 50 percent chance of decaying. It does not imply that the substance will show no activity after a period of two half-lives. Actually, it is a probabilistic notion used in cases where it is impossible to determine accurately how long a substance will be active.

14 This was filmed by Michael Mardsen in Onkalo's geological waste storage site, Finland.

15 Compare Peter Galison and Robb Moss's documentary film *Containment* (2015; http://www.containmentmovie.com).

16 Scientists, however, are well aware that the comparison between Oklo and nuclear reactors and burial of radioactive material cannot be pushed too far. Naudet argues that this is an exceptional phenomenon because the window required for this phenomenon to occur was very narrow. As physicist Bertrand Barré says, "one cannot push the comparison to the end. To use teenager vocabulary, *Oklo c'est trop* ('that's going too far'). It would be great if we could find million year old fossil fuel reactors [. . .] But it gives us confidence in our capacity to meet the conditions, significantly more simple, of a secure geological storage" (Barré 2005, 36).

17 See, for example, Latour (2004).

18 Thank you to Philippe Bardy for the translation of this manuscript.

References

Adam, B. and Groves, C. 2007. *Future Matters: Action, Knowledge, Ethics*. Leiden: Brill.

Arendt, H. 1958. *The Human Condition*. Chicago: University of Chicago Press.

Barré, B. 2005. 'Les réacteurs nucléaires naturels d'Oklo (Gabon): 2 milliards d'années avant Fermi!' *Sciences* 3 (3): 31–36.

Barthe, Y. 2006. *Le pouvoir d'indécision. La mise en politique des déchets nucléaires*. Paris: Economica.

Commission particulière du débat public Cigeo. 2013, http://www.debatpublic-cigeo.org/debat/cndp-cpdp.html (accessed July 16, 2016).

Douglas, M. 1966. *Purity and Danger: An Analysis of Concepts of Pollution and Taboo*. New York: Routledge.

Galison, P. 2010. 'Underground Future.' In *Ecological Urbanism*, edited by Mostafavi, M. and Doherty, G., 304–305. Baden: Lars Müller Publishers.

Gras, A. and Poirot-Delpech, S. 1993. *Grandeur et dépendance. Sociologie des macro-systèmes techniques*. Paris: Presses Universitaires de France.

Halbwachs, M. 1992. *On Collective Memory*. Chicago: Chicago University Press.

Hughes, T. P. 1983. *Networks of Power*. Baltimore, MD: Johns Hopkins University Press.

Latour, B. 2004. *Politics of Nature: How to Bring the Sciences into Democracy*. Cambridge, MA: Harvard University Press.

Les Amis de la Terre. 1978. *L'escroquerie nucléaire. Danger*. Paris: Stock.

Monsaingeon, B. 2014. 'Les déchets durables.' PhD dissertation, University Paris 1 Panthéon-Sorbonne, Paris.

Naudet, R. 1991. *Oklo: des réacteurs nucléaires fossiles, étude physique*. Paris: Eyrolles.

Noualhat, L. 2009. *Déchets, le cauchemar du nucléaire*. Paris: Le Seuil/Arte Éditions.

Olivier, L. 2008. *Le sombre abîme du temps. Mémoire et archéologie*. Paris: Le Seuil.

Poirot-Delpech, S. and Raineau, L. 2016. *Science and Engineering Ethics* (22): 1813. doi: 10.1007/s11948-015-9739-9.

Proctor, R. N. and Schiebinger, L. L. 2008. *Agnotology: The Making and Unmaking of Ignorance*, edited by Proctor, R. N. and Schiebinger, L. L. Palo Alto, CA: Stanford University.

Raineau, L. 2012. 'Regard socio-anthropologique sur les choix énergétiques d'aujourd'hui: la question du nucléaire civil.' In *Repenser la nature. Dialogue philosophique, Europe, Asie, Amériques*, edited by Parizeau, M.-H. and Pierron, J.-P., 431–449. Laval: Presses de l'Université de Laval.

Sandia National Laboratories. 1991. 'Expert Judgement on Markers to Deter Inadvertent Human Intrusion into the Waste Isolation Pilot Plant,' Sandia National Laboratories Report, Albuquerque, NM, and Livermore, CA: Sandia National Laboratories.

Serres, M. 1995. *The Natural Contract*. Ann Harbor, MI: University of Michigan Press.

———. 2008. *Le mal proper*. Paris: Le Pommier.

Topçu, S. 2013. *La France nucléaire. L'art de gouverner une technologie contestée*. Paris: Le Seuil.

Wetherill, G. W. 1953. 'Spontaneous fission yields from uranium and thorium.' *Physical Review* 92 (4): 907–923.

16 Biography of a 'sand heap'

Staging the beginnings of nature

Astrid Schwarz

There is no doubt that it is a story with great dramatic potential: the metamorphosis of a desolate, gigantic dried-up hole in the ground into a greening hill now abundant with water, of dead rocks into living earth, of a lunar-like landscape into some kind of Garden of Eden – which, in a sense, the 'sand heap' *Hühnerwasser* (Chicken Creek) is for its scientists.[1] This place, to them, breathes fecundity and exudes potential for new research questions; it offers tantalizing prospects for the development of innovative methods and the creation of new phenomena. From the perspective of a philosophy of technoscience, though, there are two main aspects of the story about Chicken Creek that lend themselves to scrutiny. The first one is its dramatic form, which underscores the importance of narrative in the construction of a particular research object.[2] Framed within this dramatic narrative, scientific action along with its individual and collective actors, its scenes, performances and props come into stark focus, acquiring their own specific contours. Drama serves to focus attention on the relationships among the actors as well as on the various interests they bring to the object in question. It prompts us to take notice of the various relations between the material and the symbolic world. This then suggests the second salient aspect of the story about Chicken Creek, namely, the question of the relationship between epistemology and ontology. 'Ontological commitment' is one way of expressing the inextricable relatedness between the two. In this respect, Chicken Creek appears as a real-world simulation that affords data and theories about the initial phase of an ecosystem while at the same time being the very material product of these theories.

Portrait of the object in question

Chicken Creek is a constructed natural site, the biggest of its kind in the world,[3] created for the purpose of conducting ecosystem research. It is situated in a former open-cast mining area in northeastern Germany, not far from the city of Cottbus. The heap in question is 6 ha in size, with a difference in height of approximately 15 m and an average longitudinal slope of 3.5 percent; at the lowest part of the site is a water basin (see Figure 16.1). The catchment is characterized by four main sections, namely, the inclined back slope area, the more

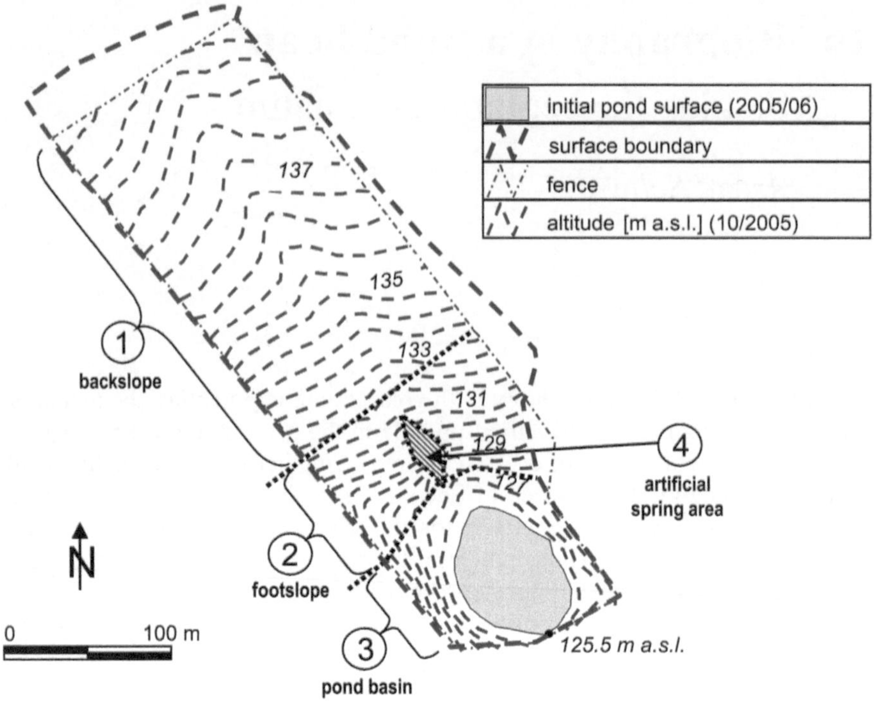

Figure 16.1 The diagram maps out the construction plan of Chicken Creek. At the same time it shows the material boundary work that serves to define the identity of the research object. Courtesy of BTU Cottbus-Senftenberg, FZLB.

Figure 16.2a and b Aerial photographs taken in 2009 and 2013 reveal the dynamic evolution of the site as a 'superdynamic system'. Courtesy of BTU Cottbus-Senftenberg, FZLB.

steeply inclined foot slope, the pond basin and a small spring area just above the basin. The construction of the hill was carried out strip-wise, a technique widely used in this mining area. The hill is situated on top of a clay layer which seals off the ecosystem from below and prevents uncontrolled influences seeping in from the subsoil. The soil in the catchment was taken from sandy accumulations, mainly to avoid the typical problems facing such sites, namely, severe acidification of soil and water due to pyrite oxidation. The whole area is surrounded by a fence, which is designed to exclude uninvited guests such as wild game, hunters, conservationists, hikers and those engaging in other recreational activities. It does not prevent smaller animals or plants from settling in the area, however. In this respect, the fence is part of the materialized rules that define what part of the ecosystem is reserved for what kind of monitoring and experimentation. These different research plots are demarcated by pathways, survey routines, and fixed instruments on top of and underneath the surface. The researchers are cautious, indeed humble, users of the sand heap who administer the ritual acts of experimentation by allowing themselves to be subject to a sophisticated set of spatio-temporal instructions: when, where, how often, and in what way it is permissible to take samples, and who is allowed to do so. It is only under these controlled conditions that the area can work as an ecosystem in its own right, and that the initiation of life-supporting structures and dynamics can be observed and measured.

The construction phase was completed in 2005. Since then the site has been performing 'the start of nature',[4] rapidly becoming an arena that is teeming with life and filled with unexpected surprises: the lake quite quickly became filled with water, the artificial spring developing differently than its construction would have suggested; the surface structures behaved differently than expected (incrustation); the course of the creek took a different direction than the one predicted; and the erosion gullies developed differently than expected. In short, Chicken Creek was set up to become a 'super dynamic system' and, sure enough, the beginnings of nature were performed with such powerfully overwhelming dynamics that, within fewer than 6 years, about 160 plant species had colonized the sand heap and made it an extremely rich system in terms of biodiversity – richer than many ecosystems that are much older and more stable.[5]

A considerable part of the following discussion is devoted to the problem of determining the point in time that constitutes a 'beginning' and thus, accordingly, to the question of whether it is possible at all to distinguish between the design of the object and its measurement. Additionally, the way this problem is handled is itself at issue, including not least the interrelatedness of the model in situ, the real-world sand heap outside in the revitalized mining area, and the model in silico, the computer simulation of Chicken Creek in the lab. However, part of the story is also devoted to the controversy over the continuation of the project, and we shall see whether the sand heap plays more of a comic or a tragic role – or possibly both – in this drama of a 'very prominent ecological research object' (what we might call an 'eco-VIP').[6]

The birth of a sand heap

The story to be told is first and foremost as a birth story. It is about the appearance of an artificial hill located in the middle of an open-cast coal mining area and dedicated to ecological research on emerging ecosystems. It is also about the creation of a 'globally unique' object, about the construction of a water catchment that is designed to be natural insofar as it enables observers to study the phenomena that occur in nature when life encounters non-living nature. In a German magazine, this was taken up in an article bearing the headline 'Everything ready for take-off' and describing how researchers from Cottbus are looking at the mechanisms by which 'life develops out of dead matter' (Willmann 2009, 1). Speaking in noticeably more moderate tones, these very researchers state that they are interested in the structures and processes that occur during the initial phase of ecosystem development. This phase starts at 'point 0', the moment of birth of the system and, as such, a fundamentally crucial moment – not only because it is quite a complex endeavor to define this 'point 0' at all, but also because entire ecosystems at the landscape level have not been studied before to this degree of completeness.

The analogy between the birth story of 'the beginning of nature' – the birth of an ecosystem – and the more common stories about 'the beginning of life' immediately suggests itself here. It serves to conjure similar expectations while simultaneously forging a link with a set of questions that are generally considered to be among the most exigent – and also perplexing – for humankind: Where do we come from and what constitutes the human condition? The human genome project sought to tell such a genesis story, and the promises of synthetic biology (particularly in the field of human enhancement) also tap into these grand themes. The stories told about the genesis of life on earth do just the same. What unites all these stories is that they are heavily laden with all kinds of values and presuppositions that are deeply rooted in our enlightenment-infused cultures and thus also in scientific culture. In order to lend this general observation more substance, it may help to illustrate it by taking a closer look at the analogies linking these birth stories – the beginning of life and the beginning of nature. Accordingly, I propose to take a brief look at what I take to be a typical geo-historical birth story of life on earth.

These stories usually begin with a swirling haze wafting over some kind of primordial soup which is then exposed to an oddly unusual atmospheric perturbation and then – given these highly dynamic, energetically heightened conditions – gives birth to the first building blocks of life, the nucleic acids. Then, we are told, the drama of life unfolds over hundreds of millions of years, with periods of stagnation and accelerated development, distinguished by different leading characters bearing names such as archaeobacteria, trilobites, dinosaurs and, most recently, the human species, ringing in the (still variously defined) Anthropocene era. Whether the genesis of life is regarded as being driven by mere chance – by a series of more or less fortunate chemical and physical accidents – or whether it is narrated as a creation myth involving divine inspiration, or is believed to be a mixture of both (or indeed something else entirely) depends on the cultural and religious context and, not least, the philosophical categories which underpin the narrative concerned.

Bearing this in mind, any decision for or against a particular line of argumentation or for or against a particular narrative is not just an epistemic issue but also an ontological one. Thus the following reflections on 'the beginning of nature' as a birth story and its ensuing drama are simultaneously a means of reflecting on this interdependence – taken to be a necessary one – between propositional knowledge and thing knowledge, between 'episteme' and 'ontos': theories and concepts *about* things are inevitably linked with a theory *of* things.

Worldly moments

This brings us to the second important implication of this birth story, which is the question of how something comes into the world, how it literally stands the reality test. It is the question of how an object is afforded, what kind of practices and technologies are involved (particularly at the moment of birth) and, finally, how this moment is to be defined.

To continue with the birth metaphor, is it the moment when the head of the baby becomes visible, when it has left its mother's body, when it takes its first breath or when the umbilical cord is cut? The long tradition of the midwife profession and its relatively recent scientization has led to this moment being well defined and, in line with the accompanying social mandate, measured.[7] As a result, most people born in Western countries know precisely to the minute when they were born. Obviously, no such tradition exists with regard to determining the 'point 0' of a water catchment (or other emerging ecosystems[8]), a situation which gave rise to a number of theoretical as well as straightforwardly practical problems in the Chicken Creek project.

Placing 'point 0', establishing a temporal order

Originally, the Lusatian landscape was formed by Pleistocene melt water streams coming from the northern glaciers and forming a hilly landscape covered with mixed oak and pine forests interspersed with small mountains and valleys. The area includes the 'Steinitz Alps' near the Chicken Creek headwater, which feeds a small stream that flows into the Spree. When coal mining excavations were intensified during the 1970s, all of this was destroyed, including the headwater, and the groundwater table was considerably lowered. As the mining company Vattenfall Europe Mining AG (VEM) is legally obliged to restore the landscape and the stream, including a functioning headwater, the newly constructed Chicken Creek ought to ensure a minimum discharge level for this emerging stream in the near future. Thus the design of the constructed natural site had to take account of the legal regulations on coal mining restoration as well as the special requirements of a research site, ensuring a maximum of oversight regarding the materials used and the overall structure of this novel object. The construction of the catchment took place between May 2004 and September 2005. Large mining machinery was used, including conveyor belt systems that involved bucket wheel excavators with combined stackers. Bit by bit, strip by strip, the excavated, detached material was redirected and recontextualized, affording what was to become a meticulously monitored sand heap. In August 2004 the construction of the sandy aquifer began, starting with the eastern area and followed over the subsequent two months by the central parts. In the second phase sandy material was dumped in the lower part of the growing hill, and in the third phase the western part of the clay layer was completed as well and covered, again, with sandy material, while in the center of the area a central trench was left open. This was the situation in which the premature sand heap entered its winter hibernation. Construction did not recommence until May 2005, when the 'abdominal cavity' was filled with the material that had been dumped in the upper and lower sections. This relocation of the material on the site itself mainly served to 'reset' the substrate in a primordial condition – in other words, to remove all traces of life such as seeds, algae, moss and the like.

The very last construction phase commenced with the sand heap being given shape, in accordance with a 'mixed design' involving research requirements and safety-relevant stipulations, to finally achieve a defined object with clear surface boundary conditions. The whole surface was then flattened once again and, in September 2005, was homogenized to remove any surface structures that had been generated by the construction work (mainly car and bulldozer tracks) and to ensure as uniform surface conditions as possible for the entire site. Finally, the area that was to become the research object Chicken Creek was fenced in completely. This step was seen as marking "the completion of the construction process", and was "therefore defined as 'point zero' of the development of the site" (Gerwin et al. 2010, 20). At this point, the first measuring devices were installed in order to monitor the initial state of the system and to observe the genesis of structures.

In one sense, the time of birth of the sand heap was September 2005; however, the instruments to be used to monitor it were not sufficiently well installed to capture this birth drama adequately from point zero. That is why, years after the literal 'point zero', a number of scenarios have been simulated – in situ – to get a more fine grained data set and an image magnification of the initial momentum. Besides this more recent reconstructive work associated with 'point zero', expectations have continually been readjusted in relation to what is notionally intended to be a natural situation of a nascent water catchment. In other words, the object Chicken Creek did not behave as the researchers expected it to. This was the case in the digital as well as in the analogue world. At the start of the project, a workshop was organized with hydrology modeling experts from all over the world. They were invited to enter a contest to devise the best computer model for predicting the development of the water catchment. None of the models won – the sand heap performed as an opaque object and served as a matter of surprise.

Other instances of surprise that gave rise to interventions on the site included lateral erosion gullies that had developed a tendency to destabilize the adjacent fence; this necessitated channelization, which reduced the extent of the erosion considerably (MacArthur and Wilson 1967, 26f.). Additionally, the border of the object had to be maintained and reconstructed several times: additional small ramparts had to be built in order to prevent potential water flux from outside the watershed; similarly, the part of the catchment outside the fenced area was leveled again to create clear catchment borders. This brings us to the ontological questions and, above all, to the connection between narratives, practices and material resistance.

Relating objects and 'relative ontology'

Willard Van Orman Quine tellingly referred to this mutual relation of the epistemic and the ontological in terms of an 'ontological commitment' that makes no general claim about the existence of a thing in reality but suggests an inextricable connection between a particular theory and the presumed ontology. "Ontology is indeed doubly relative [. . .] We cannot know what something is without knowing how it is marked off other things. Identity is thus of a piece with ontology. Accordingly it is involved in the same relativity" (Quine 1969, 54f.). This concept of ontological relativity is helpful when it comes to analyzing objects such as Chicken Creek that are at once malleable and recalcitrant. In addition, the notion of ontological relativity serves to differentiate the theoretical entities and identities that are encountered in interdisciplinary projects. Here, hydrologists 'see' a different sand heap than botanists or limnologists do.[9] Accordingly, one of the most interesting challenges for the scientists involved in the project was to establish a common, shared object of scientific investigation. The challenge for the philosopher doing her participatory observation is then to follow this language game and to conceptualize its dynamics. Interestingly enough, the deliberation about a common identity happened (and still happens) for the most part, first, by means of Chicken Creek in silico, a computer model of the site, and, second, through regulating practices and feedback control on the real-world sand heap.

Quine states quite explicitly that "it is the very facts about meaning, not the entities meant, that must be construed in terms of behaviour" (Quine 1969, 27f.), thereby referring to John Dewey's notion that language is a mode of interaction. Accordingly, Quine identifies two distinct aspects to knowing a word, the phonetic and the semantic aspect. The first requires an element of relatedness: learning takes place by observing and imitating our neighbors. That is, the phonetic aspect of the word (talking to one another) establishes relationships among people. The second aspect relates to the world of objects: we have to see the same object as other people do. It is not possible "to say in absolute terms just what the objects are",[10] but – to introduce yet another notion of relatedness – it is possible to say what they are in terms of a specific (ecological) theory. Having identified and individualized the relevant parts of the object, a more local theory can be placed into a broader general theory. This latter eventually encompasses a number of specific theories and thus allows the theorist to move subtly from one fragment of theory to another. Appreciating these different identities and their related ontologies allows for a likewise comprehensive and inclusive theorizing, particularly of assemblages such as 'Chicken Creek'.

It must be pointed out that the ontological commitment in the Quinian sense remains limited to the realm of propositional knowledge and is blind to scientific practice. In spite of this emphasis on the contingency of meaning with regard to things and the importance of both behavior and relational semantics, Quine never refers explicitly to the role of experimental practice in creating scientific knowledge – for example, the phenomenotechnical approach adopted by Hans-Jörg Rheinberger and drawing on Gaston Bachelard and Georges Canguilhem, or 'thing knowledge' in the sense of Davis Baird (Bachelard 1938; Canguilhem 1970; Rheinberger 2001; Baird 2004).

Tinkering and re-adjusting – the nature of experimental work

The overall hypothesis of the Chicken Creek project is that the initial phase of ecosystem development determines the later stages an ecosystem goes through. The water catchment Chicken Creek is an experimental system that is expected to enable a closer analysis of the role of physical, chemical and biological parameters and of how these develop and interrelate in different stages of a system. Chicken Creek is an artificial system that was designed to study the characteristics of natural systems. Therefore this system is a specific kind of field experiment that eliminates the carefully maintained spatial separation between an experimental system and the natural system, features of which it is supposed to represent. The artificial water catchment exhibits its own performance parameters and invites the discovery of causal dependencies between different parameters. It does not require an understanding of how it represents 'real' water catchments, since Chicken Creek is quite real enough to substitute for any real system with which it shares dynamic properties. Chicken Creek is first and foremost a 'real-world simulation' and only becomes a simulation in silico at a subsequent stage, when the system is represented in a computer model.

Ontological commitment and 'the real'

Here, the concept of ontological relativity[11] might prove helpful in highlighting the fact that the debate about what 'nature' (in the singular) 'really is' has insistently naturalistic overtones. One variation of this straightforward realism is, namely, that this real nature in question ought to be restricted to the Aristotelian concept of a nature that has 'become' (*gewordene Natur*), untouched by human agency. Another, perhaps closely related, idea is that 'science' is limited to the formulation of 'laws' that hold true in every place in the universe (or in no place), and that ultimately philosophy of science should be concerned primarily with this kind of epistemology. It is not surprising, then, that practically nothing – apart from physics – turns out to be 'science' in this sense. As Hilary Putnam grumbled (with more than a hint of impatience), "Why on earth (should) we expect the sciences to have more than a family resemblance to one another?" (Putnam 1994, 471), immediately answering, "There is no set of 'essential' properties that all the sciences have in common" (Putnam 1994, 472). He concludes finally, "the time has come to recognize that scientific theories are of different 'types' and that informative philosophizing has to descend to a more 'local' and less 'global' level" (Putnam 1994,

478). Thus, the plurality of sciences needs to be addressed by means of more local epistemologies, a task which appears to be all the more urgent given that, in the geography of the sciences, quantum mechanics and relativity theory[12] constitute a minority over against a landscape dominated overwhelmingly by biology and chemistry, pharmacy, geo-, agro-, hydro- and systems sciences. These are classified, far from adequately, as being merely part of the realm of *technè* or as applied sciences that simply bear on the conceptual tools of theoretical sciences – a belief often underlined by the still common credo (not just among scientists but also philosophers of science) that on the path toward truth theory always precedes experiment. All of these topics have attracted some attention in recent debates on the role and identity of science, including the relationship between epistemology and ontology, and a number of conceptualizations have been developed that have helped us to re-assess established narratives regarding the order of the sciences and their epistemic fruit, that is, knowledge that changes the world. These conceptions include the experimental or entity realism put forward by Ian Hacking (1983); the emphasis placed on the work of purification done in the theoretical sciences, accompanied by positive allusions to the 'impure sciences' (Latour 1994; Shapin 2010); the emphasis on local theory production in the applied sciences (Radder 2003; Wilholt 2005), and its 'application-related innovativeness' (*Anwendungsinnovativität*) (Carrier 2006); the assertion of an ontological difference between the twins that are science and technoscience (Carrier 2010) or proposals to look at them as complementary ideal conceptions (Bensaude Vincent et al. 2011) that create some space to focus attention on 'specific material articulations' (Bensaude Vincent et al. 2011, 371) and thus on the numerous muddy 'in-betweens'.[13]

In the following, two main issues will be addressed: first, the concept of affordance, which must be scrutinized particularly in relation to the concept of disposition; and second, the relationship between simulation in the lab and the real-world simulation of the sand heap, along with consideration of the different types of experimentation undertaken in lab and field research.

Affording a starting point

What James Gibson, who first established the concept of affordance (1997), sought to suggest by it is a particular perspective on the furniture of the world. In the process of doing so, he also came up with a form of expression that illustrates this relational view of real-world things. For example, a vertical, flat, extended and rigid surface such as a wall or a cliff face affords falling – it is fall-off-able – but might also become climb-up-able if we are attentive, and so on. Thus when scientists do research in the fenced-off environment of a sand heap – thereby defining what is the research site Chicken Creek and what is not – this object makes ecological research attract-able, the beginning of living nature represent-able, subterranean water flows measur-able, complex data sets comput-able, the in silico model account-able and so on. The concept of real-world simulation is especially interesting in that it makes it possible to name perceived threats and promises of things without defining them as fixed properties of those things.[14] The concept of affordance plays a key role here, in that it points to a relationship and not primarily to a quality. Affordances accord something to an organism in its environment that would not otherwise be possible.

Gibson's 'theory of affordance' became the core element of his ecological psychology. The concept is derived from the overall thrust of concepts such as valence, invitation character and demand character.[15] However, there is one crucial difference, as Gibson himself pointed out: the affordance of something does not change as the need of the observer changes. What his new term was intended to describe was the services the environment offers to an animal: the affordances of the environment are what the environment provides, either for good or for ill. Affordance implies the complementarity of the animal and its environment.

This is far from being a radically novel idea. Gibson refers explicitly to the concept of the ecological niche that describes the processes of exchange between an animal and its environment: the animal is influenced and even shaped by its environment. Conversely, the environment is altered by the animal. He reformulates the niche as a set of affordances, as the expression of an animal's way of life. However, this complementarity of animal and environment as well as the plasticity of this natural relation has its limits, according to Gibson: the environment existed prior to the animal and has to follow the laws of conservation; otherwise it cannot be 'invariant' for animals. It is the environment that provides recalcitrance for biological evolution. At the same time the environment is described as having 'unlimited possibilities' by virtue of confronting organisms with ever new challenges to adapt their ways of life. This motif of transgression is again counterbalanced by the claim that there are conditions of the environment and that these are the same for all animals. There is only one world – however diverse – and all animals live in it, including humans. Neither does Gibson accept the distinction between a 'new environment', the anthropogenically modified world, and the 'old' natural one. For him, changes of shape and substance in the environment are merely the consequences of what it affords to human beings. The reference to ecology provides him with a concept of an oikos that entails a kind of familiarity while creating natural relations to humans and non-humans. Affordances are not subjective, phenomenal or mind-made, as are values and meanings. It is media such as substances and surfaces that afford different features and possibilities; they differ in their sets of affordances. What we perceive when we look at objects (with a relational interest) are their affordances – not their qualities, which are nonetheless present. Affordances, then, involve active perception as well as the potential of moving around or manipulating things, of acting in a world with 'limitless opportunities'.

'Affordance' in the context of philosophy of science

In the philosophy of science, the concept of affordance is a controversial one. However, there is an interesting conception that at least partly picks up on the Gibsonian idea about relational qualities, namely, the "apparatus/World complex", coined by Rom Harré. The connection works via the idea that we are involved in "an engagement, of a sort, with Nature" (Harré 1988) when we are conducting an experiment. To characterize the experimental situation – that is, the interplay between experimental setting and what we customarily call nature, in which there is a reliance on the distinction between the artificial and the natural – Harré proposes that we reframe this relationship as one between the domesticated and the wild. By isolating fragments of nature, we domesticate them, making them (and 'nature') available for material manipulation. The laboratory is like a farm; it is neither wholly artifactual nor wholly wild. The material setup has been tamed rather than represented or caged: "domestication permits strong back inference to the wild, since the same kind of material systems and phenomena occur in the wild and in domestication. An apparatus of this sort is a piece of nature in the laboratory."

The apparatus/World complex is a conception of relatedness that draws heavily on the material (real) dimension where the 'powers of Nature' are accepted as given – that is, explanatory power is attributed fully to the world (and not to the social context of laboratory conditions). An apparatus is different from instruments: it serves as a model, a simplified version of a naturally occurring material setup. In contrast, instruments such as thermo-/ or hygrometers, "must be interpreted as analogous to states of the natural system being modelled." Harré points out that "the apparatus/World complex displays the appropriate surface appearances, just in so far as the World has the power to afford them when conjoined with apparatus. [. . .] It is apparatus/World complexes that afford perceptible phenomena" (Harré 1988, 376). As long as the "apparatus is locked into a system with Nature," we are able to glean knowledge about the piece of nature shared within that system. In this sense the apparatus/World complex comes close to the concept of real-world simulation being used in this analysis.

Real-world simulation

The real-world simulation is a conception that originates in the acknowledgement of a paradox deriving from two ideals of experimentation. The laboratory ideal involves the design of well-controlled experimental systems within a – literally – closed space, and its aim is to produce general propositions. It allows for the production of failures without taking risks that would affect society or the environment. By contrast, the field ideal acknowledges the complexity and blurred boundaries of interventions and the unpredictable responses to them that may occur. This raises the following dilemma: On the one hand the field experiment is supposed to apply solid, proven knowledge and thus to yield reliable technology. On the other hand, the very application of such knowledge usually implies new technological designs and scales, environmental conditions and organizational settings which have not been previously explored and may therefore lead to surprises. The notion of 'society as a laboratory' (Krohn and Weyer 1994) was conceptualized in relation to technological innovation, such as the implementation of technological prototypes in society (e.g., nuclear power plants). The concept 'real-world experiment' was subsequently developed to capture the highly specific character of innovative experiments as well as the fact that scientists nevertheless expect their results to be generalizable and transferable to other projects (e.g., to restoration projects). The knowledge acquired in such projects often takes the form of a kind of expertise that merges scientific background knowledge with experience gathered by observing particular cases. Both real-world experiments and real-world simulations combine features of the lab ideal and the field ideal, and in this way they establish a connection between instances of generalization and instances of individualization and value-ladenness.

The sand heap simulation in silico and the real-world simulation become entangled

It was stated previously that the artificial water catchment exhibits its own performance parameters and invites the discovery of causal dependencies between different parameters. In this sense Chicken Creek is first and foremost a 'real-world simulation' and only becomes a simulation in silico subsequently, when the system is represented in a computer model. However, the elements required to build the in silico model also influenced the design of the in situ model. It was necessary to perform some experiments in the field in order to better describe the initial conditions – after the first initial phase of Chicken Creek had already passed. In addition, the positioning and number of the measuring points in the field was pre-determined by the in silico model in order to construct the 3D grid; the same applies to the frequency and quantity of samples.[16] Conversely, it was necessary to find a consensual design for the in silico model in order to meet the requirements of qualitatively different data sets as well as different scientific cultures. The previously mentioned ontological relativity is a promising concept in the attempt to describe the dynamics of these deliberations and the creation of a shared identity (at least to some extent) through the practices and theories developed on the object Chicken Creek (such as the measurement of subterranean water streams, temperature and solar radiation, the quantity of bacteria and invertebrates in the soil, the quantity of plants on the ground and so on). However, at least from the perspective of the reviewers of the first project phase, this requirement was insufficiently met: the topics seemed to be too fragmented and there was no 'cohesive big picture' (*geschlossenes Gesamtbild*). This also applied – quite literally – to the lacking power of the in silico model.

What kind of drama is the Chicken Creek story?

We have seen that the Chicken Creek project is a birth story in several respects: the artificial catchment literally evolved and was declared as being born at a given moment – the 'point zero' of the object Chicken Creek. The artificial water catchment was then used to do research on the development of early-stage ecosystems which, again, told a birth story of a becoming ecosystem or, as it had been dubbed by the media, a take-off story about the beginning of nature. But how does the drama of this eco-VIP sand heap continue? Is it more of a tragic drama or a comedy drama? And what is its status in the research landscape?

In 2011 the project's funding was cut almost completely. It was no longer a 'special research area' (SFB) supported mainly by the German Research Council, but became a local and regional project supported by mixed financing from BTU Cottbus, Vattenfall and the Brandenburg Ministry for Research and Education. This local consortium guarantees the continuation of long term observation, thereby enabling monitoring and data collection activities to be maintained. More recently, Chicken Creek is set to become part of the 'Terenoworld',[17] an observation network involving different terrestrial ecosystems (TERENO = TERrestrial ENviromental Observatoria, organized by UFZ Leipzig). One interesting question might be how this will change expectations of Chicken Creek and the way its performance is handled, and how its identity and that of the researchers will be influenced and altered.

Concluding remarks

Chicken Creek can be seen as a paradigmatic case of conceptual merging and mixed practices, as described in detail in the previous analysis. Despite having been designed artificially, the artificial water catchment system is treated as a natural system. It exhibits its own parameters and in this sense is considered a real-world simulation. It is an attractive object for scientists, and it is unique. It pursues the lab ideal of total experimental control in the context of a field experiment and is ultimately a high-tech object. Whereas science secures its objects through its representation of facts, technoscience affords things through assemblage. Chicken Creek is a paradigmatic example of this in that it is simultaneously assembled as a real-world object (in situ) and recreated in virtual reality (in silico). Step by step, researchers have been piecing together a plethora of separate results to form a three-dimensional image of the research area which grew – and continues to grow – in parallel to the development of the real-life watershed. This birth story takes the iconic form of a genesis narrative that brings forth this technoscientific object and its open-ended potentiality. Chicken Creek is a technoscientific object in the proper sense: it gathers together theoretical knowledge, instruments, skills and purposes and in doing so affords the proof of concept of the interlocking nature of the epistemic and the ontological: theories and concepts *about* things are inevitably linked with a theory *of* things.

Notes

1 I wish to thank Walter Gerwin and Michael Mutz for extensive conversations about the assemblage that is Chicken Creek and, not least, for having facilitated a direct encounter with the very real recalcitrance of this real-world simulation.

2 In social anthropology the power of drama has been acknowledged in that it not only shapes experience but also, as a reiterated form, 'makes theory fact'. "[T]he road to discovering what we assert in asserting [. . .] lies less through postulating forces and measuring them than through noting expressions and inspecting them" (Geertz 1983, 30, 34). Accordingly, the decision about what textual form is to be adopted (in this case, drama) is at once also a decision about construing semiotic connections: What kind of

relations, between what and whom, become visible and thus the subject of analysis? It has been suggested by a number of authors that the drama of objects should be narrated in the literary form of a biography; this was seen as an apt method to reveal relationships between people and objects, see in Appadurai (1986).

3 "With its 6 ha area and tremendous potential for interdisciplinary ecological research, the site offers globally unique conditions" (Gerwin et al. 2010, 3).

4 "The Start of Nature" was the title of a series of film clips on the Chicken Creek project that were produced on behalf of the DFG (German Research Foundation) (http://www.dfg-science-tv.de/en/projects/the-start-of-nature; accessed 3.7.2016).

5 Conversation with W. Gerwin, 24.8.2012.

6 In 2009, Chicken Creek became "Ausgewählter Ort 2009" (selected place 2009), awarded by the initiative "Germany – country of ideas" under the auspices of the German Federal President.

7 A German midwifery handbook (AHT 2000) defines the moment of birth as the release of the placenta. This date is then to be written down on a ribbon, together with the family name, the calendar date, and a 'designation of sex', and fixed to the baby's wrist. Together, these data define the baby's identity. It is interesting to note that the date of birth is defined differently in different handbooks of midwifery, even within European cultures.

8 However, some work has been done in population ecology, specifically a study of recolonisation after a volcanic event (e.g., MacArthur and Wilson 1967).

9 To be sure, the concept of relative ontology does not hold that the identities marked off are fixed and irremovable. Instead Quine points out that one ontology may be reduced to another "with help of a *proxy function*: a function mapping the one universe into part or all of the other" (Quine 1969, 55).

10 Quine continues, "we are sometimes unable even to distinguish objectively between referential qualifications and a substitutional counterfeit" (Quine 1969, 67), this being of particular relevance with respect to a distinction between a simulation of the real world and experimenting in it.

11 In the foregoing section I pointed out that 'relativity' in the 'ontological commitment' refers first to relative as opposed to absolute and second to a necessary 'relatedness' between *episteme* and *ontos*. In the following it refers mainly to the first aspect.

12 These well-established models of physics are the references for not a few philosophers of science. However, even modern physics turns its back on a monolithic image, as Peter Galison shows in the present volume. He describes the evolution from the popular image of a pyramid of matter (with strongly realist connotations) toward a complex of decentered theories – 'the ring' – which are ontologically indifferent.

13 This is just as the existence of good and bad angels points to the possibility of being a perfect and pure human being and simultaneously relieves us of the burden of trying to be one.

14 Affordances have been also characterized as emergent properties, particularly in Stoffregen (2003).

15 'Affordance' is a neologism, but the concept had some forerunners, developed mainly by gestalt psychologists. As Kurt Koffka emphasized, things have a 'demand character': each thing says what it is. Thus, a fruit says 'eat me', water says 'drink me', and a woman says – according to Koffka – 'love me'. Kurt Lewin coined the term '*Aufforderungscharakter*', which has been translated variously as 'invitation character' or 'valence', with the latter term entering widespread usage. For Lewin 'valences' have corresponding 'vectors', which can be represented as arrows pushing the observer toward or away from an object.

16 A good example is the borehole investigation in 2008–2009; see Gerwin et al. (2010, 24).

17 Interview with project manager R. Hüttl, 4.9.2012.

References

AHT (Aachener Hebammen-Team). 2000. *Handbuch für die Hebammen*. Stuttgart: Hippokrates.

Appadurai, A. (ed.). 1986. *The Social Life of Things: Commodities in Cultural Perspective*. Cambridge: Cambridge University Press.

Bachelard, G. 1938. *La formation de l'esprit scientifique: contribution à une psychanalyse de la connaissance objective*. Paris: Vrin.

Baird, D. 2004. *Thing Knowledge: A Philosophy of Scientific Instruments*. Berkeley: California Press.

Bensaude Vincent, B., Loeve, S., Nordmann, A. and Schwarz, A. 2011. 'Matters of interest: The objects of research in science and technoscience.' *Journal for General Philosophy of Science* 42: 365–383.

Canguilhem, G. 1970. *Etudes d'Histoire et de Philosophie des Sciences*. Paris: Vrin.

Carrier, M. 2006. 'The Challenge of Practice: Einstein, Technological Development and Conceptual Innovation.' In *Special Relativity: Will It Survive the Next 101 Years?*, edited by Ehlers, J. and Lämmerzahl, C., 20–28. Berlin: Springer.

———. 2010. 'Knowledge, Politics, and Commerce: Science under the Pressure of Practice.' In *Science in the Context of Application: Methodological Change, Conceptual Transformation, Cultural Reorientation*, edited by Carrier, M. and Nordmann, A., 11–30. Boston Studies in the Philosophy of Science, 274. Dordrecht: Springer.

Geertz, C. 1983. 'Blurred Genres: The Refiguration of Social Thought.' In *Local Knowledge*, edited by Geertz, C., 19–35. New York: Basic Books.

Gerwin, W., Schaaf, W., Biemelt, D., Elmer, M., Maurer, T. and Schneider, A. 2010. 'The artificial catchment "Hühnerwasser" (Chicken Creek): Construction and initial properties.' *Ecosystem Development* 1, 1–56.

Gibson, J. J. 1997. *The Ecological Approach to Visual Perception*. Boston: Houghton Mifflin Company.

Hacking, I. 1983. *Representing and Intervening*. Cambridge: Cambridge University Press.

Harré, R. 1988. 'Recovering the experiment.' *Philosophy* 73: 353–377.

Krohn, W. and Weyer, J. 1994. 'Real-life experiments: Society as a laboratory: The social risks of experimental research.' *Science and Public Policy* 21: 173–183.

Latour, B. 1994. *Nous n'avons jamais été modernes*. Paris: Edition La Decouverte.

MacArthur, R. H. and Wilson, E. O. 1967. *The Theory of Island Biogeography*. New York: Princeton University Press.

Putnam, H. 1994. *Words and Life*. Cambridge, MA: Harvard University Press.

Quine, W. V. 1969. *Ontological Relativity and Other Essays*. New York: Columbia University Press.

Radder, H. 2003. 'Toward a More Developed Philosophy of Scientific Experimentation.' In *The Philosophy of Scientific Experimentation*, edited by Radder, H., 1–18. Pittsburgh: University of Pittsburgh Press.

Rheinberger, H.-J. 2001. 'Introductory remarks: Conference "experimental cultures": Configurations between science, art, and technology, 1830–1950.' *MPI Berlin Preprint* 213: 3–12.

Shapin, S. 2010. *Never Pure: Historical Studies of Science as If It Was Produced by People with Bodies, Situated in Time, Space, Culture, and Society, and Struggling for Credibility and Authority*. Baltimore: The Johns Hopkins University Press.

Stoffregen, T. A. 2003. 'Affordances as properties of the animal – environment system.' *Ecological Psychology* 15 (2): 115–134.

Wilholt, T. 2005. 'Bedingungen wissenschaftlicher Innovation unter der Vorherrschaft von Anwendungsinteressen. Freiheit und Komplexität.' In *Sektionsbeiträge des XX. Deutschen Kongresses für Philosophie*, edited by Abel, G., 2 vols, vol. 2, 377–388. Hamburg: Meiner (in Kommission für den Universitätsverlag TU Berlin).

Willmann, U. 2009. 'Alles auf Start. Auf einem Versuchsfeld im Kohletagebau verfolgen Cottbuser Forscher, wie sich auf toter Materie Leben entwickelt', *Zeit Online*, 10 (5): 1–6. http://pdf.zeit.de/2009/20/U-Boden-Cottbus.pdf.

Index

Note: Figures and tables are indicated with the page number in *italics*.

α6β4 transmembrane protein 32

Abbott, Edwin 167
adenosine triphosphate (ATP) 126–8, 194
AdS/CFT 23–4
adult-to-kid transmogrifier 169
Aeneid (Virgil) 189
Aerosil® 67n4
affordance, concept of 253–4
Against Method (Feyerabend) 118
Agence nationale pour la gestion des déchets radioactifs 232
Age of Cardboard 169
agribusiness *see* transgenics (TGs)
agro-ecology 153, 157, 159–60, 162, 163
agro-ecosystems 153, 157, 163, 164
agrotoxics 154, 163
Alice robots 42–3
Alice's Adventures in Wonderland 31
allotropes of coal 186–8, 190, 192, 196
altruism, of robots 43–4
American Language (Mencken) 139
Amonton, Guillaume 61
Amonton's Laws of static friction 61, 64, 65
Anderson, Philip 16–17
Animal Laborans 240
anontological 18
Anonymous, hactivist collective 114–18
Anselmino, Otto 95
Anslinger, Harry J. 97
Antarctica: Dronning Maud Land 218, 221–2; ice cores 215, 221–2, 226; Vostok research station 221
anthracene 190
anthropic principle 22–3
Anthropocene 169, 228, 248
anti-ontology 17–18
Apollo program 80–1
Appadurai, Arjun 91
apparatus/World complex 254–5
aquifers 136, 250
archaeobacteria 248
archaeology 238–9
Arendt, Hannah 240
Aristotle 191, 252
Arrhenius, Svante 219
asphalt 191
aspirin 89–90
assemblage 15; of cardboard 169; Chicken Creek as 251, 257; robots as 51–2
assessments, ecological 143–4, 147
Atlantic Oceanographic and Meteorological Laboratory (AOML) 208
Atomic Energy Commission (CEA) 233
atomic weight 190
ATP (adenosine triphosphate) 126–8
attractive technoscientific objects 1–3
Audubon Society 138
Aurelius, Marcus 92
Austro-Prussian War of 1866 93–4
automobile industry 81–2
Avogadro's number 190

Bachelard, Gaston 2, 11, 252
Bacon, Francis Thomas 77–8, *79*, 80
Bacon cell 78, *79,* 80
Baird, Davis 252
banking of wetlands, environmental mitigation 141–4, 145, 147
Barthe, Yannick 237
Barthes, Roland 210
Bataille, Christian 237
batteries: gas 75–8, 81; voltaic 73–4
Bauhaus 169
Baur, Emil 77
Bayer, Herbert 169
Bayer (Friedrich Bayer & Co.) 89, 94, 95

Becquerel, Antoine-César 73–4
Becton Dickinson Electronics 29
Belgium, and heroin 101
Bensaude Vincent, Bernadette ix, 10, 185–200
Berthollet, Claude-Louis 73
Big Science 217
biodiversity 153–4, 205, 247
biography of Kabuki syndrome 123–35; geneticist interview 129–31; parent interviews 127–9, 131–2; pediatric neurologist interviews 125–7, 129; translational medicine 124–5, 132–3; whole exome sequencing 124–5; Wittgensteinian approach 123
biotechnology, and transgenics 151, 158, 164
birth story 248–9, 257
bitumen *185*, 191
Bonhoeffer, Karl 95
boxes, cardboard *see* cardboard
Box Manufacturer's Certificate 173
Brandenburg Ministry for Research and Education 257
British Petroleum (BP) 81
British Ropes 81
Brookhaven 24
Brooks, Michael 118
Brooks, Rodney 44
BTU Cottbus *246*, 257
Bunyan, John 137
Bure, Lorraine 232, 236–7
Burst Strength Test (BST) 173
Bush, George 141, 145, 147

Caine's Arcade 169
caloric 73
cancer stem cells (CSCs) 27–41; assumptions of theory 30–1; division of 28, 32; leukemic and hematopoietic 28–9; metaphysical views of stemness 34–6; properties: self-renewal and differentiation 31–4; targeting therapy *vs.* niche-targeting therapy 27, 35–6
cancer stem cell theory 30–1
Canguilhem, Georges 252
capitalism, cardboard's role in commerce and 168, 169–70, 173–5, 177
carbon 185–200; as equivalent 194–5; as fire 193–4; heteronyms 187–9, 195; as material 191–2; as memory 193–4; as mephitis 189; multiple modes of existence, allotropes 186–8, 190, 192, 196; ontography of 187, 195–7; as skeleton or backbone 192; as a standard 189–90; as substance 191; as surface 192–3
carbon, forms of: anthracene 190; asphalt 191; bitumen 191; bituminous coal *185*; carbon fibers 191–2; carbon nanotubes 192, 193; charcoal 187, 194; coal 188, 191, 194; diamond 188, 190, 194; fullerenes 192, 197; graphene 192, 193, 197; graphite 188, 192, 193; oil 187, 194
"Carbon Democracy" (Mitchell) 194
carbon dioxyde 71, 189, 194; polar ice cores and temperature variations 216, 218, 221
carbon dioxyde equivalence 194
carbon economy 71–2
carbon equivalent 187
carbon fibers 191–2
carbon nanotubes 192, 193
carbon trading 187, 195
carcinogenesis 27
cardboard 166–82; commerce and industry 168, 169–70, 173–5, 177; era of cardboard-humanity 168–9; packaging science 170–5, 180; paper foundations of literacy 175–7; paperless office 166–7; strength stress tests 173; as technoscientific object 178–80; types of boxes 167
Carnap, Rudolf 17
categorical property of stemness 34–5
Cavendish, Henry 72
Cell Broadband Engine (Cell processor) 108
cell state hypothesis, of stemness 34
Centre industriel de stockage géologique 232
CGIAR 158, 161, 162
Chang, Hasok 187
charcoal 187, 194
Chicago wetlands banking market 143
Chicken Creek 245–60; affordance and apparatus/World complex 254–5; affording a starting point 253–4; birth of a sand heap 248–51, 257; construction of 245–7; experimental work 252–5; ontological relativity 251–2; placing 'point 0' 249–51, 256; real-world simulation 255–7
climate change 70–1; ice core studies 216, 218, 220–1, 223–4, 227–8; *see also* Dansgaard/Oeschger events
cloning 28–9, 32

clonogenic 28–9
coal 188, 191, 194
coal mine, and ecosystem research 245, 248, 249
coal tree *185*
cocaine 95–6, 97
Cold War 78, 171; space race 78
commodification of seeds 153–5
commodity fetishism 174
complexity 46–7, 160
concept of affordance 6–7
conductance, theory of 58–9, 62
Consortium of International Agricultural Research Centers (CGIAR) 158, 161, 162
Container Corporation of America *172*, 175, *179*
container ship spills 207–8
containment policy 171
Conté, Nicolas Jacques 188
context-sensitive strategies (CSs) 163
Continuous Flow Analysis (CFA) 223
Copernican Revolution 2
Copernicus 17
Coriolis effect 207
corrugated boxes 170, 173, 175; *see also* cardboard
Cottbus, Germany 245, 248, 257
CSC *see* cancer stem cells (CSCs)
Cypress swamps 138

Dansgaard, Willi 218, 219
Dansgaard/Oeschger events 220–1, 224, 225, 227
Dante 137
Datta, Supriyo 58–9, 62–3
Davis, Jake ('Topiary') 118
DDoS attacks 106, 114–15, 117–18
debris in oceans *see* garbage patches in oceans
decontextualizing strategies (DSs) 162–4
Deleuze, Gilles 42, 48, 50
de novo acquisition 33
Descartes, René 159
Descola, Philippe 18
devices: inertial device 59–60; nanoelectronic device 59, 62; superlubric device 65–6
Dewey, John 251
diamond 188, 190, 194
differentiation property of stem cells 31–3
Dimopoulos, Savas 22
Dioscurides 92
disorder, Kabuki syndrome 124
dispositional ontology of stemness 34–5
Distant Early Warning (DEW) line 219, 220, 222
distributed denial-of-service (DDoS) attacks 106, 114–15, 117–18
distributed green hydrogen economy 70
Divine Comedy (Dante) 137
DNA 192
domes, for ice cores 216, 217, 219, 221–2
Douglas, Mary 235
drama 245, 256
Dreser, Heinrich 95
drifters *see* garbage patches in oceans
Dronning Maud Land, Antarctica 218, 221–2
drug cartels 102
Duisberg, Carl 95
Dumas, Jean-Baptiste 194

Ebbesmeyer, Curtis 201, 207
ecological assessments 143–4, 147
ecological niche 254
ecosystem research *see* Chicken Creek
ecosystem services 142–3
eco-VIP 247
Edge Crush Test (ECT) 173
Edison, Thomas 192
Edmonds, David 123
EEG 125–6
Egyptians, ancient 191
Eidinow, John 123
Einstein, Albert 18, 21
Ekman transport 207
electrical dynamo 70
electrical generators 73, 75, 76, 78
electrical networks 70
electrical power, from heat engines 70–2
Electrical Research Association (ERA) 77–8
electrochemistry 73–5
electrolysis of water 74, 76
electron transport 56–7, 58–9, 62–3, 67
elements, chemical 189–90
Elliot, Kevin ix, 9, 136–49
Elzinga, Aant 2, 10, 215–31
embryonic stem cells (ES) 27, 31
Energy4All 128, 132
Energy Conversion Ltd 81
energy sources 70
energy technology: compared to information technology 82–3
energy transition 71
engineering 19, 25
entity theory, of stemness 34

environmental banking of wetlands 141–4, 145, 147
epistemology 1, 245
Ernst, Joël 97–8
ether 17
European Parliament 70
European Program for Ice Coring in Antarctica (EPICA) 219, 221, 223, 226
European Union (EU) 70–1
evolutionary biology, and robots *see* robots, in study of evolution
evolutionary theory 42, 44
exobiology 29
explanation 15

FACS technologies (Fluorescence-Activated Cell Sorting) 29
Faraday, Michael 73
Farbenfabriken 89, 94
Federal Bureau of Narcotics 97
fens 137
Fermilab 16
Feyerabend, Paul 118
field experiment 252, 255, 257
final waste 235
Flatland 167, 178
Floreano, Dario 43, 48–9, 51
floristic biodiversity 143
floristic quality assessment (FQA) 143–4
flotsam drift studies 207–8
Folding@home 108–10, 112, 118
Forman, Paul 3
de Fourcroy, Antoine 73
France *see* nuclear waste
Frankel, Fritz 97–8
Frederick II (Kaiser) 99
Free Radicals: The Secret Anarchy of Science (Brooks) 118
French Academy of Science 73–4
friction: by grime 61–2, 64; theory of 61–3, 65; vanishing friction events 60, 62, 65, 67, 68n9
frictionless surfaces 56–69; frictional force and duality 61–2, 63–6; idealized or simple 62–3; molecular electronics 56–7, 58–9, 67; nanoelectronics 56–7, 58–9, 62–3, 67; scientific theory of physics of conductance 58–9, 62
Friedrich Bayer & Co. 89, 94, 95
Fuchs, Harald 68n9
fuel cells: Bacon invention 78, *79*, 80; current usage 81–3; expectations of 72
fullerenes 192, 197
furniture of the world 6

Gabrys, Jennifer ix, 10, 201–14
Gair, Robert 175–6, 180
Galenius 92
Galison, Peter ix, 6, 8, 15–26
gamers *see* PlayStation Network (PSN)
garbage patches in oceans 201–14; geomythologizing 201–3, 209–11; locating the patches 202–6; studying ocean circulation 206–9; as technoscientific objects 203, 206, 209–11
gas battery 75–8, 81
gauge theory 21
Gell-Mann, Murray 17
gene, history of 2, 7
genesis: birth of sand heap 248; of garbage patch 203, 210; parody of hydrogen world 72–82; of wetlands as technoscientific objects 139–41
genetically modified organisms (GMOs) *see* transgenics
geneticist 124, 129–31
genetic medicine 124
genetics, and transgenics 151, 164
Genome Institute of Singapore 32
genomic microarray 125
geomythology, of ocean garbage patch 201–3, 209–11
Georgi, Howard 21
German Research Council 257
Germany: Federal Environmental Protection Agency 168; heroin history in 89, 91, 93–100, 102; Pulp and Paper Association 166
germ-line stem cells 31
Gibson, James J. 6, 253–4
Glashow, Sheldon 21
Glauber, Johann Rudolph 92
Global Drifter Program 203, 208–9
Global Earth Observation System of Systems (GEOSS) 208
global warming 70–1
global warming potential (GWP) 194
glyphosate 152, 155
Goldberg, David 44
golden rice 150
Google Earth 202, 204
Gosselink, James 140–1, 145
GOTO project 10
Gould, Stephen Jay 47
Grand Unified Theory 16
graphene 192, 193, 197
graphite 188, 192, 193
Gras, Alain 194

Great Britain: heroin history in 90, 101
Great Depression: wetland destruction 138
Great Pacific Garbage Patch 201; *see also* garbage patches in oceans
greenhouse gases 70–1, 187, 194–5, 216, 227
Greenland: Camp Century 219; ice cores 215, 218, 220–1, 225, 227
Greenland Icecore Program (GRIP) 220, 225
Greenland Ice Sheet Program (GISP) 220, 225
Grendel 137
Gross, David 22–3
Grove, William 74–5
Guest Keen and Nettlefolds 81
Guyton de Morveau, Louis Bernard 73
gyres 201, 204, 206–9; *see also* garbage patches in oceans

hackers, and Sony 105, 112–18
Hacking, Ian 18, 171, 253
hactivism 117
Hague, The 233
Halbwachs, Maurice 238, 239
Hamburger, Adolf 95
Hamilton, William D. 52n1
Hamilton's rule 42–3, 50, 51, 52n1
Hanford Site 242n11
Haraway, Donna 4, 6, 143, 171
Harré, Rom 254–5
Harrison, Phil 113
Harvard University 32
HAVL (long-lived nuclear waste) 233, 238
heat engine 70–2; combustion engine 70, 81–2, 194
Heidegger, Martin 6, 11
hematopoietic stem cells 28–9
hepatitis 100–1
herbicide resistance 152–3, 158
heroin 89–104; doctors lose control of 97–102; invention of 89–90, 94–5; morphine injections and rise of morphinism 93–4; opium tamed in Middle Ages 92–3; prohibitions against 96–102; quest to obtain 91; theriac and opium historical use 92
Higgs particle 15
HIV epidemic 100–1, 102
Hoffmann, Felix 89, 94
Holland, John 44
holographic principle 23–4
Home, for PlayStation Network 108
homo depictor 57
homo faber 57, 240
Horowitz, Jules 241
Hottois, Gilbert 3–4
Hotz, George (GeoHot) 113, 115–16
Hough, Palmer 142
Huang, Sui 36
Hughes, Thomas 241
human condition 248
human genome project 248
Husebö, Stein 101
hybridoma technology 29
hybrids 6, 154
hydric soils 145
hydrocarbons: carbon as memory 193–4; as energy sources 70
hydrogen: discovery of 72–3; H_2 and gas batteries 76
hydrogen economy 70–2
hydrogen world 70–86; energy *vs.* information technology 82–3; heat engines 70–2; parody of genesis of 72–82

Ice Age 169, 219
idealization 62–3
identity of cancer stem cells 28
Ihde, Don 171
Illston, Susan 113
immunodeficient mice 29
immunologists 29
indulgences, of Catholic Church 195
Industrial Centre for Geological Storage (CIGEO) 232
industrial revolutions 70, 187, 228
information technology: compared to energy technology 82–3
insect toxicity 152–3, 158
integrin-alpha-6 (ITGA6) gene 32
intellectual property rights (IPR): of transgenics 152, 154, 158, 159, 160–1
Intergovernmental Panel on Climate Change (IPCC) 70–1, 223, 224, 227
International Geophysical Year 218
International Opium Conference 96
International Polar Year 215
International Rice Research Institute 150
International Thermonuclear Experimental Reactor 242n8
International Union of Pure and Applied Chemistry (IUPAC) 190
International Union of Pure and Applied Physics (IUPAP) 190
Into Eternity (documentary film) 238
IPCC (Intergovernmental Panel on Climate Change) 70–1, 223, 224, 227

iPS cells technology 32–3
isotopes of carbon 190
Is Sex Necessary? Or Why You Feel the Way You Do (White) 174
ISTUK drill 217

Jacob, Walter 95
Jefferson Physics Laboratory 21
Jonas, Hans 238
Jouzel, Jean 226

Kabuki syndrome 124–6, 129–32; *see also* biography of Kabuki syndrome
Kant, Immanuel 11
Kavli Institute for Theoretical Physics 22
Kavli Prize in Nanoscience 192
Keller, Laurent 43, 48–9, 51
Kelty, Christopher 8, 42–55
kin selection 42, 52n1
Körner, Hans Harald 102
Krabbenborg, Lotte 125
Kraepelin, Emil 94
Krömeke, Franz 94
Kuhn, Thomas 18
Kutaragi, Ken 108
Kyoto Protocol 71

Lacan, Jacques 111
Lacey, Hugh x, 9, 150–65
Landauer, Rolf 58, 67n5
Lander, A. D. 33–4
Langer, Carl 76–7
Langton, Christopher 44
language, scientific 133
language games 123
Langway, Chester 219
Laplane, Lucie x, 8, 27–41
Latour, Bruno 4, 67, 107, 168, 171, 187, 241
Lavoisier, Antoine Laurent de 73–4, 194, 196
laws: Bataille Law (1991, France) 232, 236–7, 240; Clean Water Act (U.S.) 138, 142, 143–4; Coastal Zone Management Act (U.S.) 138; Computer Crimes Act 113; Digital Millennium Copyright Act (DMCA, U.S.) 113; Endangered Species Act (U.S.) 138; National Environmental Policy Act (U.S.) 138; Opium Act (1920, Germany) 97, 98, 99; Smoking Opium Exclusion Act (1919) 96
Leesona-Moos 80
Lemischka, Ihor 32
Leno, Jay 105
les trente glorieuses 233
leukemia 29; chronic lymphoid 28; myeloid 28
leukemic stem cells 28–9
Levinstein, Eduard 93
Lewin, Louis 97
Leyden jars 73
Life with Playstation (PS3) 109
Lim, Bing 32
Linde, Andre 22
literacy, paper foundations of 175–7
Loeve, Sacha x, 10, 185–200
Lorentz, Hendrik 17
Lorius, Claude 218, 219
lowlands 137, 139
Low Orbit Ion Cannon (LOIC) 114
LulzSec 117–18
Lunar Rover Vehicle 81
lymphocytes 31

Macquer, Pierre Joseph 194
macroplastics 202, 205
Maldecena, Juan 22
Marcoule, France 234
Marcus, George E. 91
Marcus Aurelius 92
marijuana 97
marine debris *see* garbage patches in oceans
Marine Pollution Bulletin 204
markers *see* time markers
markets: carbon 195; green 194
marriage, boxed-in 174
marsh 139, 189
Maryland, wetlands banking 145
Matteucci, Carlo 74
Maxwell, James C. 17
McGreevy, John 23–4
McLuhan, Marshall 176
medicine: genetic 124; heroin as 89–90, 95–102; molecular 124; translational 124–5, 132–3
Melton, Douglas 32
Memorandum of Agreement, on wetlands (1990) 142
Mencken, Henry Louis 139
Mendeleev, Dmitiri 190, 191
mephitic air 186, 189
Mephitis 189, 194
mercury 235
message in a bottle (MIB) 207
methane 216, 218
Meusnier de la Place, Jean-Baptiste 73
microplastics 204–6
Microsoft's Xbox Live 108

Midway Atoll 202
Milburn, Colin x, 9, 105–22
military-industrial complex 171
Mitchell, Timothy 194
Mithridates (King) 92
mitigation banking of wetlands 141–4, 145, 147
mitochondrial dysfunction 126, 128, 131–2
Mitsch, William 140–1, 145
mixed oxides (MOX) 234
mixed practices 257
model in silico 256
model in situ 256
mole (Avogadro's number) 190
molecular biology, and transgenics 151, 158, 164
molecular electronics 56–7, 58–9, 67
molecular medicine 124
Mond, Ludwig 76–7
Monte Carlo method 19–20
Montgolfier brothers 73
Moore, Charles 201, 204
Moore's law 82
morphine 89, 92–8, 102; *see also* heroin
Morvan, France 235
Moss, David 139–40
motion: Newton's first law of 59–60; perpetual motion 60, 65–7
MRI 125
Mullen Test 173
Müller, Lothar 176–7
Mumford, Lewis 170, 173–4, 175, 180
Musée d'histoire naturelle 73
Müser, Martin 61
Myth of the Paperless Office, The 167

nanoelectronics 56–7, 58–9, 62–3, 67
nanofacture 19
nanoscience 19
nanotribology 65–7
nanoworld of carbon 186, 192
narcotics *see* heroin
National Aeronautics and Space Administration (NASA) 78, 80–1
National Agency for Management of Radioactive Waste (ANDRA) 232, 236, 238
National Cancer Institute 27
National Museum of Play 169
National Research and Development Corporation (NRDC) 78, 81
National Research Council 139–41
National Wetlands Policy Forum 141
Native Americans: use of wetlands 137
Nature 219, 221
Naudet, Roger 241
neoliberal approaches 142
neurological disorders *see* biography of Kabuki syndrome
neutrinos 16
Newton, Isaac 18, 59–60, 63
New World modernism 170
nitrogen 73
Nixon, Richard 27, 80
Nobel Prize, and carbon allotropes 192
NOC-SCID mice 29
no-go areas 233
no net loss 136, 138, 141
non-stem cancer cells 33
Nordmann, Alfred x, 8, 56–69
North Atlantic Subtropical Gyre 207
North Greenland Eemian Ice Drilling (NEEM) 227
North Greenland Icecore Program (NGRIP) 221
North Pacific Gyre 201
n-tuple packages 155, 159–60, 162, 164
nuclear power 28, 234, 241
nuclear waste 232–44; in Bure, Lorraine, France 232, 236–7; deep geological burial 235–7; memory and markers for the future 233, 237–9; natural analogues 240–1; reprocessing of used nuclear fuels 233–4; as reverse salient 241; storage and reversibility 237–8; as technoscientific object 239–40; toxic final waste 235

ocean garbage patches *see* garbage patches in oceans
Ocean Surface Current Simulator (OSCUR) 207
Oeschger, Hans 218, 219
oikos 254
oil 187, 194
oil crisis 81
Oklo, Gabon 240–1
Olivier, Laurent 239
Omega Minus 19
oncology 27
oncomouse 6
ontography of carbon 187, 195–7
ontological commitment 245, 251, 252
ontological indifference 8–9
ontological relativity 251, 252, 256
ontology 6, 7, 8
oocytes 32
Operation Sony 114

opium 90, 92, 102
Opium Act (1920, Germany) 97, 98, 99
Opium Convention 96
Osiander, Andreas 17
OtherOS (PlayStation 3) 113
Ötzi the Iceman 188
Out of the Box 169
oxygen in fuel cells 72–8

Pacific Garbage Patch 201–2, 203–4, 206
packaging science of cardboard 170–5, 178
Pande, Vijay 109–10
Paneth, Friedrich 191
paper *see* cardboard
participation, robots in evolutionary biology 42, 45–6, 49–50
particle physics 15–16
patient organization 128, 132–3
pediatric neurology *see* biography of Kabuki syndrome
pencils, graphite 188
persistent organic pollutants (POP) 205
Pessoa, Fernando 188–9, 196
pharmakon 189
phenomenotechnical 252
phenotypic 29, 31
philosophy: of experiment 75; of technoscience 3
phlogiston 17, 73
Phoenicians 191
physics, ontology of pyramid and ring 15–26; anti-ontology 17–18; Monte Carlo methods 19–20; particle physics 15–17; string theory 21–5
Pierrelate, France 240
Pilgrim's Progress (Bunyan) 137
plastic bags 168
plasticity: of plastics in ocean garbage patches 206; of wetlands 144–6, 147
plastics in oceans *see* garbage patches in oceans
Plato 47–8; inverted Platonism of technoscience 60, 63, 65, 66; Platonism of science 60
PlayStation 2 112
PlayStation 3 (PS3) 106, 108–10
PlayStation Network (PSN) 105–22; devotion of gamers 110–12; epic fail 106–7; Folding@home 108–10, 112, 118; hacking attacks 112–18; resurrection 118–19
PlayStation Portable (PSP) 108
PLOS Biology 42
plutonium 233–4
Poincaré, Henri 17
Poirot-Delpech, Sophie x–xi, 10, 232–44
polar ice 2–3
polar ice cores 215–31; Antarctica 221–2; as climate change messengers 215–16; drilling back into time 219–22, 228; drilling techniques 216–17; engaging with ice core 222–4; Greenland 220–1; ice as marker of time 218–19; as political issue 227–8, 229; recalcitrance of 225–6
pollution, debris in oceans 202, 204–5
Popper, Karl 123
Porter, Steven G. 97
positrons 16
postmodernism 3–4
power plant (for Apollo program) 80–1
Pratt and Whitney 80, 81
Pravaz, Charles-Gabriel 93
Princeton University 32
principle of presumed legitimacy (PLT) 156–7
Puritans 137
Putnam, Hilary 252
pyramid, physics as *see* physics, ontology of pyramid and ring
pyrite oxidation 247

quantum electrodynamics 15
quantum mechanics 16, 18, 253
quarks 16, 17
quasi-objects 196
Queen of Hearts *30*, 31
Quine, Willard V. O. 251–2

Radboud University Medical Center 124
radiation and cancer stem cells 28
rapid assessment methods (RAMs) 143–4, 147
rapid ecological assessment 144
rare neurological disorders 124, 130
real-world simulation 253, 255–7
recalcitrance: of ice cores 225–6, 229; of wetlands 144–6, 147
recycling logo 175
recycling paper and cardboard 166, 168, 175
Reed, Mark 56–7, 59
regular slotted container (RSC) 170
relational ontology of stemness 34–6
relational property 32, 35
relativity theory 253
research, development & innovation (R&D&I) 151, 152, 155–7, 160–1
reverse salient 241

Rheinberger, Hans-Jörg 252
rice fields, in South Carolina 138
Ricoeur, Paul 238
Rimon, Noam 109
ring *see* physics, ontology of pyramid and ring
Robertson, Morgan 141–6
robots, in study of evolution 42–55; embodiment of 49–50; a life of their own 51–2; physical *vs.* simulated robots 48–9; scale of situatedness 45–7; simulacrum 47–8, 52; simulation and participation of 42, 45–6, 49–52; social behavior 50; using robots 42–4
Royal Society of London 75, 76, 80
Rule 41, for cardboard boxes 173
Rushdie, Salman 76

Sabille, Eric van 208
Salmeron, Miquel 63
sand heap *see* Chicken Creek
San Francisco Exploratorium 169
Sanger sequencing 125, 130
scale of situatedness 45–7
Schiappa, Edward 145
Schirmeisen, André 64
Schönbein, Christian 74
Schwarz, Astrid xi, 10, 245–60
Science 219
science: compared to technoscience 3, 57, 66–7
science and technology studies (STS) 4
science trench, for ice cores 217, 222
self-maintaining system hypothesis 28
self-renewal property of stem cells 31–3
Serres, Michel 107, 171, 196
Sertürner, Friedrich Wilhelm 93, 94
Shenker, Steve 23
Simondon, Gilbert 11, 80, 82
simulacrum 47–8, 52
simulated robots 42, 46, 48–9
simulation: real-world 253, 255–7; robots in evolutionary biology 42, 45, 51–2
simulations, in physics 19–20
single-gene defects 29
Smalley, Richard 197
Smith, H. Allen 175–6, 180
social behavior, robot simulations 50
Soentgen, Jens xi, 9, 89–104
Sony *see* PlayStation Network (PSN)
Sony Pictures 117, 118
Souriau, Etienne 187
South Carolina, wetland conversion 138
Sownage 117–18
space of agricultural options 151–2, 153, 155, 157, 161–2, 163
Sputnik 78
Standard Model 15
standard risk assessments (SRA) 155–6, 157, 158, 160, 161, 162
Stanford University 108–9
Stauffer, Bernhard 219
stem cells, types of 27, 31; *see also* cancer stem cells
stem cells hypothesis 28
stemness 27–8, 32–6
stemness signature 31, 35
stress tests for cardboard 173
string theory 21–5
Strominger, Andy 22–3
Subtropical Convergence Zone 204
Superconducting Supercollider 16–17
super dynamic system *246*, 247
superlubricity 62–6
superweeds 155
Susskind, Leonard 22
swamps 137–9
Swan, Sir Joseph 192
syndrome *see* biography of Kabuki syndrome
system-level property 34
system ontology of stemness 34–5, 36

technoscience: compared to science 57, 66–7; defined 3–4; inverted Platonism of 60, 63, 66
technoscientific objects 1, 8–11; attractive objects 1–3; carbon as 195–6; cardboard as 166, 178–80; fuel cells 82; ice core as 218, 222, 228; nuclear waste as 239–40; ocean garbage patches as 203, 206, 209–11; transgenics 151
technoscientific research 5
technoscientific theory 57–9, 66–7
technoscientific world, furniture of 5–8
Teissier, Pierre xi, 9, 70–86
Tennant, Smithson 194
Terenoworld 257
TERrestrial ENviromental Observatoria 257
theriac 92
Thoreau, Henry David 138
thorium 240
time markers: ice cores as 218, 224, 228; for nuclear waste 237–9
Tobler, Jakob 77
Tour, James 56–7, 59
Townsend, Kilpatrick 114
Traceable Drifter Unit (TDU) 207

transgenics (TGs) 150–65; agribusiness of 150, 152–5, 158, 160–1; controversies over 156; critical narrative 156–7, 159; crop seeds as renewable regenerative resources 153–5, 160, 161; dualist ontology 159; legitimating narrative 157–8, 159; risk assessments for 155–6; safety of 160; science backing of 162–3; seeds in n-tuple packages 155, 159–60, 162, 164; space of agricultural options 151–2, 153, 155, 157, 161–2, 163; as technoscientific objects 151, 163–4; values of capital and the market 160–2, 164
translational medicine 124–5, 132–3
Treaty of Versailles 96–7
tribology 65–7
tuberculosis 101
tumorigenic cells 27
Twitter: from the ocean 208; in Sony hacking incident 114, 117

Ugly Duckling, The 169
unified atomic mass unit (UMA) 190
unified electroweak theory 19
unified field theory 15
United Nations: Single Convention on Narcotic Drugs 99
United Technologies Corporation 71
uranium 233–4, 236, 240
U.S.: cardboard trash and recycling 166, 175; history of wetlands 136–8; ice coring projects 217; war on drugs 96–7, 100–2
U.S. Army Corps of Engineers (Corps) 138, 141, 142
U.S. Copyright Office 113
U.S. Environmental Protection Agency (EPA) 138, 142
U.S. Fish and Wildlife Service (FWS) 139, 144
U.S. National Marine Fisheries Service (NMFS) 142
U.S. National Oceanic and Atmospheric Administration (NOAA) 202

van den Burgh, Simone xi, 9, 123–35
van Helmont, Joan Baptista 189
Vattenfall Europe Mining AG (VEM) 249, 257
Versailles Peace Treaty 97
Vienna Circle 17
Vileisis, Ann 137, 145
Virgil 189
Volta, Alessandro 73–4
voltaic battery 73–5
Vostok research station, Antarctica 221

Walker, Richard 139
water: definition as H_2O 187; electrolysis of 74
water catchment *see* Chicken Creek
Watt, James 70
weak interaction theory 15
Web of Science database 31
Weinberg, Steven 15, 21
wetlands 136–49; genesis of concept 139–41; history of, in U.S. 136–8; mitigation banking 141–4, 147; plasticity and recalcitrance of 137, 144–6, 147
White, E. B. 174
White Magic, The Age of Paper (Müller) 176
White Planet, The 215
whole exome sequencing (WES) 124–5, 129–33
Wikipedia 168, 170
Winsberg, Eric 45, 51
Winthrop, John 137
Witten, Ed 22
Wittgenstein, Ludwig 123
Wittgenstein's Poker (Edmonds and Eidinow) 123
wonder drugs 89–90
Wood, Alexander 93
World War I: heroin and 94, 95, 96, 97
World War II: wetland destruction 138
Written Declaration 16/2007 70
Wunderkammer 5

Xbox Live 108
xeno-transplantations 29
Xylander, Cheryce von xii, 9, 166–82

yellow-cakes 234
YouTube, in Sony hacking incident 113

zombies 119